IET MATERIALS, CIRCUITS AND DEVICES SERIES 80

Physical Biometrics for Hardware Security of DSP and Machine Learning Co-processors

Other volumes in this series:

Physical Biometrics for Hardware Security of DSP and Machine Learning Co-processors

Anirban Sengupta

The Institution of Engineering and Technology

Published by The Institution of Engineering and Technology, London, United Kingdom

The Institution of Engineering and Technology is registered as a Charity in England & Wales (no. 211014) and Scotland (no. SC038698).

© The Institution of Engineering and Technology 2023

First published 2023

The Institution of Engineering and Technology
Futures Place
Kings Way, Stevenage
Hertfordshire SG1 2UA, United Kingdom

www.theiet.org

British Library Cataloguing in Publication Data
A catalogue record for this product is available from the British Library

ISBN 978-1-83953-821-6 (hardback)
ISBN 978-1-83953-822-3 (PDF)

Typeset in India by MPS Limited

Cover Image: Paper Boat Creative/Stone via Getty Images

Printed and bound by CPI Group (UK) Ltd, Croydon, CR0 4YY

Contents

Acknowledgements

I would like to thank my family and friends for the support and encouragement throughout the execution of the book project. I would also like to thank Indian Institute of Technology (IIT) Indore for the support in executing this work.

Preface

This book on "**Hardware Security of DSP and Machine Learning Coprocessors using Biometrics**" presents state-of-the art explanations for securing and protecting digital signal processing (DSP) and machine-learning coprocessors (hardware intellectual property (IP) cores) against hardware threats. DSP coprocessors such as FIR filters, image processing filters, discrete Fourier transform, and JPEG compression hardware are extensively utilized in several real-life applications. Further machine-learning coprocessors such as convolutional neural network (CNN) hardware IP core can play a vital role in several applications such as face recognition, medical imaging, autonomous driving, and biometric authentication. Thus security/protection of these hardware coprocessors against hardware threats such as IP abuse/misuse that includes fraud claim of IP ownership and IP piracy becomes extremely vital. This book presents state-of-the art hardware security solutions for such DSP and machine learning coprocessors using biometric as well as other techniques.

Broadly the theme of this book includes the following:

- **Chapter 1** presents an "**Introduction: secured co-processors for machine-learning and DSP applications using biometrics**". This chapter discusses background on securing hardware coprocessors including its hardware threats. It also discusses the role of behavioral synthesis design process in the security and IP core protection of hardware coprocessors. It further highlights basic details of biometric security used for machine learning and DSP coprocessors.
- **Chapter 2** presents "**Integrated defense using structural obfuscation and encrypted DNA-based biometric for hardware security**". The significant features of this chapter include highlighting the background on DNA/genome sequencing, fundamentals of hardware watermarking, hardware steganography and hash-based digital signature. It also discusses in details the process of integrated defense using structural obfuscation and encrypted DNA-based biometric for hardware security.
- **Chapter 3** presents "**Facial signature-based biometrics for hardware security and IP core protection**". The significant features of this chapter include discussion on the importance of high-level synthesis for designing-secured DSP coprocessors., advantages of employing facial biometric over fingerprint biometric for IP core protection of hardware coprocessors. Finally, it discusses the detailed methodology of employing facial biometric for securing DSP coprocessors.

- **Chapter 4** presents "**Secured convolutional layer hardware co-processor in convolutional neural network (CNN) using facial biometric**". The significant features of this chapter include discussion on why designing CNN convolutional layer IP core is important. It also discusses the detailed approach for designing a secured CNN convolutional coprocessor IP core using facial biometric.
- **Chapter 5** presents "**Handling symmetrical IP core protection and IP protection (IPP) of Trojan-secured designs in HLS using physical biometrics**". The significant features of this chapter include a discussion on HLS-based symmetrical IP core protection using IP buyer fingerprint biometric and IP seller facial biometric. It also discusses the process of employing facial biometric for protecting Trojan-secured system-on-chip design against piracy.
- **Chapter 6** presents "**Palmprint biometrics vs. fingerprint biometrics vs. digital signature using encrypted hash: qualitative and quantitative comparison for security of DSP coprocessors**". The significant features of this chapter include overview on palmprint biometric based IP core protection, fingerprint biometric-based IP core protection as well as digital signature-based IP core protection. It also provides a qualitative and quantitative comparison between palmprint biometric, fingerprint biometric and digital signature in terms of IP core protection and hardware security.
- **Chapter 7** presents "**Secured design flow using palmprint biometrics, steganography and PSO for DSP coprocessors**". This chapter highlights the low-cost steganography based hardware security design flow using palmprint biometric and steganography. It also provides analysis on various case studies in terms of design cost analysis and security.
- **Chapter 8** presents "**Methodology for exploration of security-design cost tradeoff for signature-based security algorithms**". The significant features of this chapter include motivation on performing security-design cost tradeoff, summary of 'signature-based security algorithms for hardware IPs' in the literature and the methodology for exploration of security-design cost tradeoff for signature-based security algorithms.
- **Chapter 9** presents "**Taxonomy of hardware security methodologies: IP core protection and obfuscation**". The significant features of this chapter include discussion on possible hardware threats and attacks in the design flow of hardware integrated circuits, taxonomy representation of IP core protection methodologies and taxonomy representation of obfuscation methodologies.

Authors believe that there is no book that presents details of hardware security of DSP and machine learning coprocessors using biometrics, under one canopy. By covering chapters under this special topic, it will empower readers to drive their borders of knowledge to plunge into some latest security and design aspects of modern hardware coprocessors, especially for DSP and machine-learning applications. The book is prepared keeping in mind that can be easily integrated to any graduate level course. Furthermore, it also serves as designer's hand-book who is eager to integrate hardware security solutions for DSP and machine learning applications.

Sincerely,
Book Author
Dr. Anirban Sengupta, Ph.D., Assoc. Professor
Fellow of IET, Fellow of British Computer Society (BCS), Fellow of IETE, Senior Member of IEEE
IEEE Distinguished Lecturer (IEEE Consumer Electronics Society)
IEEE Distinguished Visitor (IEEE Computer Society)
Former Board Member and Former Chair, IEEE CTSoc Security and Privacy of CE Hardware and Software Systems (SPC) Technical Committee (TC)
Former Chair, IEEE Computer Society Technical Committee on VLSI
Founder & Former Chair, IEEE Consumer Technology Society Bombay Chapter (Now MP Chapter)
Deputy Editor-in-Chief, IET Computers & Digital Techniques,
Editor-in-Chief, IEEE VLSI Circuits and Systems Letter
Awardee, IEEE Chester Sall Memorial Consumer Electronics Award (IEEE CE Society)
Associate Editor – IEEE Transactions on VLSI Systems, IEEE Transactions on Consumer Electronics
Former Editorial Board Member – IEEE Transactions on Aerospace and Electronic Systems, IEEE Access, IEEE Consumer Electronics Magazine, IET Computers and Digital Techniques, IEEE Letters of the Computer Society, IEEE Canadian Journal of Electrical and Computer Engineering, Elsevier Microelectronics Journal
General Chair, 37th IEEE International Conference on Consumer Electronics (ICCE), Las Vegas
General Chair, 23rd International Symposium on VLSI Design and Test (VDAT-2019), India
Executive Committee, IEEE International Conference on Consumer Electronics (ICCE) – Berlin and Las Vegas
IEEE Distinguished Lecturer Nominations Committee, IEEE CE Society
Computer Science and Engineering
Indian Institute of Technology Indore
Email: asengupt@iiti.ac.in
Web: http://www.anirban-sengupta.com

Authors' Biography

 **Anirban Sengupta** is an associate professor in the Discipline of Computer Science and Engineering at Indian Institute of Technology (IIT) Indore. He has around 270 publications and patents, 50 book chapters and 5 books.

His is a recipient of awards/honors such as Fellow of IET, Fellow of British Computer Society, Fellow of IETE, IEEE Chester Sall Memorial Consumer Electronics Award, IEEE Distinguished Lecturer, IEEE Distinguished Visitor, IEEE CESoc Outstanding Editor Award, IEEE CESoc Best Research Award from CEM, Best Research paper Award in IEEE ICCE 2019, IEEE Computer Society TCVLSI Outstanding Editor Award and IEEE TCVLSI Best Paper Award in IEEE iNIS 2017. He held/holds around 17 Editorial positions in IEEE/IET Journals. He is the Editor-in-Chief of IEEE VCAL (Computer Society TCVLSI), Deputy EiC of IET Computers & Digital Techniques and General Chair of 37th IEEE Int'l Conference on Consumer Electronics (ICCE) 2019, Las Vegas. He is consistently ranked in Stanford University's Top 2% Scientists globally across all domains. Complete details available at: http://www.anirban-sengupta.com/index.php

Professional leadership role in scientific community – editors

- Deputy Editor-in-Chief: IET Computers and Digital Techniques (2018–Present)
- Editor-in-Chief: IEEE VLSI Circuits & Systems Letter, IEEE CS-TC on VLSI (2017–2020)
- Associate Editor: IEEE Transactions on VLSI Systems (TVLSI) (2018–Present)
- Associate Editor: IEEE Transactions on Aerospace & Electronics Systems (TAES) (2016–2020)
- Associate Editor: IEEE Transactions on Consumer Electronics (TCE) (2019–Present)
- Guest Editor: IEEE Transactions on VLSI Systems (TVLSI) (2016–2017)
- Guest Editor: IEEE Transactions on CAD of Integrated Circuits & Systems (TCAD) (2019–2020)
- Associate Editor, IEEE Letters of the Computer Society (LOCS) (2019–2020)
- Associate Editor: IET Computers and Digital Techniques (CDT) (2015–2018)
- Senior Editor: IEEE Consumer Electronics Magazine (CEM) (2017–2019)
- Associate/Executive Editor: IEEE Consumer Electronics Magazine (CEM) (2016–2017)
- Associate Editor: IEEE Canadian Journal of Electrical and Computer Engineering (2018–2020)
- Associate Editor: IEEE Access (2015–2018)
- Associate Editor: IEEE VLSI Circuits & Systems Letter (2015–2017)
- Guest Editor: IEEE Access Journal (2016–2017)
- Guest Editor: IET Computers and Digital Techniques (CDT) (2017–2018)
- Editor: Elsevier Microelectronics Journal (2016–2018)

Chapter 1

Introduction: secured co-processors for machine learning and DSP applications using biometrics

Anirban Sengupta[1] and Mahendra Rathor[2]

The chapter gives an introduction on security requirements of co-processors for machine learning (ML) and digital signal processing (DSP) applications and the role of biometrics in securing them. This introduction of the book tries to build interest in readers about the various DSP and ML co-processors; behavioral synthesis design process for generating secured DSP and ML co-processors and importance of biometric security for hardware authentication.

The chapter is organized as follows: Section 1.1 introduces about the co-processors, different hardware threats, and conventional security solutions; Section 1.2 highlights the significance of behavioral synthesis in designing and securing co-processors; Section 1.3 introduces about the co-processors for ML applications, why ML co-processors need to be secured, and how behavioral synthesis plays a crucial role in securing ML co-processors; Section 1.4 introduces about the behavioral synthesis perspective in designing and securing DSP co-processors; Section 1.5 introduces about the biometric security based on fingerprint, face, and palmprint for ML and DSP co-processors.

1.1 Security of co-processors: an introduction, hardware threats, and conventional security solutions

Modern human life is enriched with a number of electronic devices and gadgets such as television, cell phones, laptops, tablets, smart wearable, and digital camera. This could be possible with the tremendous advancement in research and technology. The technological advancement has not only made our day-to-day life sophisticated and comfortable but also played a pivotal role in the life critical systems such as healthcare, military, and defense. Be it a consumer field, banking, or any critical infrastructure, co-processors used in the electronic systems have proved themselves

[1]Department of Computer Science and Engineering, Indian Institute of Technology Indore, India
[2]Software Innovation Center, Indian Institute of Technology BHU, India

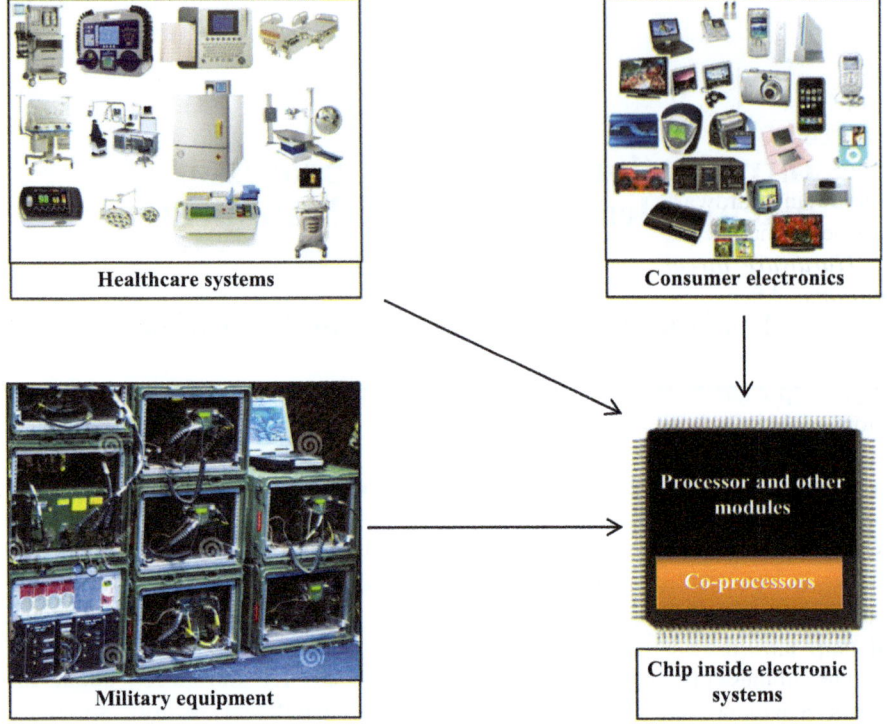

Figure 1.1   Co-processors used in various consumer and critical systems

to be very useful (Pilato *et al.*, 2018; Sengupta *et al.*, 2017a). Figure 1.1 shows some typical applications of co-processors in consumer and life critical systems.

A co-processor can be defined as a dedicated hardware unit used for processing some specific computations or executing intended functionalities. Examples of some co-processors are as follows:

- Digital signal processing (DSP) co-processors are used for processing DSP algorithms such as infinite impulse response (IIR) filter, finite impulse response (FIR) filter, discrete cosine transform (DCT), inverse discrete cosine transform (IDCT), and discrete wavelet transform (DWT) (Mahdiany *et al.*, 2001; Sengupta *et al.*, 2017a).
- Multimedia co-processors are used to execute dedicated multimedia applications such as image compression/decompression, audio encoding/decoding, and video encoding/decoding (Sengupta *et al.*, 2018).
- ML co-processors are used to execute dedicated ML applications such as artificial neural network (ANN), convolutional neural network (CNN), linear regression (LR), and support vector machine (SVM) (Lemley *et al.*, 2017; Sengupta and Chaurasia, 2022; Spencer *et al.*, 2019; Struharik *et al.*, 2018; Zhao *et al.*, 2017).

The co-processors mentioned above are widely employed in modern system-on-chip (SoC) designs to cater a wide variety of DSP, multimedia, and ML applications. The use of co-processors in the modern systems offers high performance by offloading the general purpose processor or central processing unit from their heavy computational load. Because of delivering acceleration in performance, the co-processors are also referred to as hardware accelerators. Though the advantage of achieving hardware acceleration is traded-off with the cost of additional area, the high-performance requirement in the modern systems overpowers this limitation.

Hardware threats to co-processors or IP cores and conventional security solutions: After giving a brief introduction on co-processor, its different types, and utility in modern systems, let us also discuss about its security perspective. The security threats to the co-processors have come into the limelight because of globalization of design supply chain. A whole process of designing co-processors or hardware accelerators or semiconductor intellectual property (IP) cores is generally not carried within a single house by a single entity. The current design supply chain involves various entities or offshore design houses and fabrication units (foundries) in the journey of obtaining a co-processor chip from its specifications. The entire very large-scale integration (VLSI) design process of obtaining an integrated circuit (IC)/chip from its specifications is distributed into various design phases namely determining chip specifications, high-level synthesis (HLS)/behavioral synthesis, logic synthesis, layout or physical synthesis, and chip fabrication (Sengupta, 2016, 2017). Some important reasons behind the execution of these different design phases in different design houses and foundry are as follows (Sengupta, 2020; Sengupta and Mohanty, 2019):

(i) Increasing design complexity of system-on-chips and time-to-market pressure demands modular design paradigm where entire system functionality is divided into different modules. The various modules, also referred to as cores, are procured from different vendors by the SoC integrator. The procurement of pre-designed and pre-verified IP cores or design modules from outside design houses reduces not only the design complexity but also the design time.
(ii) Cost of building and maintaining a fabrication facility for the advanced technology nodes is excessively high, which is not favorable for most of the SoC design houses. Hence, IC design houses found outsourcing the fabrication of chips to offshore foundries economically beneficial and thereby they became fabless.

Though the globalization of design supply chain offers benefits in terms of reduction in design complexity, shorter design cycle/time and saving in the cost of managing an advanced fabrication unit, nevertheless its dark side cannot be overlooked. Involvement of various entities in the globalized design supply chain gives rise to various kinds of hardware security threats such as IP piracy or cloning, IP/IC counterfeiting, IP/IC illegal reverse engineering (RE), hardware Trojan (malicious logic) insertion, and IC overbuilding. In the IP piracy or cloning threat, potential rogue elements in the SoC design house or foundry may steal the IP of original vendor and sell under his/her own brand name to earn illegal income. This may immensely harm the revenue of genuine IP vendor/designer/owner. Whereas in IP/IC counterfeiting threat, an adversary may sell poor quality or fake IPs or refurbished ICs under the brand name of the genuine supplier. Thereby, the counterfeiting threat not only hits the original supplier's revenue but also his/her brand value. In the case of false claim of IP ownership threat, an adversary fraudulently claims the ownership of the IP in the court and could be successful in the absence of producing proper IP ownership evidence by the genuine owner. Further, the RE attacks on IP or IC are such malicious efforts by a potential adversary which aim to back track the design stages with the intention of stealing design intents or inserting malicious logic or hardware Trojan. A potential hardware Trojan hidden inside the system chips can be catastrophic in the case of critical applications such as healthcare and defense. Some adverse effects of hardware Trojan insertion are as follows (Sengupta *et al.*, 2017b; Sengupta and Rathor, 2019a): (i) denial of service, (ii) leakage of confidential or sensitive data from the critical systems, (iii) password theft, (iv) corrupting computational output value, (v) battery explosion, and so on. Furthermore, the IC overbuilding threat results into illegal over production of ICs by a fabrication unit without the knowledge of the design owner. The IC overbuilding threat may also severely affect the revenue of the original designer. Hence, ensuring security of co-processors during their design process is vital not only from designer's perspective but also from end user's perspective.

There are some traditional approaches for securing the co-processor or IP core designs against the aforementioned threats. Figure 1.2 depicts the security mechanisms deployed for different kinds of hardware threats (Sengupta, 2020). As shown in the figure, popular security approaches against IP piracy or cloning are logic locking (or functional obfuscation) (Roy *et al.*, 2008; Sengupta *et al.*, 2019a), IP watermarking (Koushanfar *et al.*, 2005; Sengupta and Bhadauria, 2016), and steganography (Sengupta and Rathor, 2019a). The logic locking technique offers preventive control against piracy whereas the IP watermarking and steganography techniques offer detective control against piracy. Moreover, the IP watermarking and steganography techniques are also capable to secure co-processors and IP cores against IP counterfeiting and false claim of ownership threats. Further, the RE attacks on hardware can be prevented using the logic locking technique. Hardware or structural obfuscation (Lao and Parhi, 2015; Sengupta *et al.*, 2017a; Sengupta and Rathor, 2019b) and layout camouflaging are the potential techniques that make the RE of an IC and its design arduous for an attacker (Sengupta and Mohanty, 2019). Moreover, the structural obfuscation and camouflaging techniques are also capable to prevent the potential hardware Trojan insertion attack in an untrustworthy design house or foundry (Sengupta, 2020; Sengupta and Mohanty, 2019). Further, various active and passive hardware metering techniques have been proposed to provide security against IC overbuilding by an untrustworthy foundry (Koushanfar, 2012; Koushanfar and Qu, 2001). Besides the conventional security mechanisms of hardware protection, recently, biometric techniques have emerged as a robust security paradigm for hardware authentication and enabling protection against IP piracy and counterfeiting (Sengupta and Rathor, 2020a, 2021; Sengupta *et al.*, 2021). We will discuss the biometric techniques for hardware security in Section 1.5.

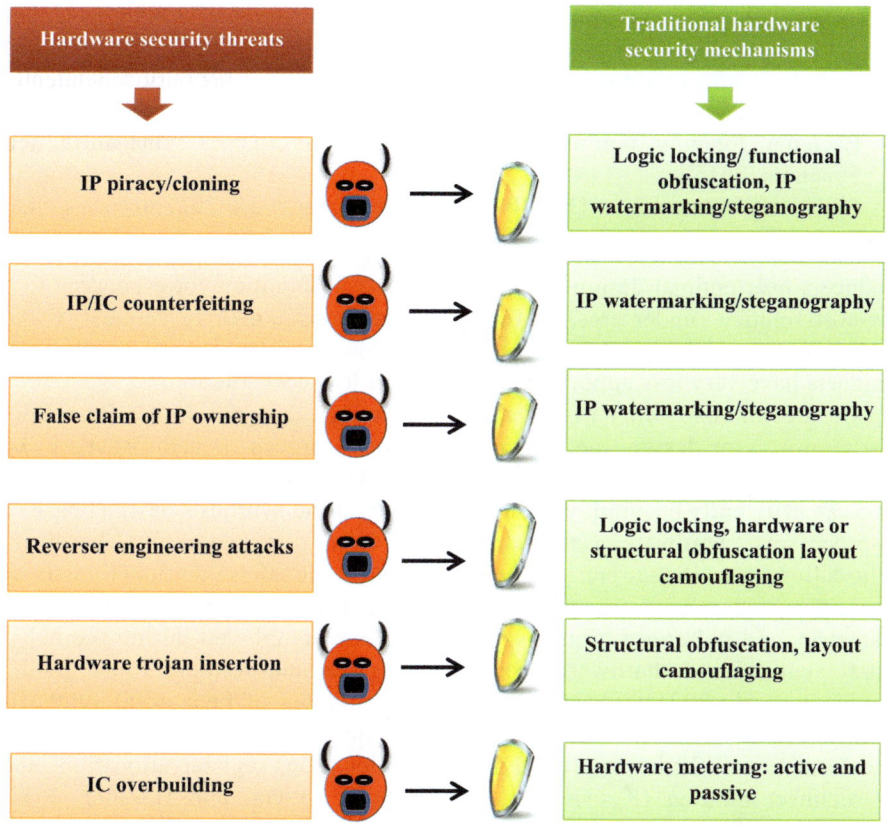

Figure 1.2   Various hardware security threats and respective security solutions

1.2   Role of behavioral synthesis design process in security of co-processors

This section discusses the importance of behavioral synthesis or HLS (McFarland *et al.*, 1988) design process in designing secured co-processors. Since the co-processors for DSP, multimedia, and ML applications are computationally intensive and have complex designs, their lower level design descriptions such as register transfer level (RTL) and gate level are not readily available. In contrast, their high or algorithmic or behavioral level functional descriptions are available in the form of mathematic equations, transfer function, or C/C++ codes. Therefore, by applying behavioral synthesis process, the high/algorithmic level functional description of a co-processor application can be converted into the next level of design abstraction i.e., the RTL (Sengupta, 2020; Sengupta *et al.*, 2010). Moreover, the HLS design process offers the benefit of exploring an optimal design architecture using a design space exploration (DSE) technique (Sengupta, 2020). Some DSE techniques used for design architecture exploration are listed below:

1. Genetic algorithm-driven DSE (GA-DSE) (Krishnan and Katkoori, 2006; Sengupta *et al.*, 2012).
2. Particle swarm optimization-driven DSE (PSO-DSE) (Mishra and Sengupta, 2014; Sengupta and Mishra, 2014).
3. Bacterial foraging optimization-driven DSE (BFO-DSE) (Bhadauria and Sengupta, 2015).
4. Firefly algorithm-driven DSE (FA-DSE) (Sengupta *et al.*, 2017c).

The above-mentioned optimization-based DSE techniques are capable to produce a near optimal design point or resource configuration corresponding to a low-cost solution (under area, power, and delay constraints). In contrast, if the co-processors are designed from a relatively lower abstraction level, then the designers have very less opportunity to explore a low-cost solution.

Besides the capability of offering an optimized design solution for a low-cost co-processor design, the behavioral synthesis process is also proved to be efficient for deploying a security mechanism. The following security mechanisms can efficiently be employed during the behavioral synthesis design phase of co-processors: hardware watermarking, hardware steganography, hardware obfuscation, etc. (Sengupta, 2020). Apart from the aforementioned security techniques, recently, biometric approaches such as fingering biometric, face biometric, and palmprint biometric have also been employed during the behavioral synthesis for hardware security (Sengupta and Rathor, 2020a, 2021; Sengupta *et al.*, 2021). The following various phases of behavioral synthesis offer the flexibility of embedding security constraints into the design: high-level transformation, scheduling, functional unit allocation, register allocation, and interconnect binding (Koushanfar *et al.*, 2005; Le Gal and Bossuet, 2012; Sengupta, 2017). Employing security mechanisms during an early design phase such as behavioral synthesis also helps perform the early design-cost trade-off and minimize the impact of embedding security constraints on overall area, power, and performance of the system. Figure 1.3 shows a typical security and

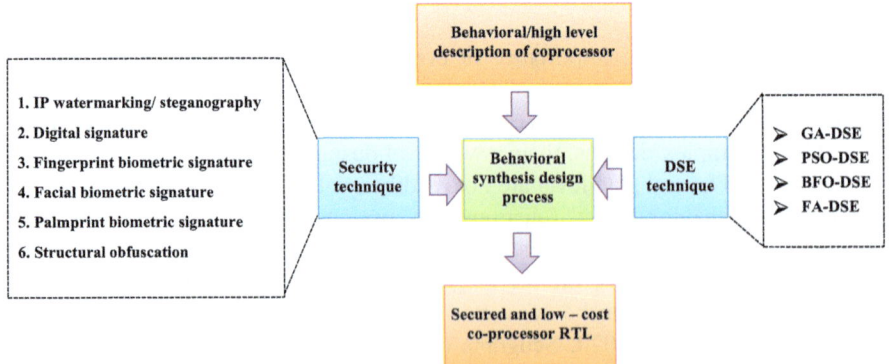

Figure 1.3   Security and cost aware behavioral synthesis design process for designing secured and low-cost co-processors

cost aware behavioral synthesis design process for designing co-processors. Additionally, adding security mechanism with the behavioral synthesis process also enables the security of coprocessor designs at subsequently lower abstraction levels such as RTL, gate, and layout level of designs (Sengupta, 2020).

1.3   Introduction to ML co-processors and their security

As discussed in Section 1.1, co-processors used in modern electronic systems can be designed for various applications such as DSP and ML (Sengupta and Chaurasia, 2022). This section gives an introduction on popular ML algorithms, their different types, and applications. Further, we discuss the need of ML co-processors, their security perspective, and the role of behavioral synthesis in designing secured ML co-processors.

1.3.1   What are ML algorithms and their co-processors

The modern consumer technology is empowered by ML because of its capability of adding intelligence to various consumer electronics (CE) and healthcare devices. The following are some popular ML algorithms that are widely employed in modern electronic industry: convolutional neural network (CNN), deep learning, linear regression (LR), support vector machine (SVM), k-means clustering, and so on. Be it Internet of Things (IoTs), smart cities, smart home, smart traffic/transportation, global positioning system (GPS) tracking, etc., the aforementioned ML algorithms are playing a crucial role. Some important applications of CNN and LR-based ML are discussed below (Bazrafkan *et al.*, 2017; Lemley *et al.*, 2017; Mahdavinejad *et al.*, 2018; Sengupta and Chaurasia, 2022; Spencer *et al.*, 2019; Struharik *et al.*, 2018; Zhao *et al.*, 2017):

- Particularly, a CNN-based ML is capable of offering high accuracy and thereby widely applied in CE applications to perform tasks such as image segmentation and classification, face recognition, object/curve detection, emotion detection, and voice analyzing. Moreover, tech-giants are commonly applying CNNs for their product recommendations, in photo search, and for automatic tagging systems. Further, the technological advancement in autonomous driving, medical diagnostics and video surveillance, etc. is also driven by CNN-based ML.
- The LR-based ML is a supervised learning technique which is applied mainly for prediction or forecasting. The energy usage and traffic load prediction in modern systems is driven through LR-based ML. A linear regression function can be employed for energy efficient usage of appliances in a smart home, temperature forecasting, prediction of traffic speed, etc. Thereby, LR-based ML finds its significance in a variety of applications in cyber physical systems such as smart home systems, intelligent transportation systems, GPS, IoT data, and also in some consumer-specific applications such as camera blur spread estimate and depth estimation from camera lens.

1.3.2   Why modern systems need ML co-processors and why to secure them

An ML algorithm is fundamentally a mathematical and probabilistic model that performs high computations on input data, whether the objective is classification or prediction. For example, a CNN framework is composed of different layers such as convolutional layer, pooling, flattening layer, and fully connected layers. Among the various different layers, the convolutional layer involves huge computations on input data. Thereby, due to high computational intensiveness, the realization of CNN-based ML as a dedicated co-processor is imperative for image centric applications (Sengupta and Chaurasia, 2022). Similarly, an LR-based ML framework functions on a huge number of data points in the training phase. Hence, it also requires high computations and huge data crunching during the training phase. This entails offloading the functionality of LR-based ML on a dedicated hardware platform. The dedicated hardware platforms such as co-processor or reusable IP core, graphics processing unit (GPU), field programmable gate array (FPGA), and application-specific integrated circuit (ASIC) are capable to efficiently handle the data and computational intensive functionality of ML. Designing a dedicated co-processor for the ML models such as CNN and LR helps satisfy the design parameter constraints, specifically performance, than a general purpose processor counterpart. In comparison to GPUs, other hardware platform such as a dedicated IP core or co-processor is a better alternative for executing ML algorithms for the modern consumer and healthcare applications. This is because many devices may not support the amount of power required to run a GPU. For example, a GPU may need power of around 450 W, including central processing unit and motherboard. In contrast, a dedicated co-processor for executing only ML computations can be designed as per the specified area power and delay constraints. Thereby, dedicated co-processors or IP cores for ML have secured their place in modern lightweight, low-power, and high-performance systems (Everything you Need to Know About Hardware Requirements for Machine Learning, 2019; Hardware Accelerators for Machine Learning, 2020; Hardware for Machine Learning, 2021).

Secured co-processor for ML: the design process of an ML co-processor incorporates multiple offshore entities such as design houses and a foundry. Because of this distributed supply chain, following scenarios may occur:

1. A malicious designer of an ML co-processor may secretly insert a hardware Trojan into the design and sell the infected designs that are to be integrated in larger systems. For example, different combinational and sequential trigger-based hardware Trojans can be placed into the design to malfunction the original functionality.
2. A rogue element in the foundry may secretly insert the Trojan during the fabrication process. For example, a hardware Trojan can be inserted by manipulating the dopant level during the fabrication process to form a side channel to facilitate the leakage of sensitive information.

3. A dishonest foundry may reverse engineer the GDS file to steal the IP of an ML co-processor and sell illegally to earn illegitimate income. The adversary may even claim the ownership of the ML co-processor IP fraudulently.
4. A dishonest foundry may breach the terms of fabricating the ML co-processor chips and may produce extra chips for personal benefits.
5. Some poor quality or fake/counterfeit ML co-processors can be sold by an IP broker to a system integrator. Such counterfeit ML co-processors may contain hidden Trojan inside them or may not be as per the desired performance specification.

The aforementioned scenarios highlight the security threats that may occur during the design and fabrication process of ML co-processors. Hence, ensuring security of ML co-processor is vital too. Recently, researchers have started focusing their attention towards designing secured ML co-processors. For instance, a CNN-based ML co-processor has been secured using a biometric signature technique to cater the threat of IP piracy and counterfeiting. Further, a LR-based ML co-processor has been secured using an obfuscation technique to cater the threat of illegal RE and hardware Trojan insertion. In the next section, we discuss how a behavioral synthesis process can be useful in designing secured ML co-processors.

1.3.3 Role of behavioral synthesis in designing and securing ML co-processors

As discussed earlier, ML co-processors are highly computationally intensive in nature as they require a number of operations to be performed on huge dataset. Further, the ML applications are generally available in the form of their algorithmic or behavioral descriptions. Hence, the behavioral synthesis process can be easily applied to generate RTL designs of ML co-processors. Furthermore, behavioral synthesis framework is amenable for embedding security features into the ML-coprocessor design. By embedding security during the early design phase such behavioral synthesis, the design of ML co-processor can have more control on design parameters to satisfy the user constraints. This is possible because of the ability of behavioral synthesis framework of integrating a DSE process that helps explore a low-cost security solution. A typical design flow of generating RTL of an ML co-processor using the behavioral synthesis process is shown in Figure 1.4. As shown, the process is accomplished in the following steps: (i) a high-level framework (such a mathematic or transfer function) is converted into corresponding data flow architecture that shows the dependency of different operations in the ML application; (ii) the data flow architecture is subjected to high-level transformation followed by scheduling, allocation, and binding steps of behavioral synthesis. During these steps, security mechanisms such as structural obfuscation and embedding signature constraints can be applied to generate a secured ML co-processor design. Moreover, a DSE process can be integrated with the behavioral synthesis process to explore a low-cost solution for the intended ML co-processor. Further, datapath and controller synthesis steps are executed to generate an optimal and secured RTL of ML co-processor.

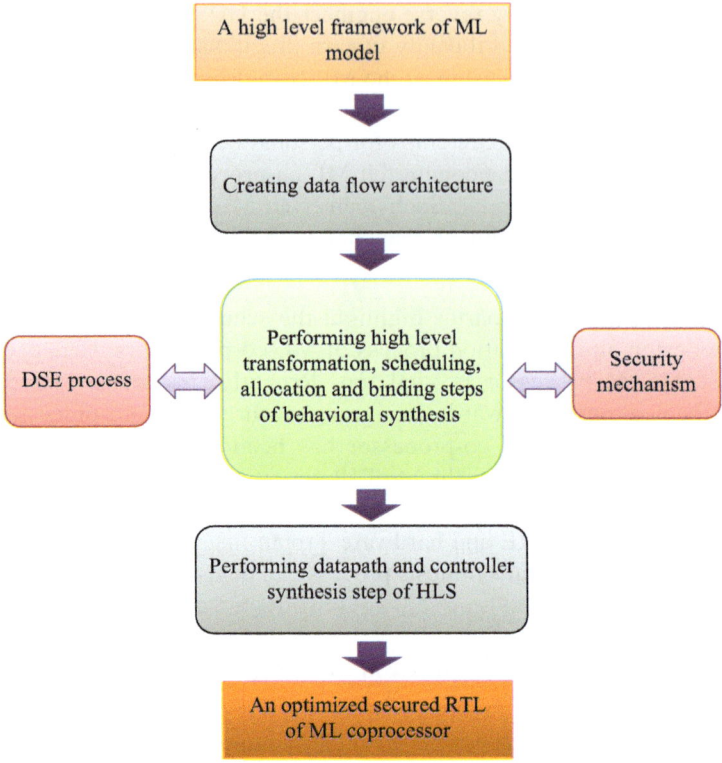

Figure 1.4   *A methodology of generating secured and low-cost RTL design of ML co-processor*

1.4   Introduction to DSP co-processors and their security: a behavioral synthesis perspective in designing and securing DSP co-processors

DSP co-processors are the dedicated application-specific processors or IP cores designed to execute computationally intensive DSP algorithms. Following are some of the most useful DSP algorithms that are widely used in modern electronic systems: IIR filter, FIR filter, DCT, IDCT, DWT, etc. These DSP algorithms facilitate tasks such as compression and decompression of images, audio and videos, and filtering out noise from digital signals. Because of the wide applications in the modern era, the market of DSPs is thriving rapidly. However, with the need of modern electronic gadgets to become lighter in weight, efficient in energy consumption, and good in performance, satisfying orthogonal design constraints of area, power, and performance has become challenging. Moreover, designing DSP co-processors beginning at RTL or logic level is not easy as their low-level descriptions are not readily available. This is where a behavioral synthesis process comes to rescue. The behavioral synthesis process paves the way of early estimation of design parameters of a DSP co-processor and hence provides the opportunity to explore such a design solution which is capable to satisfy given design constraints or generate a low-cost solution. Further, easy availability of algorithmic description of DSP applications makes the generation of their RTL design using the behavioral synthesis design process easier (Sengupta, 2020).

Security of DSP co-processors: with the growth in market place of DSP co-processors, paying attention on their security has become vital. Similar to the other IP cores, the following are the major security threats to the IP cores of DSP co-processors: hardware Trojan insertion by a rogue DSP designer and a foundry, piracy or cloning of DSP cores by SoC integrators or foundry, counterfeiting of DSP cores by rival IP designers, RE attack by foundry or a design house to steal IP or insert malicious logic. Recently, a number of research works have been proposed in the literature which primarily focused on security of DSP co-processors. Koushanfar *et al.* (2005), Le Gal and Bossuet (2012), and Sengupta and Bhadauria (2016) proposed watermarking techniques for securing DSP cores against the threat of IP piracy and securing IP ownership rights. These authors leveraged behavioral synthesis framework to secure the DSP cores. Koushanfar *et al.* (2005) and Sengupta and Bhadauria (2016) leveraged the register allocation and binding phase, whereas Le Gal and Bossuet (2012) leveraged datapath synthesis phase of behavioral synthesis. Further, Sengupta *et al.* (2018) proposed a more robust triple phase watermarking technique where three phases such as scheduling, register allocation, and functional unit allocation phases of behavioral synthesis were leveraged to secure DSP cores. A physical level watermarking technique for securing DSP cores was also proposed in the literature (Sengupta and Rathor, 2020b). Furthermore, IP core steganography techniques (Rathor and Sengupta, 2020; Sengupta and Rathor, 2019a) have also been proposed in the literature to secure DSP cores. These techniques embed vendor's secret stego-constraints during the behavioral synthesis framework of DSP cores. Additionally, logic locking techniques for offering preventing control against piracy and RE attacks to DSP cores were proposed by Sengupta *et al.* (2019a) and Rathor and Sengupta (2019). The logic locking technique integrated IP core locking blocks (ILBs) into the gate level design to generate an encrypted DSP core. For securing DSP cores against RE and Trojan insertion attacks, structural obfuscation techniques were proposed by Lao and Parhi (2015) and Sengupta *et al.* (2017a). These techniques used high-level transformation phase of behavioral synthesis to structurally obfuscate the design architecture of DSP co-processors. Moreover, a multi-key-based structural obfuscation technique was also proposed by Sengupta and Rathor (2020b) to secure the DSP cores. Interestingly, most of the security techniques for DSP cores leveraged behavioral synthesis process to embed the security features. This is because of the capability of behavioral synthesis process of offering (i) different phases such as high-level transformation, scheduling, allocation, binding, and datapath synthesis for embedding security features; (ii) opportunity of integrating DSE process for finding a low-cost security solution; (iii) security to the subsequent design levels of DSP cores.

In the next section, beyond the conventional IP security/authentication techniques, this chapter provides an introduction to some emerging biometric techniques of securing ML and DSP co-processors.

1.5   Biometric security for ML and DSP co-processors

So far, traditional approaches such as hardware watermarking/steganography have been prevalent for securing or authenticating co-processors/IP cores. However, recently, some biometric-based hardware IP core authentication approaches have gained attention of researches because of their natural ability to offer a unique signature. This section of the chapter provides an insight about the following: (i) how the biometric security for hardware authentication is different than a user authentication process; (ii) advantages of biometric-based security techniques over traditional approaches; (iii) different physical biometric-based hardware security techniques proposed in the literature namely fingerprint biometric, facial biometric, and palmprint biometric. A thematic representation of biometric-based hardware security approach is depicted in Figure 1.5.

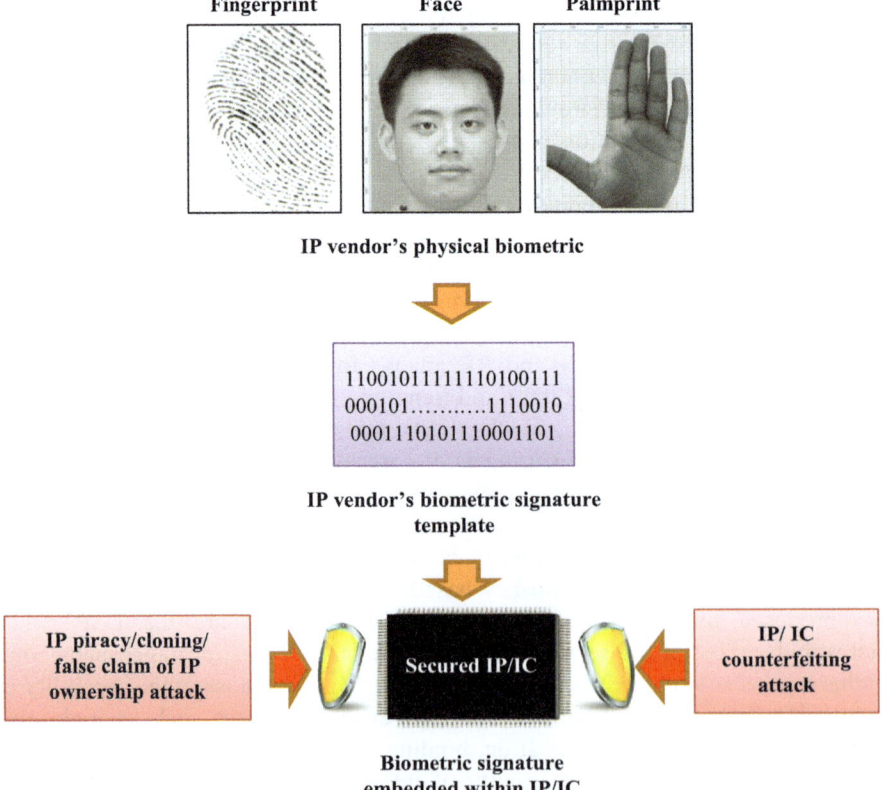

Figure 1.5   Thematic representation of biometric security approach of securing IPs

1.5.1   How biometric security for hardware authentication is different than a user authentication

Biometrics are biological or physical or behavioral characteristics that are leveraged to identify individuals. Following are the different biometrics:

*   Morphological or physical biometrics uses the physical traits such as fingerprint, shape of the face, palm veins, and irises etc.
*   Biological biometrics uses traits at a genetic and molecular level such as DNA/chromosome features.
*   Behavioral biometrics uses a unique behavioral pattern of an individual such as voice pattern and handwritten signature.

Conventionally, biometric security has been applied for user authentication in major organizations/enterprises. The biometric security systems help recognize people using their physiological or biological or behavioral characteristics. In these systems, biometric of the user or person to be authenticated needs to be captured live. Hence, biometric of the user is re-captured during the verification process. On the contrary, the hardware authentication using biometric traits is independent of re-capturing the biometric during the IP counterfeit detection or authentication process. Rather, a pre-stored biometric of the concerned individual is used to reproduce the biometric constraints embedded into the design. More explicitly, the biometric captured for the constraints generation and embedding process is stored in a tamper-proof memory and the same is used to regenerate the biometric constraints during the verification of the author and authenticating the hardware IP design. The existing biometric-based hardware security techniques in the literature have utilized the pre-storage biometric such as facial image and palmprint image during the verification process.

1.5.2   Why biometric security is required for hardware protection: advantages over traditional security mechanisms

The traditional hardware IP authentication techniques (watermarking and steganography) are based on creating vendor's secret information and embedding into the design, which is followed by the verification of secret constraints into the design during the detection process. However, the secret information associated with the vendor's watermark or stego-mark is not unique as it does not represent vendor's natural identity. In such a case, if the secret watermark or stego-mark is compromised by an adversary then justifying the secret-mark for proving IP ownership and detecting IP cloning may become challenging for the original IP designer. Additionally, rogue IP supplier can sell the counterfeit IPs pretending them to be secured with the stolen watermark or stego-mark. This is how the adversary can misuse a stolen secret mark to claim the IP ownership fraudulently or escape the IP counterfeit detection process. The following reasons highlight the vulnerability or replicability aspect of a watermark and a stego-mark (Sengupta and Rathor, 2020a).

- The overall privacy or robustness of a watermark relies on the following factors: (i) types of signature literals selected; (ii) size of the signature i.e. total number of literals; (iii) encoding of signature literals into hardware security constraints.
- Similarly, the secrecy of a stego-mark depends on the following factors: (i) secret design data, (ii) secret stego-key, and (iii) encoding of stego-constraints (secret design constraints) into hardware security constraints.

If the above-mentioned secret information and the secret-mark generation algorithm are known to an adversary then s/he can have the opportunity to replicate the secret-mark and misuse it. This may nullify the objective of hardware watermarking or steganography. Further, an RSA encryption and hashing-based digital signature approach (Sengupta *et al.*, 2019b) has also been employed to enhance the IP core security. However, this approach is also vulnerable to forging with signature because of its dependency on encryption key. Hence, keeping in mind the above-mentioned limitations of traditional hardware IP authentication mechanisms, the biometric-based mechanisms have recently come into limelight. The biometric-based IP authentication techniques are capable to offer the following advantage. In the biometric-based techniques, vendor's natural identity is mapped with the hardware security constraints to be implanted into the designs. Because of the natural uniqueness of an individual's biometric information, the attacker can never replicate or copy the vendor's biometric signature and misuse for false IP ownership claim or false authentication. Thereby, implanting a signature generated from the unique biometric traits of an individual into the ML or DSP co-processor designs provides a seamless authentication or counterfeit detection of IP cores.

1.5.3   Types of different physical biometric-based mechanisms for hardware security

Having discussed the significance of biometric-based security mechanisms for IP core protection, let us provide some highlights on the following recently published biometric security mechanisms: (i) fingerprint biometric for hardware security (Sengupta and Rathor, 2020a, 2020c); (ii) face biometric for hardware security (Sengupta and Rathor, 2021); (iii) palmprint biometric for hardware security (Sengupta *et al.*, 2021).

(i)   Fingerprint biometric for hardware security

The fingerprint biometric-based hardware security approach was first introduced by Sengupta and Rathor (2020a) to secure hardware IP cores against the IP piracy, counterfeiting, and false claim of IP ownership threats. The approach of fingerprint biometric-based hardware security leverages some important minutiae features such as ridge bifurcations and ridge endings to create a fingerprint signature template and enable the detection of the vendor's biometric fingerprint in the IP core designs.

Since the minutiae points namely ridge bifurcations and ridge endings on a fingertip are unique for each individual and hence offers the opportunity to distinctly identify the IP vendor's authentic designs based on his/her fingerprint embedded. A generic flow of the fingerprint biometric-based approach for hardware IP authentication/counterfeit detection is depicted in Figure 1.6. As highlighted in the figure, the approach is divided into the following different phases: (1) fingerprint biometric constraints generation; (2) biometric constraints embedding; (3) fingerprint constraints detection. In the fingerprint biometric constraints generation phase, biometric fingerprint image of the IP vendor is first subjected to quality enhancement through a Fourier transform (FFT) process.

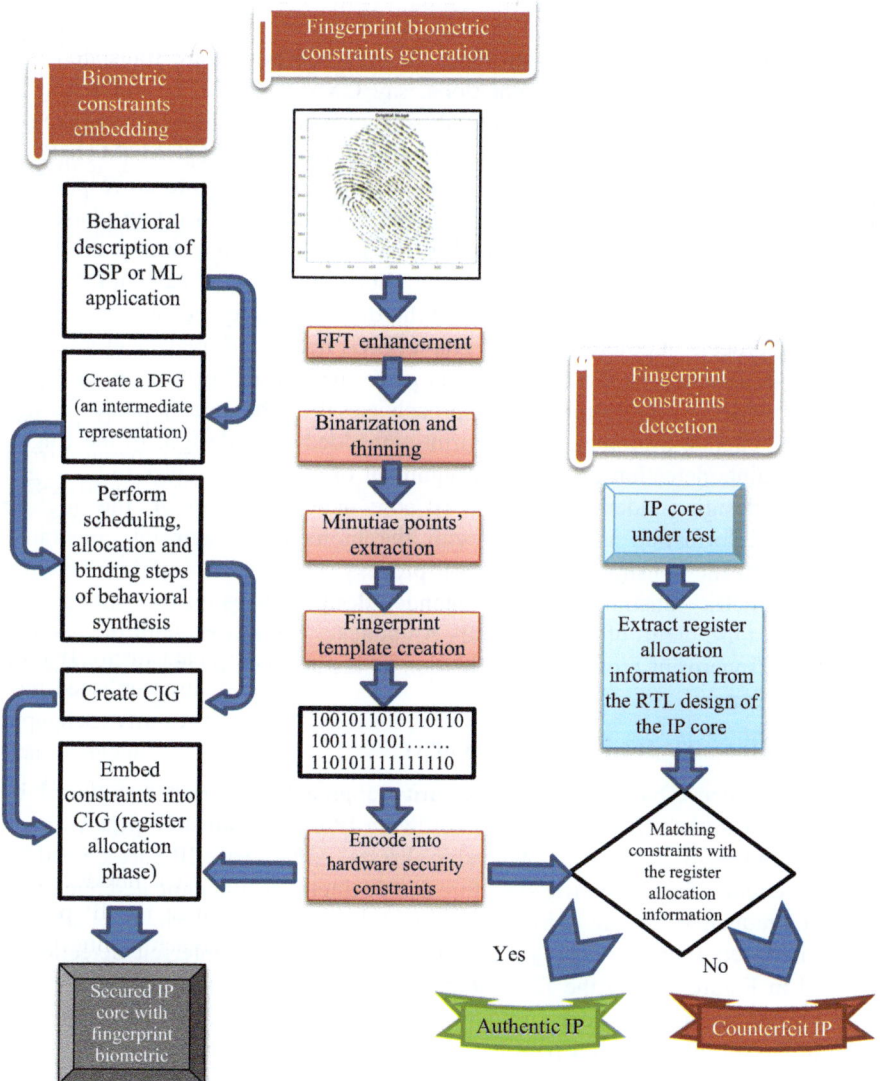

Figure 1.6   Fingerprint biometric-based approach for hardware IP authentication/counterfeit detection

The FFT process enhances the quality in terms of fine separation of the ridge lines and reconnecting the broken ridge lines, etc. Further, the fingerprint image is converted into a binarized image which is then subjected to minutiae extraction process. The following four attributes are used to characterize each minutiae point:

1. CX: *x*-coordinate
2. CY: *y*-coordinate
3. MT: minutiae type (bifurcation or ending)
4. RA: ridge angle

Next, the above-mentioned attributes of minutiae points are translated into their corresponding binary representations, say CXb, CYb, MTb, and RAb. To generate the fingerprint digital template, the binary values of different attributes of each minutia are concatenated in the following manner: CXb||CXb||MTb||RAb. Once the fingerprint digital template is obtained, it is mapped into corresponding hardware security constraints. In the biometric constraints embedding phase, the hardware security constraints are added to the intended design during the behavioral synthesis design process. The register allocation phase of behavioral synthesis is used to enable the embedding of hardware security constraints. In the embedding process, a colored interval graph (CIG) framework plays a crucial role where the embedding of the constraints is performed in the form of extra edges. Thus, the biometric fingerprint-based approach uses the behavioral synthesis design process to generate a biometric fingerprint-embedded RTL design. In the fingerprint constraints detection phase, the assignment of storage variables to the registers of the design is identified in the IP core under-test. Further, this register assignment information is matched with the biometric fingerprint constraints obtained from the constraints' generation process. The presence of vendor's fingerprint biometric constraints into the intended design proves the authenticity and nullifies the false claim of IP ownership. An attacker can never claim the genuine IP owner's fingerprint biometric information as his/her own owing to its inherent uniqueness feature.

(ii) **Face biometric for hardware security**

The face biometric-based hardware security approach was first introduced by Sengupta and Rathor (2021) to secure hardware IP cores against IP piracy, counterfeiting and false claim of IP ownership threats. A hardware IP core embedded with the vendor's face biometric signature can distinctly be authenticated due to the inherent uniqueness of an individual's facial features. A generic flow of face biometric-based approach for hardware IP authentication/counterfeit detection is depicted in Figure 1.7. In the face biometric signature generation phase, the following facial features have been used: (1) height of forehead; (2) height of face; (3) width of nasal ridge; (4) inter pupillary distance; (5) ocular breadth; (6) bio-ocular breadth; (7) inter ocular breadth; (8) width of face; (9) width of nasal base; (10) nasal breadth; (11) oral commissure width. Once the features are selected, their dimensions are evaluated using the Manhattan distance. Further, the binary values of features dimension are concatenated to generate the intended facial signature

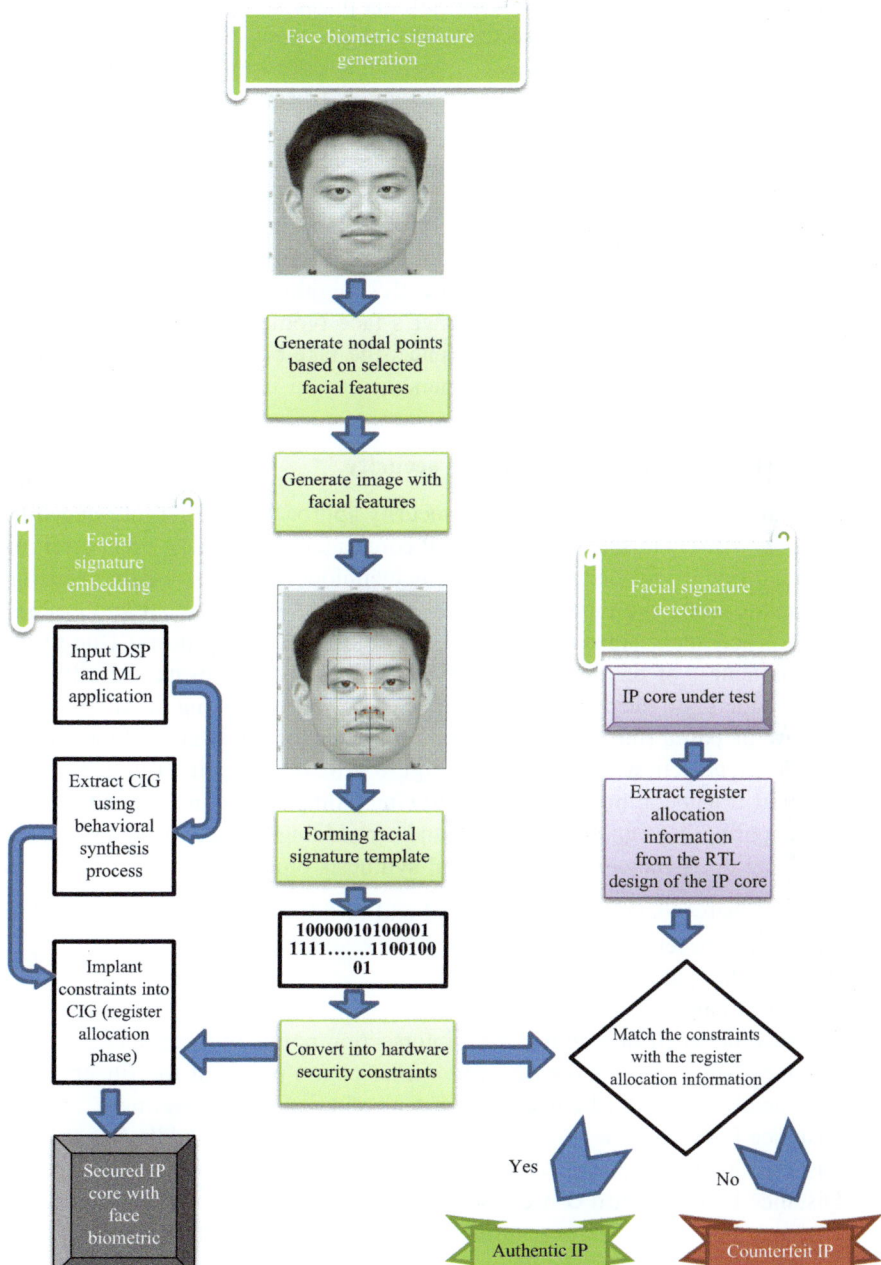

Figure 1.7   Facial biometric-based approach for hardware IP authentication/ counterfeit detection

digital template. In the facial signature embedding phase, the signature is converted into the corresponding hardware security constraints using the designer's developed mapping rules. Thus, obtained constraints corresponding to the facial biometric are embedded into the hardware design of IP cores such as DSP and ML co-processors. The embedding of constraints is performed during the register allocation phase of behavioral synthesis process. Thus, a secured DSP and ML co-processor design carrying the facial signature constraints can be generated. In the facial signature detection phase, a pre-stored facial image with the designer's selected facial features is used, thus making the face biometric-based authentication approach a contact-less technique. Figure 1.7 highlights the facial signature detection process. The presence of IP owner's face biometric constraints into the intended design proves its authenticity. Hence, the face biometric-based approach disables an attacker claiming the IP ownership due to its capability of offering inherent uniqueness.

(iii) **Palmprint biometric for hardware security**

The palmprint biometric-based hardware security approach was first introduced by Sengupta *et al.* (2021) to secure hardware IP cores against IP piracy, counterfeiting and false claim of IP ownership threats. This technique generates a palmprint signature template using the unique palmprint features of an individual and embeds into the co-processor designs during the behavioral synthesis process. Due to the inherent uniqueness of palmprint biometric, its respective hardware security constraints embedded into the designs are capable of distinctly proving the IP owner. A generic flow of palmprint biometric-based approach for hardware IP authentication/counterfeit detection is depicted in Figure 1.8. In the palmprint biometric-based secured IP generation phase, the following palmprint features are chosen for creating the palmprint signature template and embedding into the design using the register allocation framework of behavioral synthesis process:

1. Distance between start of life line and end of life line
2. Distance between datum points of head line and life line
3. Width of the palm
4. Length of palm
5. Distance between first consecutive intersection points of forefinger
6. Distance between second consecutive intersection points of forefinger
7. Distance between third consecutive intersection points of forefinger
8. Distance between first consecutive intersection points of middle finger
9. Distance between second consecutive intersection points of middle finger
10. Distance between third consecutive intersection points of middle finger
11. Distance between first consecutive intersection points of ring finger
12. Distance between second consecutive intersection points of ring finger
13. Distance between third consecutive intersection points of ring finger
14. Distance between first consecutive intersection points of little finger
15. Distance between second consecutive intersection points of little finger
16. Distance between third consecutive intersection points of little finger

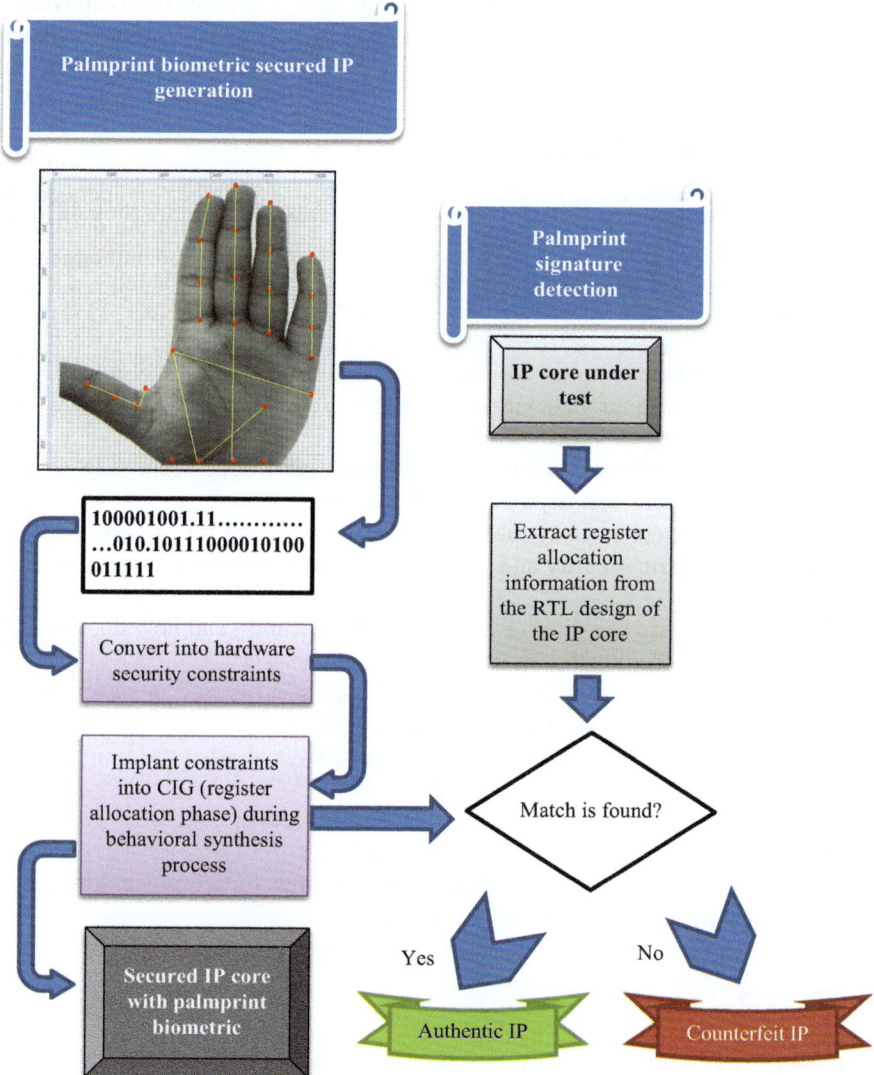

Figure 1.8 Palmprint biometric-based approach for hardware IP authentication/ counterfeit detection

17. Distance between first consecutive intersection points of thumb finger
18. Distance between second consecutive intersection points of thumb finger
19. Distance between starburst point and third intersection point of thumb.

In the palmprint biometric constraints detection phase, authenticity of the intended IP is verified using the process shown in Figure 1.8. Like the face biometric approach, the palmprint biometric is also a contact-less verification approach where a pre-stored palmprint image with the designer's selected features is used.

1.6   Questions and exercise

1. What are co-processors used in electronic systems and how are they different from general purpose processors?
2. What are DSP co-processors, their different types, and applications?
3. What are machine-learning co-processors, their different types, and applications?
4. Why are machine-learning co-processors required in modern systems?
5. What is the role of behavioral synthesis process in designing DSP and machine learning co-processors?
6. What is PSO-DSE and how is it important for generating secured IP cores?
7. What are different security threats to hardware IP cores?
8. What are different security mechanisms to protect hardware IP cores?
9. What is biometric security and how is it useful for personal or enterprise level security?
10. What is biometric security for hardware authentication and how does it differ from a user authentication system?
11. What is fingerprint biometric-based hardware security and what are unique fingerprint attributes used for hardware security?
12. What is face biometric-based hardware security and what are unique facial features used for hardware security?
13. What is palmprint biometric-based hardware security and what are unique palmprint features used for hardware security?
14. How is the biometric information embedded into the co-processor designs verified during the authentication process?

References

Bazrafkan, S., T. Nedelcu, P. Filipczuk, and P. Corcoran (2017), 'Deep learning for facial expression recognition: a step closer to a smartphone that knows your moods', in: *Proceedings of the ICCE*, Las Vegas, pp. 217–220.

Bhadauria, S. and A. Sengupta (2015), 'Adaptive bacterial foraging driven datapath optimization: exploring power-performance tradeoff in high level synthesis'. *Applied Mathematics and Computation*, vol. 269, pp. 265–278.

Everything you Need to Know About Hardware Requirements for Machine Learning (2019), https://www.einfochips.com/blog/everything-you-need-to-know-about-hardware-requirements-for-machine-learning/, last accessed on October 2022.

Hardware Accelerators for Machine Learning (2020), https://cs217.stanford.edu/, last accessed on October 2022.

Hardware for Machine Learning (2021), https://inst.eecs.berkeley.edu/~ee290-2/ sp21/, last accessed on October 2022.

Koushanfar, F. (2012), 'Hardware metering: a survey', in: *Introduction to Hardware Security and Trust*, New York, NY: Springer, https://doi.org/ 10.1007/978-1-4419-8080-9_5.

Koushanfar, F. and G. Qu (2001), 'Hardware metering', in: *Proceedings of the 38th Design Automation Conference*, pp. 490–493.

Koushanfar, F., I. Hong, and M. Potkonjak (2005), 'Behavioral synthesis techniques for intellectual property protection', *ACM Transactions on Design Automation of Electronic Systems*, vol. 10, no. 3, pp. 523–545.

Krishnan, V. and S. Katkoori (2006), 'A genetic algorithm for the design space exploration of datapaths during high-level synthesis', *IEEE Transactions on Evolutionary Computation*, vol. 10, no. 3, pp. 213–229.

Lao, Y. and K. K. Parhi (2015), 'Obfuscating DSP circuits via high-level transformations', *IEEE Transactions on Very Large Scale Integration (VLSI) Systems*, vol. 23, no. 5, pp. 819–830.

Le Gal, B. and L. Bossuet (2012), 'Automatic low-cost IP watermarking technique based on output mark insertions', *Design Automation for Embedded Systems*, vol. 16, no. 2, pp. 71–92.

Lemley, J., S. Bazrafkan, and P. Corcoran (2017), 'Deep learning for consumer devices and services: pushing the limits for machine learning, artificial intelligence, and computer vision', *IEEE Consumer Electronics Magazine*, vol. 6, no. 2, pp. 48–56.

Mahdavinejad, M. S., M. Rezvan, M. Barekatain, *et al.* (2018), 'Machine learning for internet of things data analysis: a survey', *Digital Communications and Networks*, vol. 4, no. 3, pp.161–175.

Mahdiany, H. R., A. Hormati and S. M. Fakhraie (2001), 'A hardware accelerator for DSP system design', in: *Proceedings of the ICM*, pp. 141–144.

McFarland, M. C., A. C. Parker, and R. Camposano (1988), 'Tutorial on high-level synthesis', in: *DAC '88 Proceedings of the 25th ACM/IEEE Design Automation*, vol. 27, no. 1, pp. 330–336.

Mishra, V. K. and A. Sengupta (2014), 'MO-PSE: adaptive multi-objective particle swarm optimization based design space exploration in architectural synthesis for application specific processor design', *Elsevier Journal on Advances in Engineering Software*, vol. 67, pp. 111–124.

Pilato, C., S. Garg, K. Wu, R. Karri and F. Regazzoni (2018), 'Securing hardware accelerators: a new challenge for high-level synthesis', *IEEE Embedded Systems Letters*, vol. 10, no. 3, pp. 77–80.

Rathor, M. and A. Sengupta (2019), 'Robust logic locking for securing reusable DSP cores', *IEEE Access*, vol. 7, pp. 120052–120064.

Rathor, M. and A. Sengupta (2020), 'IP core steganography using switch based key-driven hash-chaining and encoding for securing DSP kernels used in CE systems', *IEEE Transactions on Consumer Electronics*, vol. 66, no. 3, pp. 251–260.

Roy, J. A., F. Koushanfar, and I. L. Markov (2008), 'EPIC: ending piracy of integrated circuits', in: *Proceedings of the DATE*, Munich, pp. 1069–074.

Sengupta, A. (2016), 'Intellectual property cores: protection designs for CE products', *IEEE Consumer Electronics Magazine*, vol. 5, no. 1, pp. 83–88.

Sengupta, A. (2017), 'Hardware security of CE devices [hardware matters]', *IEEE Consumer Electronics Magazine*, vol. 6, no.1, pp. 130–133.

Sengupta, A. (2020), *Frontiers in Securing IP Cores – Forensic Detective Control and Obfuscation Techniques*, The Institute of Engineering and Technology (IET), ISBN-10: 1-83953-031-6, ISBN-13: 978-1-83953-031-9.

Sengupta, A. and S. Bhadauria (2016), 'Exploring low cost optimal watermark for reusable IP cores during high level synthesis', *IEEE Access*, vol. 4, pp. 2198–2215.

Sengupta, A. and R. Chaurasia (2022), 'Secured convolutional layer IP core in convolutional neural network using facial biometric', *IEEE Transactions on Consumer Electronics (TCE)*, vol. 68, no. 3, pp. 291–306.

Sengupta, A. and V. K. Mishra (2014), 'Automated exploration of datapath and unrolling factor during power–performance tradeoff in architectural synthesis using multi-dimensional PSO algorithm', *Expert Systems with Applications*, vol. 41, no. 10, pp. 4691–4703.

Sengupta, A. and S. P. Mohanty (2019), *IP Core Protection and Hardware-Assisted Security for Consumer Electronics*, The Institute of Engineering and Technology (IET), ISBN: 978-1-78561-799-7, e-ISBN: 978-1-78561-800-0.

Sengupta, A. and M. Rathor (2019a), 'IP core steganography for protecting DSP kernels used in CE systems', *IEEE Transactions on Consumer Electronics*, vol. 65, no. 4, pp. 506–515.

Sengupta, A. and M. Rathor (2019b), 'Protecting DSP kernels using robust hologram-based obfuscation', *IEEE Transactions on Consumer Electronics*, vol. 65, no. 1, pp. 99–108.

Sengupta, A. and M. Rathor (2020a), 'Securing hardware accelerators for CE systems using biometric fingerprinting', *IEEE Transactions on Very Large Scale Integration Systems*, vol. 28, no. 9, pp. 1979–1992.

Sengupta, A. and M. Rathor (2020b), 'Enhanced security of DSP circuits using multi-key based structural obfuscation and physical-level watermarking for consumer electronics systems', *IEEE Transactions on Consumer Electronics*, vol. 66, pp. 163–172, doi: 10.1109/TCE.2020.2972808.

Sengupta, A. and M. Rathor (2020c), 'HLS based IP protection of reusable cores using biometric fingerprint', *IEEE Letters of the Computer Society*, vol. 3, no. 2, pp. 42–45.

Sengupta, A. and M. Rathor (2021), 'Facial biometric for securing hardware accelerators', *IEEE Transactions on Very Large Scale Integration Systems (TVLSI)*, vol. 29, no. 1, pp. 112–123.

Sengupta, A., R. Sedaghat, and Z. Zeng (2010), 'A high level synthesis design flow with a novel approach for efficient design space exploration in case of multi-parametric optimization objective', *Microelectronics Reliability*, vol. 50, no. 3, pp. 424–437.

Sengupta, A., R. Sedaghat, and P. Sarkar (2012), 'A multi structure genetic algorithm for integrated design space exploration of scheduling and allocation in high level synthesis for DSP kernels', *Elsevier Journal of Swarm and Evolutionary Computation*, vol. 7, pp. 35–46.

Sengupta, A., D. Roy, S. P. Mohanty, and P. Corcoran (2017a), 'DSP design protection in CE through algorithmic transformation based structural obfuscation', *IEEE Transactions on Consumer Electronics*, vol. 63, no. 4, pp. 467–476.

Sengupta, A., S. Bhadauria, and S. P. Mohanty (2017b), 'TL-HLS: methodology for low cost hardware trojan security aware scheduling with optimal loop unrolling factor during high level synthesis', *IEEE Transactions on CAD of Integrated Circuits and Systems*, vol. 36, no. 4, pp. 655–668.

Sengupta, A., S. Rathlavat, and M. K. Naskar (2017c), 'A firefly algorithm driven approach for high level synthesis', in: *Proceedings of the IEEE International Symposium on Nanoelectronic and Information Systems*, pp. 15–19.

Sengupta, A., D. Roy, and S. P. Mohanty (2018a), 'Triple-phase watermarking for reusable IP core protection during architecture synthesis', *IEEE Transactions on Computer Aided Design of Integrated Circuits & Systems*, vol. 37, no. 4, pp. 742–755.

Sengupta, A., D. Roy, S. P. Mohanty, and P. Corcoran (2018b), 'Low-cost obfuscated JPEG CODEC IP core for secure CE hardware', *IEEE Transactions on Consumer Electronics*, vol. 64, no. 3, pp. 365–374.

Sengupta, A., D. Kachave, and D. Roy (2019a), 'Low cost functional obfuscation of reusable IP cores used in CE hardware through robust locking', *IEEE Transactions on Computer Aided Design of Integrated Circuits & Systems*, vol. 38, no. 4, pp. 604–616.

Sengupta, A., E. R. Kumar, and N. P. Chandra (2019b), 'Embedding digital signature using encrypted-hashing for protection of DSP cores in CE', *IEEE Transactions on Consumer Electronics, vol.* 3, pp. 398–407.

Sengupta, A., R. Chaurasia, and T. Reddy (2021), 'Contact-less palmprint biometric for securing DSP coprocessors used in CE systems', *IEEE Transactions on Consumer Electronics*, vol. 67, no. 3, pp. 202–213.

Spencer, B., O. Alfandi, and F. Al-Obeidat, (2019), 'Forecasting temperature in a smart home with segmented linear regression', *Procedia Computer Science*, vol. 155, pp. 511–518.

Struharik, R., B. Vukobratović, A. Erdeljan, and D. Rakanović (2018), 'CoNNA – compressed CNN hardware accelerator', in: *Proceedings of the DSD*, Prague, pp. 365–372.

Zhao, R., W. Luk, X. Niu, H. Shi, and H. Wang (2017), 'Hardware acceleration for machine learning', in: *Proceedings of the ISVLSI*, Bochum, 2017, pp. 645–650.

Chapter 2

Integrated defense using structural obfuscation and encrypted DNA-based biometric for hardware security

Anirban Sengupta[1] and Rahul Chaurasia[1]

This chapter describes a robust hardware security methodology capable of providing integrated defense using multi-layered structural obfuscation and encrypted deoxyribonucleic acid (DNA)-based biometric security (Sengupta and Chaurasia, 2022). The presented security methodology in this chapter enables the defense against the threats of register transfer (RT) level design modification by performing structural obfuscation. Additionally, it also provides detective defense control against intellectual property (IP) piracy using integrated encrypted DNA biometric security.

The organization of the chapter is as follows: Section 2.1 provides the introduction of the chapter; Section 2.2 highlights the background details of deoxyribonucleic acid (DNA)/genome sequencing; Section 2.3 presents the discussion and analysis of some of the major state of the art approaches; Section 2.4 explains the encryption process to encrypt DNA-based biometric signature as well as presents the integrated defense using structural obfuscation and encrypted DNA for hardware security; Section 2.5 shows the detection and validation of embedded encrypted DNA signature corresponding to the target register transfer level hardware design; Section 2.6 discusses the security properties and design cost of the discussed approaches; and Section 2.7 concludes the chapter.

2.1 Introduction

In the present technological era, remarkable advancements and innovations have led to manifestations in the form of smart/portable consumer electronics (CE) devices, smart cities, computing devices, smart health care systems, etc. All these innovations are playing a pivotal role in providing an end consumer with easy-to-use interface and adaptability of technology in almost every activity in our daily life. Therefore, to analyze all the aspects of these devices, it becomes crucial to understand their designing, security vulnerabilities, and reliability concerns.

[1]Department of Computer Science and Engineering, Indian Institute of Technology Indore, India

First, this chapter discusses the underlying hardware in these devices and their importance in terms of applications they perform. For example, in consumer electronics and computing devices, the underlying hardware is responsible for performing several crucial tasks ranging from image processing to audio/video processing. Furthermore, their usages in the field of health care, robotics, Internet of Things (IoT), and in mission critical applications cannot be overlooked. In all these aforementioned applications, underlying digital signal processing (DSP) hardware coprocessors play a crucial role. For example, joint picture expert group compression and decompression (JPEG Codec), discrete Fourier transform (DFT), fast Fourier transform (FFT), and discrete cosine transform (DCT) are used in tasks such as image and video compression, radar applications, digital video broadcasting, and audio and video compression. Moreover, digital filters such as finite impulse response filter (FIR), infinite impulse response filter (IIR), and image processing filters are used in audio processing devices, robotics vision, biometrics, and medical imagery. It is interesting to realize that all these applications are required to perform computationally intensive and data-intensive tasks. Therefore, it is realistic to design them as dedicated hardware coprocessors or reusable IP cores to accelerate the device performance with higher efficacy.

So far, we have conferred the underlying DSP hardware from the perspective of application and the need of designing them as dedicated hardware coprocessor. Next, we discuss design aspects of these hardware coprocessors. As a designer, to design these coprocessors, several orthogonal aspects are to be taken care of. For example, from the designer's perspective, design process should be less complex resulting into lesser turnaround time. Further, from the designer's point of view, the design cost should be cheaper but without compromising important functionality. Additionally, the deployed hardware design must be reliable and should not lead to any safety and security concerns. However, from the system integrator's perspective, it is a tiresome task to manage the complete design process of all hardware coprocessors single-handedly within a single company. This is because the design process offers too much complexity, extensive turnaround time, design cost, and time to market, if done all within a single company. Therefore, outsourcing of these hardware coprocessors to the third-party vendors is the common and acceptable practice in the industry. This is where, to accelerate the design turn-around time, it opens up several security vulnerabilities in the design chain as these third-party vendors may not be trust-worthy. Therefore, using secured (and genuine) intellectual property cores or hardware coprocessors is crucial.

Now, we look into the possible security threats during the design process of such dedicated hardware coprocessors. First, we discuss the main entities involved in the design cycle and their role during the design process; one is the IP vendor or designer, who is responsible to design the coprocessor. Second is the system on chip (SoC) integrator, responsible to integrate the designs. Third entity is the foundry or manufacturing houses responsible for creating fabricated chip of the final hardware design. Let us have a look at the possible security threats around all the three levels of design process. From a system integrators perspective, he/she needs to ensure that the imported IP core or coprocessor (from third party vendor), before being integrated in the system, is authentic. This is because fake/pirated IP cores may be unsafe and unreliable as they do not undergo rigorous quality checks. Therefore, the detection and isolation of such pirated IP core designs is very crucial to restrict their integration into SoCs. On the other hand, from an IP vendor's perspective, a SoC integrator may fraudulently claim the ownership of the IP core design supplied by the third-party vendor. Therefore, the protection of IP vendor's right is also necessary against such threat. Another aspect is an adversary in the foundry or fab who may overproduce the design without the consent (or knowledge) of system integrator. It is also possible that an adversary in the foundry (untrustworthy entity) may pirate the design in terms of counterfeiting and/or cloning. Therefore, it is evident there are several security vulnerabilities that exist in the hardware design chain.

This chapter mainly focuses on safeguarding the hardware IP or coprocessors used in underlying consumer electronics and computing systems, against the threats of reverse engineering and IP piracy. To achieve robust security against both these threats, coprocessor design structure is first made unobvious in terms of RT level structure through multi-level obfuscation process. This hinders an adversary in reverse engineering the design by identifying its design functionality and hardware architectural details. Subsequently, the generated encrypted DNA signature of an IP vendor is implanted into the IP core design during high-level synthesis (HLS) process. Using such DNA signature as a secret authentic mark, it ensures detective control against pirated hardware coprocessors before being integrated into the system (Sengupta and Chaurasia, 2022).

2.2   Background on DNA/genome sequencing

Deoxyribonucleic acid, also known as genome, is a molecule that contains the unique biological information that makes each species unique in the sense of characteristics and identification. DNA is comprised of chemical building blocks called nucleotides and each nucleotide is composed of three different components such as sugar, phosphate groups, and nitrogen bases. The sugar and phosphate groups link the nucleotides together to form each strand of DNA. The four chemical elements thymine, adenine, guanine, and cytosine are the four types of nitrogen bases. The nucleotides are joined together by covalent bonds between the phosphate of one nucleotide and the sugar of the next, forming a phosphate–sugar backbone "S" from which the nitrogenous bases protrude. The configuration of the DNA molecule is highly stable, allowing it to act as a template. In general, a DNA sequence comprises of two base pairs (BP): base pair (BP)-1, consisting of two chemical base elements: thymine as "T" and adenine as "A" and in base pair (BP)-2 the elements are guanine as "G" and cytosine as "C." The final structure contains sugar–phosphate backbone as leading and lagging strands. The order or sequence of these chemical elements is used for determining the instructions that are contained in a strand of DNA. Thus, different sequence orders represent different and unique information. This chapter discusses how a DNA sequence of an IP vendor can be exploited to act as a secret authentic mark for providing detective control against pirated IP core versions and nullifying false claim of IP ownership (Sengupta and Chaurasia, 2022).

2.3   State-of-the-art: discussion and analysis

In the literature, several techniques have been presented for securing hardware coprocessors or IP cores. Some of the important security techniques employ non-signature and signature-based hardware security. Among the signature-based approaches, some of them are non-biometric-based and some on bio-metric traits. Each approach has their own distinctive characteristics and security features. Hardware steganography is non-signature-based approach. Further, IP watermarking and hash-based digital signature are the non-biometric-based approaches. Additionally, biometric-based approaches exploit fingerprint and facial biometric features for hardware security. Now, we discuss upon each state-of-the-art approach and analyze their security strengths. Subsequently, we discuss how encrypted DNA-based biometric advances the security strength of hardware coprocessors compared to other hardware security approaches.

2.3.1   *Hardware steganography*

Sengupta and Rathor (2019) presented hardware steganography-based approach for detecting pirated DSP versions before being integrated into CE systems. This approach creates stego-constraints and implant as IP vendor's covert evidence into the hardware. However, the strength of steganography approach relies only on covert design data, stego-encoder, and encoding rules that may be compromised by an attacker. Furthermore, the produced stego-information (stego-keys) is usually not large and robust. Additionally, Sengupta and Rathor (2019) presented a crypto-based dual phase hardware steganography for securing IP cores. In this approach, first the stego constraints are generated using crypto modules and then they are implanted into the design covertly during register allocation phase and hardware allocation phase of high-level synthesis process. Both these approaches are dependent on auxiliary security parameters for providing security against piracy. Moreover, they do not provide security against reverse engineering attack. On the other hand, the encrypted DNA biometric approach (Sengupta and Chaurasia, 2022) presented in this chapter is not reliant on such exterior covert data that is subjected to compromise, but rather depends on encrypted DNA signature using large secret key, shift function, Feistel cipher iterations, information of round functions, genome sequence and strength, count/ordering of the inserted poly-nucleotides, nature of sequencing of DNA base pair, dual encoding and substitution box (S-Box) type. Further, the encrypted DNA signature approach (Sengupta and Chaurasia, 2022) integrates multi-level structural obfuscation to thwart reverse engineering, thereby ensuring the double line of defense in terms of IP piracy, fraud claim of IP ownership and reverse engineering-based attack (aiming to alter/tamper RT level description).

2.3.2   Hardware watermarking

Koushanfar *et al.* (2005) presented watermarking approach for intellectual property protection. Further, Le Gal and Bossuet (2012) presented an IP watermarking included in high-level synthesis. In this approach, watermark is based on mathematical relationships between numeric values as inputs and outputs. Furthermore, Sengupta and Bhadauria (2016) presented watermarking-based security that comprises of four different variables from IP vendor. Further, Sengupta *et al.* presented another watermarking technique with seven security variables. Roy and Sengupta (2019) presented a watermarking scheme combining different design abstraction levels such as high level and RT level. Additionally, Sengupta and Rathor (2020) presented a physical design-level-based watermarking and multi-key-based structural obfuscation approach for defense of DSP hardware. In this approach, watermarking comprises of three security variables and is embedded during floor-planning at physical level of the IP core design. In these hardware watermarking-based techniques, an attacker can compromise the signature information such as its variables, their combination/size, and encoding rules without significant efforts. However, the encrypted genome/DNA signature-based hardware security approach (Sengupta and Chaurasia, 2022) inserts unique robust chromosomal DNA impression (post-encryption) consisting of large security constraints in the IP core design. Hence, replicating and regenerating the encrypted chromosomal DNA mark is infeasible for an attacker owing to requiring several complex secret information such as shift functions, a large secret key, number of iterations of Feistel cipher, information of round functions, DNA base pair sequencing type, sequence and size of chromosomal DNA, count/ordering of the polynucleotide introduced, dual encoding rules, and type of S-Box. Therefore, the original chromosomal DNA impression cannot be regenerated by an attacker for evading piracy/counterfeiting detection process and is considered more secured than watermarking and steganography approaches.

2.3.3   Hash-based digital signature

Sengupta *et al.* (2019) presented a hardware security approach with embedded digital signature generated using encrypted-hashing for enabling the security of DSP cores for enabling their secured integration into consumer electronics systems. This security technique relies on secure hash algorithm (SHA)-512 hash block and uses the RSA private key value (512 bits) for producing the digitally signed hash value. This is subsequently inserted into the target design during architectural synthesis. The security provided by the encrypted hash-based digital signature approach may get compromised if encoding and RSA encryption key is compromised. Therefore, an attacker can regenerate the digital signature to evade the piracy detection process. On the contrary, Sengupta and Chaurasia (2022) present a methodology to embed unique encrypted DNA security mark of IP vendor into the structurally obfuscated design during high-level synthesis. Therefore, encrypted DNA-based digital signature (Sengupta and Chaurasia, 2022) is more secured and robust than normal hash-based technique (Sengupta *et al.*, 2019).

2.4   Integrated defense using structural obfuscation and encrypted DNA-based biometric for hardware security

In this sub-section, we discuss the process for generating encrypted DNA biometric secret signature for hardware security. As shown in Figure 2.1, the input block comprises of DNA biometric of IP vendor body sample, secret key (N-bit) for IP vendor selected encryption rounds. The output block generates encrypted genome/ DNA signature (of IP vendor selected strength) at the end of encryption process. First, we discuss how to exploit DNA sample of IP vendor for generating DNA

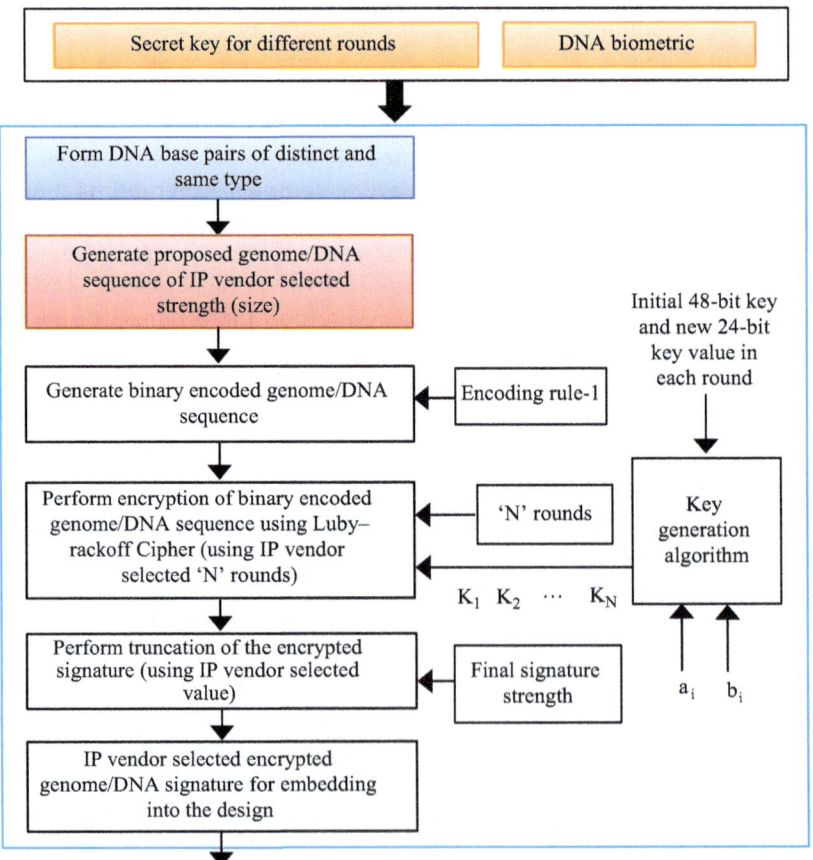

Figure 2.1   Process for generating encrypted binarized DNA signature

signature. Next, we understand the encryption mechanism to generate encrypted DNA signature. Subsequently, we highlight, how the IP vendor selected DNA signature is transformed into secret hardware security constraints followed by multi-level of encoding for embedding into hardware coprocessor design. Further, the encrypted DNA signature approach integrates multi-level structural obfuscation to thwart reverse engineering, thereby ensuring the double line of defense in terms of IP piracy, fraud claim of IP ownership and reverse engineering-based attack (aiming to alter/tamper RT level description). As shown in Figure 2.2, the input block comprises of module library, sample DSP coprocessor and designer-specified resource constraints. The output block generates structurally obfuscated scheduled hardware design followed by hardware allocation and binding. Finally, post embedding encrypted DNA constraints, secure RT level design corresponding to hardware coprocessor is obtained using high-level synthesis (Sengupta and Chaurasia, 2022). The overview of hardware security methodology with integrated defense using structural obfuscation and encrypted DNA signature is shown in Figure 2.3. This offers integrated defense using two security layers (1) security against RT-Level alteration by obfuscating the structure of design (without causing a change in functionality) and (2) detective control against IP piracy using encrypted DNA signature of IP designer. The corresponding details of hardware

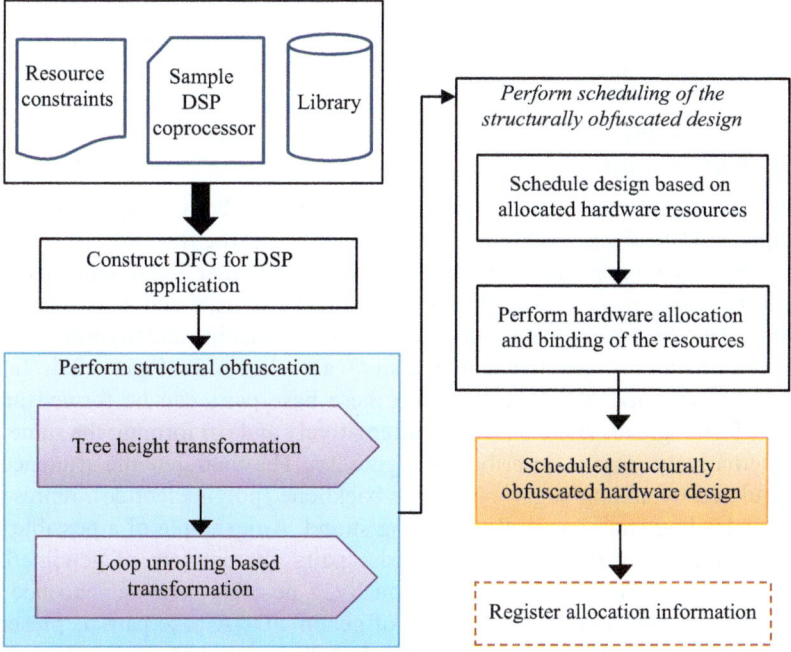

Figure 2.2   Process for generating structurally obfuscated scheduled design using HLS

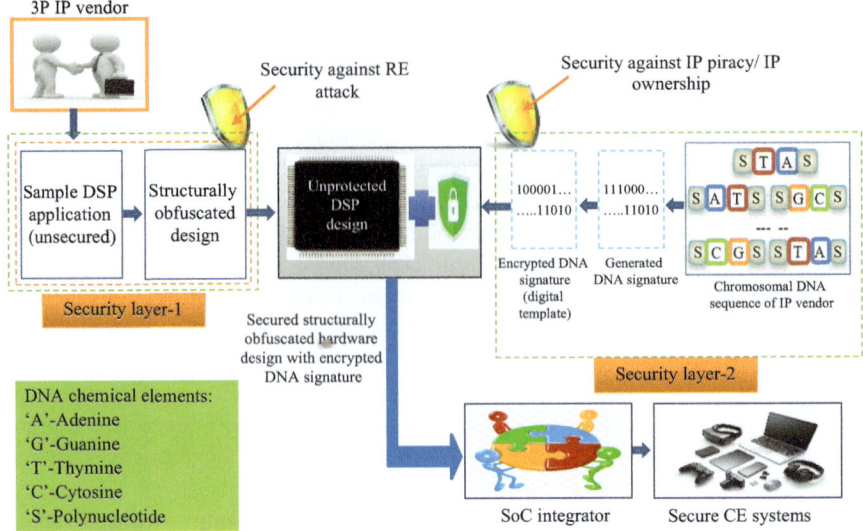

Figure 2.3   Overview of integrated defense for securing hardware coprocessors using structural obfuscation and encryption DNA-based biometric methodology

security methodology (Sengupta and Chaurasia, 2022) have been discussed in subsequent subsections.

2.4.1   Extracting DNA signature from IP vendor body sample

The process of extracting DNA signature from IP vendor body sample is as follows (Sengupta and Chaurasia, 2022): as shown in Figure 2.1, the first block is meant for formulating DNA base pairs. As we discussed earlier that DNA sequence comprises of two base pairs of four different chemical elements and sugar–phosphate backbone. Two base pairs (base pair-1 comprising of adenine and thymine and base pair-2 comprising of guanine and cytosine) are shown in Figure 2.4. In this approach, a genome/DNA sequence with these base pairs can be formed in two ways: (a) forming distinctive base pairs alternatively and (b) forming the same base pairs alternatively. This is described in Figure 2.5. The final genome sequence can be formulated by adding sugar–phosphate backbone (polynucleotide), represented as "S," as leading strand as well as lagging strand. An example of a possible final genome sequence by taking distinctive base pairs alternatively of genome/DNA base pairs is shown in Figure 2.6. Additionally, a possible genome sequence with alternative base pairing of the same type of genome/DNA base pairs is presented, as shown in Figure 2.7. Therefore, several different and unique genome/DNA sequences are possible. Thus, a final genome/DNA sequence of IP vendor chosen size (strength) is selected after the execution of the second block, for generating

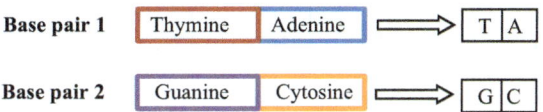

Figure 2.4   DNA base pair generation

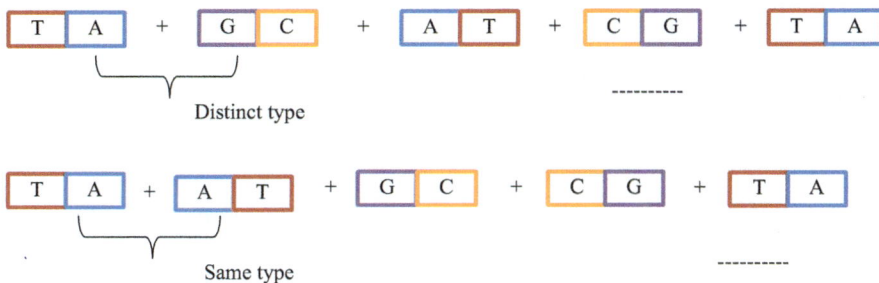

Figure 2.5   Type of DNA base pairs

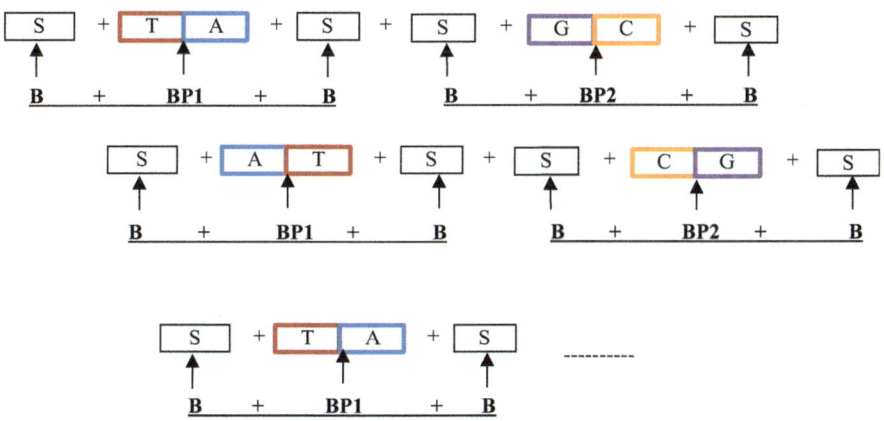

Figure 2.6   Sample genome/DNA sequence by forming the base pairs of distinct type and polynucleostride alternatively

binarized DNA signature in the next block. Next, the third block is accountable for producing the binary-encoded genome/DNA sequence (using IP vendor-specified encoding rule-1). Let us assume that DNA sequence from IP vendor body sample, followed by IP vendor-selected final genome/DNA sequence (by considering DNA base pairs of distinct type) is as follows:

STAS-SGCS-SATS-SCGS-STAS-SGCS-SATS-SCGS-STAS-SGCS-SATS-SCGS

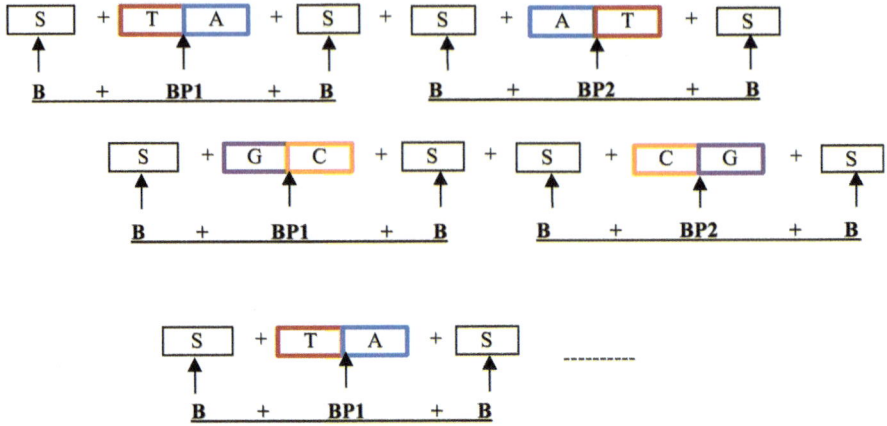

Figure 2.7   Sample genome/DNA sequence by forming the base pairs of the same type and polynucleostride alternatively

Table 2.1   Encoding rule-1

Encoding element	Alphabet value	Encoded binary representation
S	19	10011
A	1	1
T	20	10100
G	7	111
C	3	11

Now, we generate its equivalent binarized signature. To derive binarized-encoded genome/DNA sequence, the methodology employs the following encoding rule (rule-1) as shown in Table 2.1. In the encoding Table 2.1, the alphabet values represent the position number of the alphabets in their alphabetic order. Since the value of alphabets A to Z is from 1 to 26, the alphabets "S," "A," "T," "G," and "C" used in the approach are encoded as 19, 1, 20, 7, and 3, respectively. The impact of using these alphabets in this hardware security approach is in terms of generating binary-encoded DNA signature. The coding rule-1 is applicable for all the chemical components (A, T, C, G, S) used in the DNA sequence. Therefore, the corresponding binary-encoded DNA signature (considering 128-bit size) formed using encoding rule -1 is shown below:

$$1001110100110011 - 100111111110011 - 1001111010010011 -$$
$$100111111110011 - 1001110100110011 - 100111111110011 - \qquad (2.1)$$
$$1001111010010011 - 100111111110011 - 1001.$$

Similarly, we may generate the DNA sequence corresponding to alternating genome/DNA base pairs of the same type. Assuming, final DNA sequence, followed by IP vendor-selected sequence length is as follows:

STAS-SATS-SGCS-SCGS-STAS-SATS-SGCS-SCGS-STAS-SATS-SGCS-SCGS

Therefore, the corresponding binary-encoded DNA impression (e.g., 128 bit), formed using encoding rule-1, is shown below:

$$1001110100110011 - 1001111010010011 - 100111111110011 -$$
$$100111111110011 - 1001110100110011 - 1001111010010011 - \qquad (2.2)$$
$$100111111110011 - 100111111110011 - 1001$$

In the next block, this binarized DNA signature is accepted as input of multi-iteration Feistel cipher process for generating encrypted DNA signature. The details of the Feistel cipher encryption to generate encrypted DNA signature are discussed in the next subsection.

The advantages of the encrypted DNA signature method (Sengupta and Chaurasia, 2022) over a random key generation are: (a) enables stronger security and (b) renders higher robustness. Further, it is more unlikely for an adversary to replicate the encrypted chromosomal DNA imprint during ownership proof and evading piracy detection phases because of requiring several complex information such as huge secret key, shift functions, Feistel cipher iterations, count of round functions, sequencing nature of DNA base pair, chromosomal DNA sequence and size, ordering/count of the polynucleotide put in, dual encoding rules, and type of S-Box. Hence, for an adversary, it becomes impossible to exactly regenerate the original chromosomal DNA mark for evading the detection of pirated IP versions and successfully claiming fraudulent ownership.

2.4.2 Encryption of the DNA signature using DES algorithm

To generate encrypted DNA signature, the generated binary-encoded genome/DNA sequence (as shown in Section 2.4.1) is encrypted using the multi-iteration Feistel cipher process based on the IP vendor chosen "N" rounds and key generation algorithm (Sengupta and Chaurasia, 2022). The key generation algorithm accepts 48-bit initial key specified by IP vendor and produces new 48-bit key in each round ("I") that internally depends on the 24-bit secret keys for the different rounds and shift function bits "a_i" and "b_i," respectively. The encryption mechanism and the key generation algorithm are shown in Figures 2.8 and 2.9 respectively. The generated binary-encoded genome/DNA sequence of 128 bits (considering (2.1)) is fed into the multi-iteration Feistel cipher based on the IP vendor's selected "N" rounds for encryption (as shown in Figure 2.8). The 128-bit genome/DNA sequence is then alienated into two 64-bit sequences and fed into the cipher in two iterations. In the initial round, the first 64-bit binarized output of genome/DNA signature has been divided into two equal parts as "L" and "R" of 32 bits each, where L and R indicate the left and the right part, respectively. Then the right part is fed into the encryption function capable of performing diffusion/permutation (using expanded P-box) and

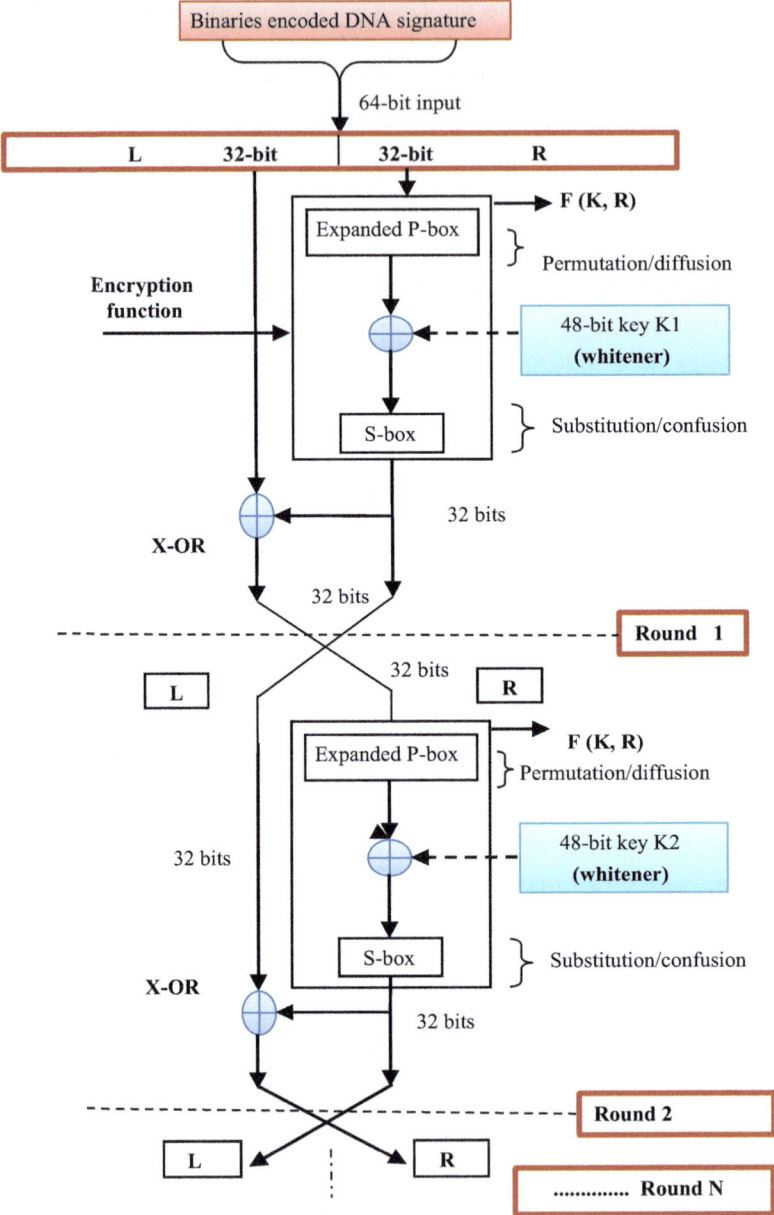

Figure 2.8   Feistel cipher process

substitution/confusion (using S-box). The structure of an instance S5 of S-boxes (S1–S8) used in this approach is shown in Figure 2.10, where "xx" indicates the outer bits and "yyyy" indicates the middle bits in the input "xyyyyx" of the S-box. Initially in the encryption function, the input "R" (32 bit) is transformed into 48-bit

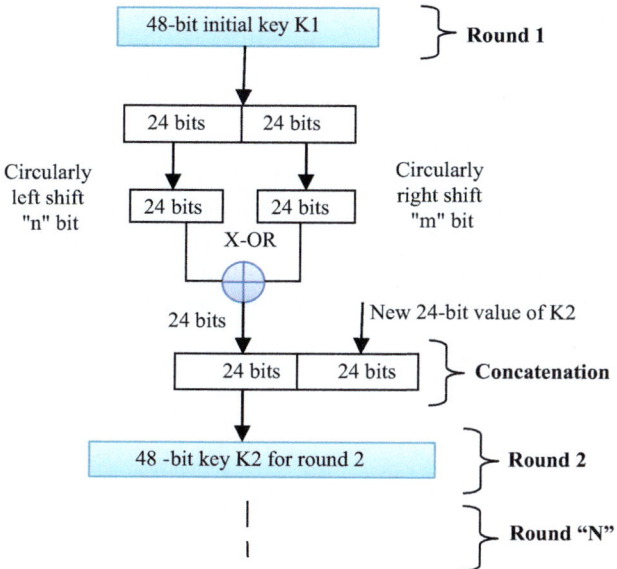

Figure 2.9   Key generation process

S₅	×00 00×	×00 01×	×00 10×	×00 11×	×01 00×	×01 01×	×01 10×	×01 11×	×10 00×	×10 01×	×10 10×	×10 11×	×11 00×	×11 01×	×11 10×	×11 11×
0yy yy0	2	12	4	1	7	10	11	6	8	5	3	15	13	0	14	9
0yy yy1	14	11	2	12	4	7	13	1	5	0	15	10	3	9	8	6
1yy yy0	4	2	1	11	10	13	7	8	15	9	12	5	6	3	0	14
1yy yy1	11	8	12	7	1	14	2	13	6	15	0	9	10	4	5	3

Figure 2.10   Structure of the S-box (S₅) used in the approach (Sengupta and Chaurasia, 2022)

output using the expanded "P"-box. Then the output 48 bits is subsequently XOR-ed with the 48-bit secret key (K1) of the initial round. Each round of the cipher generates the encrypted output based on a 48-bit key created by the IP vendor. Similarly based on the rounds (N) chosen by the IP vendor, different 48-bit keys

(K1, K2 . . . KN) will be used. The key generation process is described in Figure 2.9. In the key generation mechanism, first, the IP vendor chosen 48-bit key value is divided into two 24-bit sequence followed by performing shift function. The shift bits ("a_i" and "b_i") chosen by an IP vendor in each round "I" are responsible for performing circular left shift and circular right shift, respectively, on each 24-bit sub-divided key value from the previous round, before being X-ORed. The output of the X-OR is a 24-bit value which is concatenated with a new 24-bit value, chosen by the IP vendor in the current round. This results into a new 48-bit key value of the next round. This process continues till IP vendor-selected round "N" of the encryption process. Figure 2.11 shows an example of the key generation

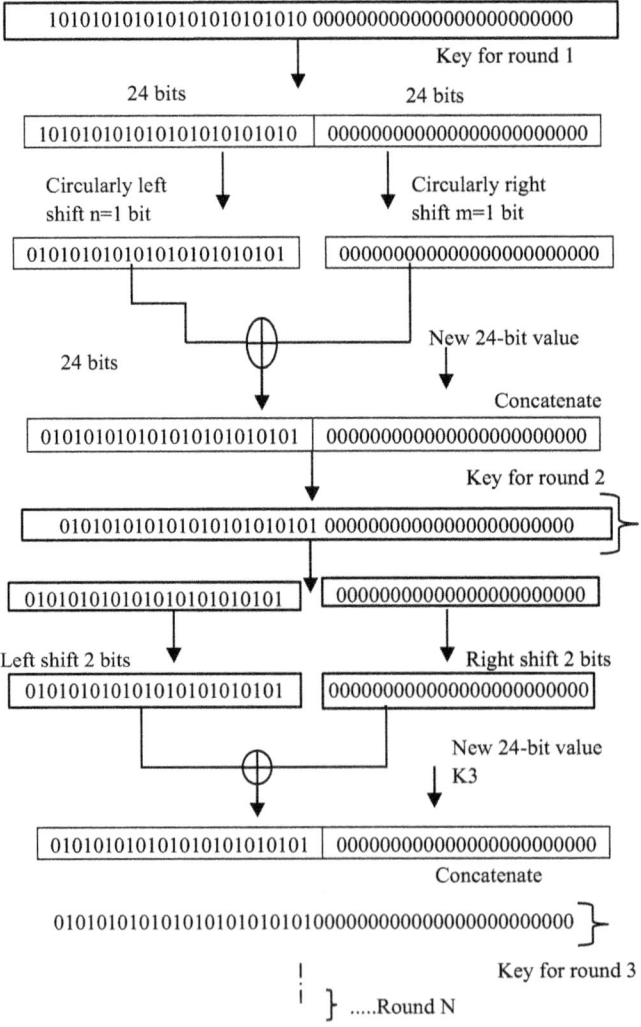

Figure 2.11   Demonstration of an example of a key generation process

process through demonstration till round 3. Additionally, a demonstrative example (input and output) of the Feistel cipher encryption process corresponding to round 3 (based on the original input of the first 64-bit DNA signature in (2.1)) is shown in Figure 2.12.

Thus, the final obtained encrypted DNA signature (considering 128-bit size corresponding to binarized DNA signature in (2.1)) is as follows:

"0010010101100000010100111101110110101110110101111011100100000110
0100100011101110000010010010101110001001010010001000100101101100
0001000100101101100"

(2.3)

The produced encrypted DNA signature (after the execution of N encryption rounds) is then fed to the next block for further truncation using the IP vendor's selected value (final signature strength). Finally, the obtained signature is used for embedding into the target design. This embedded secret encrypted DNA signature acts as digital evidence during piracy detection process (Sengupta and Chaurasia, 2022).

Now, we discuss the security features of Feistel cipher encryption. The security features of the cipher integrated into design are as follows:

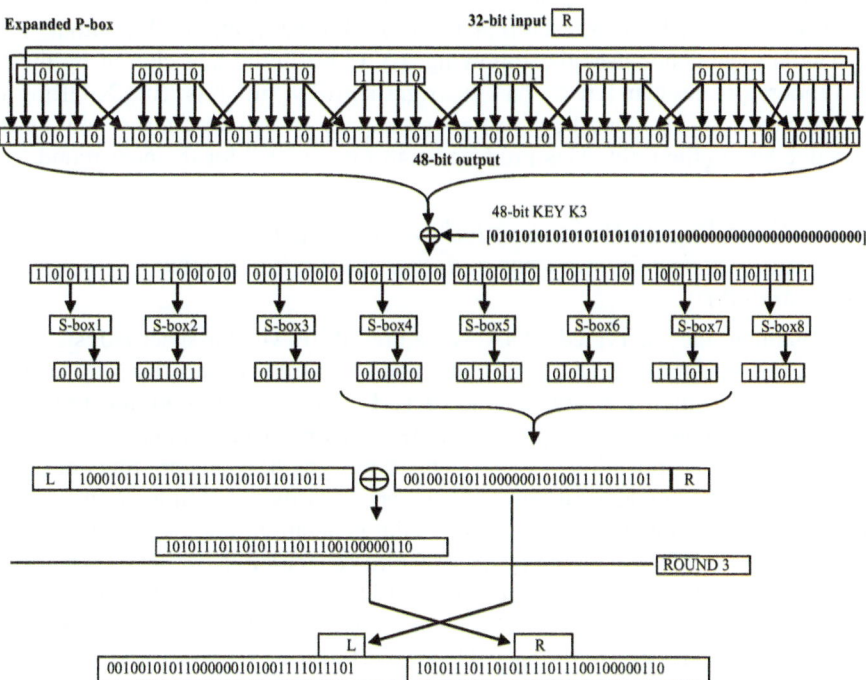

Figure 2.12   Demonstration (input and output) of the Feistel cipher encryption process in round 3 based on the original input of the first 64-bit DNA signature

1.   The encryption process exploits the Feistel structure comprising of multiple processing rounds of the plaintext (input binary encoded DNA signature), where each round consists of a substitution step followed by a permutation step. The diffusion process at the completion of each round swaps between the modified "L" and the original "R." Therefore, it results the "R" of the current round as "L" for the next round and "L" of the current round as "R" for the next round. This confusion and diffusion steps correspond to a "round." Further, the repetition of rounds is decided by the designer. After the completion of last round, the both sub-blocks, "R" and "L" are concatenated for formulating the Cipher data block. These two properties make the cipher robust against: (i) avalanche effect—a small change in plaintext results into sever modification in the Cipher data (encrypted DNA signature); (ii) completeness—each bit of Cipher data depends on many bits of the plaintext (binary-encoded DNA signature).

2.   The round function of the cipher uses the following: (i) expansion of diffusion box—since "R" is 32-bit and size of round key is a 48-bit, therefor first expansion of right input to 48 bits is performed; (ii) XOR (whitener)—after the diffusion, XOR operation on the expanded right section and the round key id performed in data encryption standard (DES) algorithm. (iii) Substitution boxes—the S-boxes carry out the real amalgamation (confusion). The number of S-boxes in DES is 8, each transforms a 6-bit input into a 4-bit output.

3.   The key generator uses the following: (i) 48-bit initial key chosen by the IP vendor; (ii) circular left by "n" bits/right shift by "m" bits; (iii) X-ORing to produce a 24-bit value key; (iv) appending an IP vendor chosen new 24-bit value key to the right. This produces a 48-bit key "K" for the next round.

2.4.3   Encoding of the encrypted DNA signature for conversion into secret constraints for hardware security

Post obtaining the encrypted DNA signature of IP vendor-specified signature strength, it is transformed into secret constraints for hardware security. The security constraints generation process (as shown in Figure 2.13) accepts the following inputs: (a) encrypted DNA signature of IP vendor selected signature strength; (b) target DSP application as its data flow graph/ transfer function; and (c) encoding rule, for generating the hardware security constraints as its output. Among these three inputs of security constraint generation process, so far, we have discussed the steps for encrypted DNA signature generation. Now, let us discuss the process of generating structurally obfuscated DFG design. To obtain structurally obfuscated DFG design, multi-level transformations based on tree height transformation (THT) and loop unrolling (LU) are performed (Sengupta and Chaurasia, 2022). The obfuscation is achieved by making the design structure unobvious or hard to interpret in terms of functionality/interconnectivity for an adversary to thwart reverse engineering. Structural obfuscation obscures the actual architecture of hardware design into non-interpretable one without compromising its original

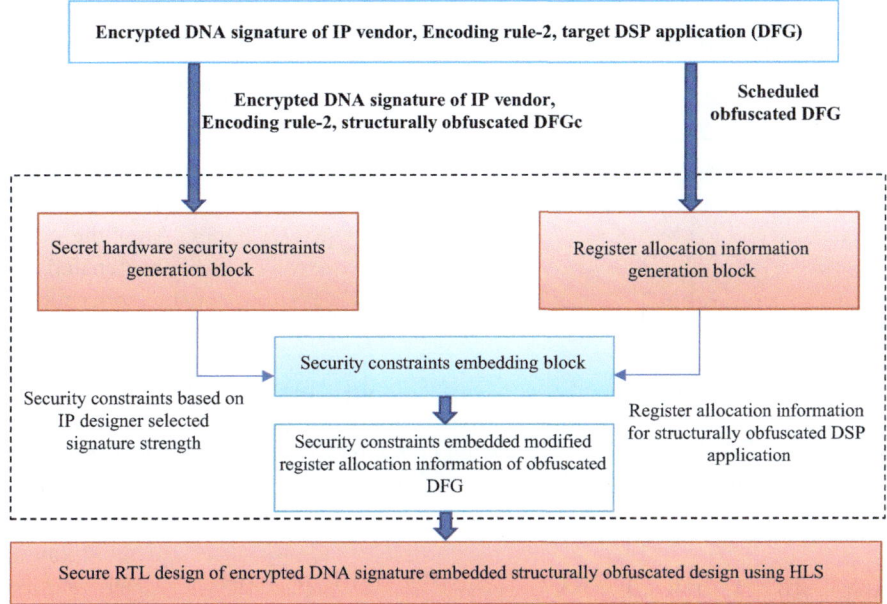

Figure 2.13   Process for generating secured RT level design post embedding encrypted DNA signature into structurally obfuscation design

functionality. For the sake of demonstration, DFG of 4-point DFT application capable of computing parallel outputs has been taken as an input to obtain structurally obfuscated design. The input DFG of 4-point DFT hardware (non-obfuscated) is shown in Figure 2.14. Next, the input DFG is exploited to perform THT-based transformation. Figure 2.15 shows the tree height-transformed design corresponding to input DFG shown in Figure 2.14. In THT transformation, sequential execution flow (leading to critical path of the design) in the graph is transformed to parallel sub-computation without compromising the original design functionality. For example, in the DFG computing first output $Y[0]$, among the sequential operations (7,9,11) of input DFG (Figure 2.14), operations 7 and 9 are transformed into parallel sub-computation as shown in Figure 2.15. Similarly, corresponding to DFG computing first output Y (Sengupta and Chaurasia, 2022), among the sequential operations (8,10,12) of input DFG (Figure 2.14), operations 8 and 10 are transformed into parallel sub-computation, as shown in Figure 2.15. Obfuscation based on THT structurally changes the interconnectivity of the RT level datapath of the DSP hardware in terms of #multiplexers and their size, #demultiplexers and their size and storage element, etc. Thus, obscuring the internal structural of the hardware design without compromising its actual func-tionality. This therefore hinders an adversary from reverse engineering the hard-ware design by identifying its true functionality and architectural details. Thus, THT-based structurally obfuscated design is obtained as shown in Figure 2.15 (Sengupta and Chaurasia, 2022).

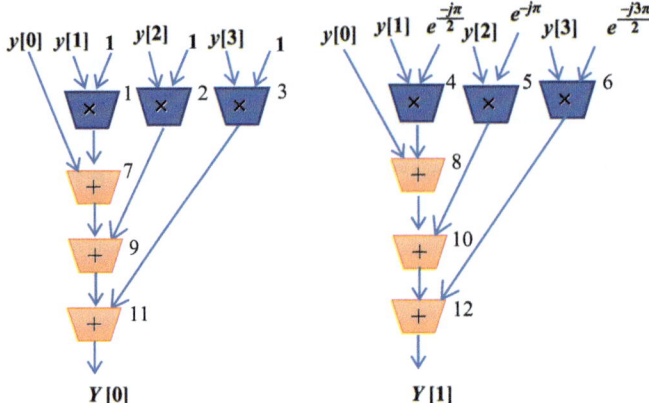

Figure 2.14   *DFG of DFT 4-point design computing two samples concurrently (Sengupta and Chaurasia, 2022)*

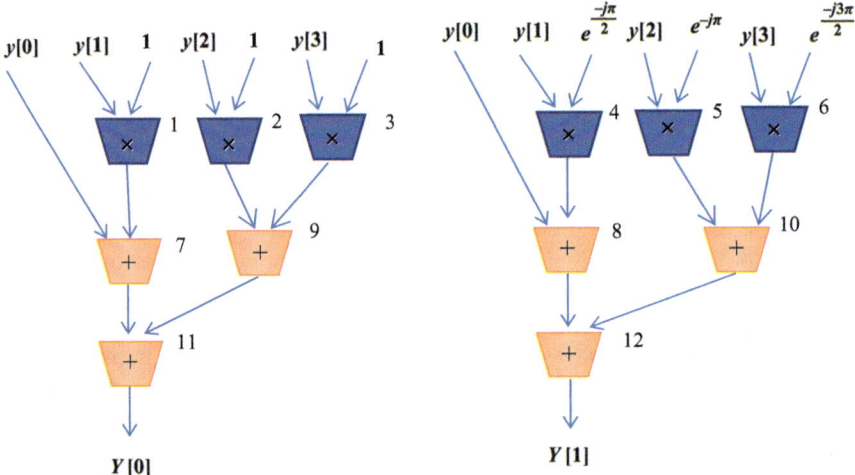

Figure 2.15   *Structurally transformed design of 4-pint DFT hardware (Sengupta and Chaurasia, 2022)*

In the loop unrolling-based transformation, the corresponding DFG is unrolled as per the value of unrolling factor. LU executes the same functionality multiple times (same as presented inside the loop). This unrolling of DFG design manifests into structurally obfuscated design. Subsequently the scheduled design of obfuscated DFG is shown in Figure 2.16. LU-based structural obfuscation introduces changes in the interconnectivity of the RT level datapath of the DSP hardware. Thus, LU enables the security of the design against the threat of reverse engineering (Sengupta and Chaurasia, 2022).

Further, IP vendor-specified encoding rules to generate the security constraints corresponding to encrypted DNA signature are shown in Figure 2.17, where a security constraint ($<Zi, Zj>$) is generated corresponding to each of the IP designer-selected encrypted DNA signature bit (0 and 1); where "i" and "j" represent the storage variables (Z1 to Z26) corresponding to scheduled structurally obfuscated scheduled design, as shown in Figure 2.16. Thus, secret hardware security constraints corresponding to IP designer selected encrypted DNA signature (128-bit size, shown in (2.3)) are generated using scheduled structurally obfuscated design (shown in Figure 2.16) and IP designer-specified encoding rules (shown in Figure 2.17). The resulting DNA constraints are shown in Tables 2.2 and 2.3, where Tables 2.2 and 2.3 present the secret constraints for hardware security corresponding to signature bit "0" and "1" respectively.

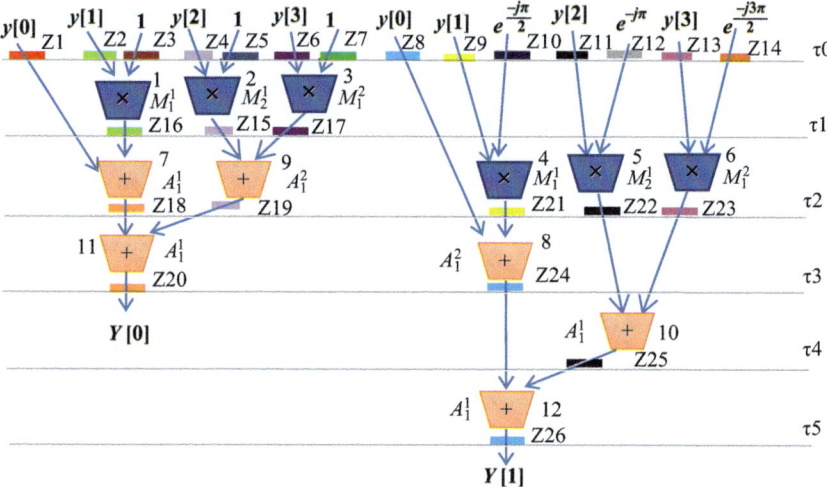

Figure 2.16   Scheduled DFG of structurally transformed DFT 4-point based on resource constraints 3 multipliers (M) and 2 adders (A) (Sengupta and Chaurasia, 2022)

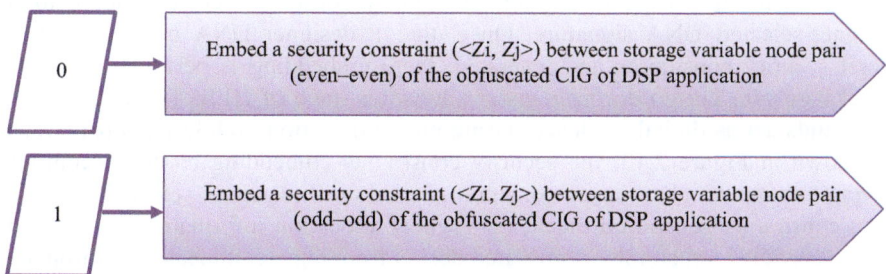

Figure 2.17   Encoding rule for hardware security constraints generation (Sengupta and Chaurasia, 2022)

Table 2.2   Encrypted DNA signature-based secret hardware security constraints corresponding to signature bit "0"

<Z2, Z4>	<Z4, Z6>	<Z6, Z10>	<Z8, Z16>	<Z10, Z24>	<Z14, Z22>
<Z2, Z6>	<Z4, Z8>	<Z6, Z12>	<Z8, Z18>	<Z10, Z26>	<Z14, Z24>
<Z2, Z8>	<Z4, Z10>	<Z6, Z14>	<Z8, Z20>	<Z12, Z14>	<Z14, Z26>
<Z2, Z10>	<Z4, Z12>	<Z6, Z16>	<Z8, Z22>	<Z12, Z16>	<Z16, Z18>
<Z2, Z12>	<Z4, Z14>	<Z6, Z18>	<Z8, Z24>	<Z12, Z18>	<Z16, Z20>
<Z2, Z14>	<Z4, Z16>	<Z6, Z20>	<Z8, Z26>	<Z12, Z20>	<Z16, Z22>
<Z2, Z16>	<Z4, Z18>	<Z6, Z22>	<Z10, Z12>	<Z12, Z22>	<Z16, Z24>
<Z2, Z18>	<Z4, Z20>	<Z6, Z24>	<Z10, Z14>	<Z12, Z24>	<Z16, Z26>
<Z2, Z20>	<Z4, Z22>	<Z6, Z26>	<Z10, Z16>	<Z12, Z26>	<Z16, Z20>
<Z2, Z22>	<Z4, Z24>	<Z8, Z10>	<Z10, Z18>	<Z14, Z16>	–
<Z2, Z24>	<Z4, Z26>	<Z8, Z12>	<Z10, Z20>	<Z14, Z18>	–
<Z2, Z26>	<Z6, Z8>	<Z8, Z14>	<Z10, Z22>	<Z14, Z20>	–

Table 2.3   Encrypted DNA signature-based secret hardware security constraints corresponding to signature bit "1"

<Z1, Z3>	<Z3, Z5>	<Z5, Z9>	<Z7, Z15>	<Z9, Z23>	–
<Z1, Z5>	<Z3, Z7>	<Z5, Z11>	<Z7, Z17>	<Z9, Z25>	–
<Z1, Z7>	<Z3, Z9>	<Z5, Z13>	<Z7, Z19>	<Z11, Z13>	–
<Z1, Z9>	<Z3, Z11>	<Z5, Z15>	<Z7, Z21>	<Z11, Z15>	–
<Z1, Z11>	<Z3, Z13>	<Z5, Z17>	<Z7, Z23>	<Z11, Z17>	–
<Z1, Z13>	<Z3, Z15>	<Z5, Z19>	<Z7, Z25>	<Z11, Z19>	–
<Z1, Z15>	<Z3, Z17>	<Z5, Z21>	<Z9, Z11>	<Z11, Z21>	–
<Z1, Z17>	<Z3, Z19>	<Z5, Z23>	<Z9, Z13>	<Z11, Z23>	–
<Z1, Z19>	<Z3, Z21>	<Z5, Z25>	<Z9, Z15>	<Z11, Z25>	–
<Z1, Z21>	<Z3, Z23>	<Z7, Z9>	<Z9, Z17>	<Z13, Z15>	–
<Z1, Z23>	<Z3, Z25>	<Z7, Z11>	<Z9, Z19>	<Z13, Z17>	–
<Z1, Z25>	<Z5, Z7>	<Z7, Z13>	<Z9, Z21>	–	–

2.4.4   Embedding of the secret DNA signature into design

So far, we discussed the generation of security constraints corresponding to IP designer-selected DNA signature. Once, the IP designer DNA biometric-driven secret security constraints are generated, there embedding is performed into the target hardware design during register allocation stage of HLS. These embedded constraints act as digital evidence during piracy detection and design verification. As shown in Figure 2.13, the security constraints embedding block accepts following two inputs: (1) security constraints based on IP designer-selected signature from constraints generation block; (2) register allocation information corresponding to scheduled structurally obfuscated design from register allocation information block. Therefore, now we discuss the process for generating the register allocation information. To do so, first, the structurally transformed design (shown in Figure 2.15) is subjected to scheduling, hardware allocation, and binding phase of

HLS process as shown in Figure 2.2. The design is scheduled on the basis of IP designer-specified resource constraints (two adders and three multipliers) and available dependency information of operational nodes. LIST scheduling algorithm has been used for the same (Sengupta and Chaurasia, 2022). This results into the scheduled structurally obfuscated design as shown in Figure 2.16. As evident from the figure, scheduled 4-point DFT design consumes total six control steps to compute parallel outputs. This scheduled and structurally obfuscated design is exploited to generate register allocation information. The extracted register allocation information corresponding to structurally obfuscated and scheduled 4-point DFT design, is shown in Table 2.4, where "Z0 to Z26" represents storage variables of the design that are assigned among the registers "R^1 to R^{14}" depending on their dependency information and liveness property. Further, "$\tau0$ to $\tau5$" are six control steps required to schedule the structurally obfuscated design based on IP designer-specified resource constraints.

So far, we discussed the process for obtaining the secret hardware security constraints and register allocation information. Now, let us discuss the implantation of security constraints into the target design (as shown in Figure 2.16). To implant the secret DNA biometric security constraints into the target design, following rule is employed: both the storage variables from the security constraint pair cannot be assigned to the same register. The following rule is useful to avoid the timing overlap of assignment of two storage variables to a single register at the same time. Further, in case if the conflict occurs, then the next available register is assigned to the conflicting variable. However, if it is not possible to accommodate a conflicting variable with any of the available register of the design, then a new register is added to the design to accommodate the conflicting storage variable. Thus, the register allocation information after embedding the secret hardware security constraints is shown in Table 2.5. Where the variables marked in red represent the local alterations performed during the embedding of security constraints. As evident from Table 2.5, no extra registers were required to embed all the available generated constraints for hardware security corresponding to encrypted DNA signature. Thus, post embedding all the encrypted DNA constraints, a secure hardware coprocessor design is generated. Next, we formulate the multiplexing and de-multiplexing scheme for functional units (adders and multiplier) and registers used in the

Table 2.4   Register allocation corresponding to structurally obfuscated 4-point DFT design (before implanting encrypted DNA constraints)

CS	R^1	R^2	R^3	R^4	R^5	R^6	R^7	R^8	R^9	R^{10}	R^{11}	R^{12}	R^{13}	R^{14}
$\tau0$	Z1	Z2	Z3	Z4	Z5	Z6	Z7	Z8	Z9	Z10	Z11	Z12	Z13	Z14
$\tau1$	Z1	Z15	–	Z16	–	Z17	–	Z8	Z9	Z10	Z11	Z12	Z13	Z14
$\tau2$	Z18	–	–	Z19	–	–	–	Z8	Z21	–	Z22	–	Z23	–
$\tau3$	Z20	–	–	–	–	–	–	Z24	–	–	Z22	–	Z23	–
$\tau4$	–	–	–	–	–	–	–	Z24	–	–	Z25	–	–	–
$\tau5$	–	–	–	–	–	–	–	Z26	–	–	–	–	–	–

Table 2.5   Register allocation corresponding to structurally obfuscated 4-point DFT design (after implanting encrypted DNA constraints)

CS	R^1	R^2	R^3	R^4	R^5	R^6	R^7	R^8	R^9	R^{10}	R^{11}	R^{12}	R^{13}	R^{14}
$\tau 0$	Z1	Z2	Z3	Z4	Z5	Z6	Z7	Z8	Z9	Z10	Z11	Z12	Z13	Z14
$\tau 1$	Z1	Z15	–	–	Z16	Z17	–	Z8	Z9	Z10	Z11	Z12	Z13	Z14
$\tau 2$	Z18	–	–	Z19	–	–	–	Z8	–	Z21	Z22	–	Z23	–
$\tau 3$	–	–	Z20	–	–	–	–	–	Z24	–	Z22	–	Z23	–
$\tau 4$	–	–	–	–	–	–	–	–	Z24	–	–	Z25	–	–
$\tau 5$	–	–	–	–	–	–	–	–	Z26	–	–	–	–	–

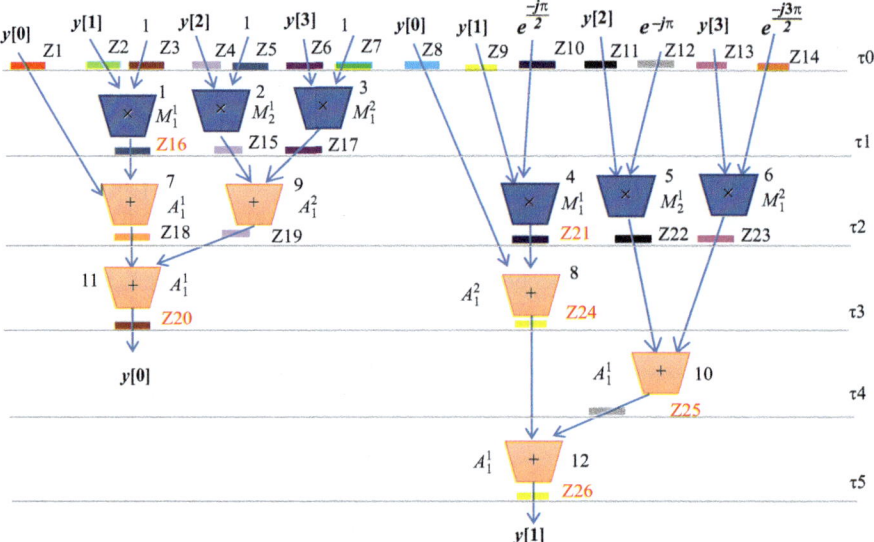

Figure 2.18   Structurally obfuscated scheduled DFG of 4-point DFT post embedding encrypted DNA signature

secure design. These multiplexing and de-multiplexing connections are derived from DNA signature embedded scheduled structurally obfuscated design as shown in Figure 2.18. Where variables marked in red represent the locally altered variables due to constraint embedding rule as discussed earlier. Finally, secure RT level data path design corresponding to target hardware coprocessor is generated followed by performing data path and controller synthesis (Sengupta and Chaurasia, 2022). The resulting secure RT level design of hardware coprocessor is capable of hindering RT level alteration and enabling the detective control against piracy through integrated defense using structural obfuscation and DNA biometric respectively.

2.5 Detection/validation of embedded encrypted DNA signature in RT level design

Encrypted DNA biometric-based methodology for hardware security provides robust detective control against counterfeited DSP coprocessors during piracy detection. The process of detection of counterfeited IP versions is shown in Figure 2.19. As apparent from the figure, the presence of authentic DNA security constraint is verified within the design (under test) to discern the authentic and counterfeited versions. For this purpose, legitimate encrypted DNA signature is regenerated by an SoC integrator, based on pre-stored encrypted DNA sequence during the piracy detection process. If the presence of encrypted DNA signature-driven security constraints is not found in the design under test, then it is probably counterfeited design. Moreover, the involvement of several complex information during encrypted DNA signature generation and implantation makes it almost impossible for an adversary to evade the counterfeit detection process. To reform the encrypted DNA signature and corresponding constraints for hardware security, following information is mandatory: (a) DNA sequence and its length; (b) dual encoding rule; (c) truncation length; (d) secret encryption key; (e) number of encryption rounds; (f) round function details; (g) S-box selection, etc. Thus, for an adversary, the exact regeneration of secret encrypted DNA biometric constraints is not possible to evade piracy detection successfully.

Further, in case if an adversary from offshore design house or at foundry, fraudulently claims IP ownership, then the integrated encrypted DNA biometric security constraints into hardware design offer seamless verification of IP

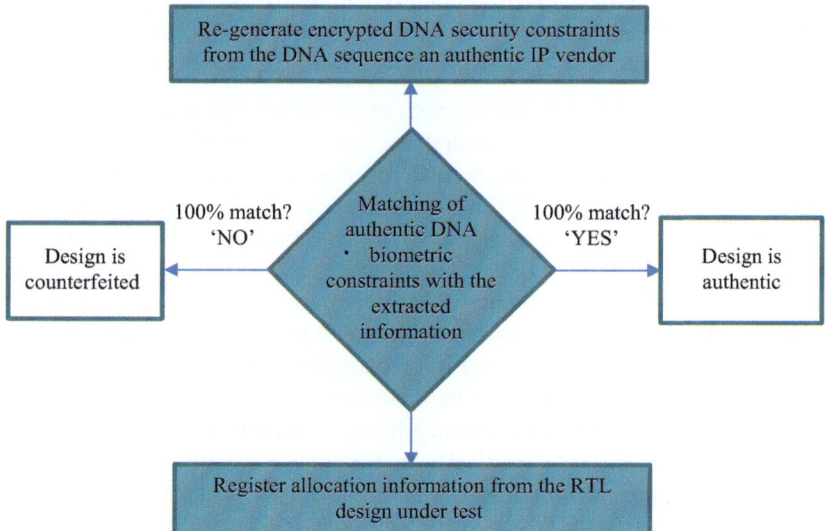

Figure 2.19   Detection of counterfeited designs

ownership. To nullify the fraudulent claim of IP ownership and awarding the IP ownership to genuine IP vendor, the positions of authentic DNA signature bits are matched bit by bit with the digital template embedded into the design under test. In case of complete matching, IP ownership is awarded to the genuine IP vendor.

Thus, the involvement of several complex information during encrypted DNA biometric signature generation and implantation makes it almost impossible for an adversary to evade counterfeit detection and fraudulently claim IP ownership successfully (Sengupta and Chaurasia, 2022).

2.6   Discussion and analysis

This section analyses the multi-level structural obfuscation and encrypted DNA biometric-based security methodology for securing hardware coprocessors in terms of offered security strength and its impact on the cost of the design (Sengupta and Chaurasia, 2022). The subsequent sub-sections discuss the security properties of encrypted DNA biometric, security analysis of multi-level structural obfuscation, security analysis of embedded encrypted DNA signature and the resulting design cost for generating a secure hardware coprocessor.

2.6.1   Security properties/parameters of encrypted DNA signature

Encrypted DNA signature offers highly robust security through its inherit security properties and crucial security parameters employed by IP designer, during DNA signature extraction and encryption. Therefore, for an adversary, it is not possible to evade piracy detection by replicating and implanting the authentic DNA signature into pirated versions and fraudulently gaining IP ownership. The regeneration of encrypted DNA signature is practically impossible for an adversary as the encrypted DNA signature offers the following security properties:

(a) *Secret encryption key* (E): The strength of the secret encryption key is $48*N*C$ bits; where "N" signifies the rounds of encryption used for a single Feistel cipher and "C" signifies the number of iterations a Feistel cipher is used depending on the length of the binary encoded DNA signature. Both these parameters (N and C) are unknown to an attacker. Additionally, binary-encoded genome/DNA signature strength depends on the genome sequence formed originally. For example, if the binary-encoded DNA signature is 128 bits, then C = 2, while if the binary-encoded genome/DNA signature is 192 bits, then C = 3 and so on. Therefore, for example, if the number of encryption round are Z= 10 and number of iterations are C= 2, then E= 960 bits and the key space is 2,960. Therefore, for an adversary, finding an exact signature from this massive key space is practically impossible.

(b) *Shift function parameters* ("n" *and* "m"): Secret encryption key (E) also depends on the shift bits "a" and "b." The exact value of shift bits "a_i" and "b_i" used in each round of the encryption key generation is unknown to an attacker. This also prevents an attacker from finding the exact signature.

(c) *DNA sequence and length*: The sequence and length of base pairs used during formation of the chromosomal DNA sequence is unknown to an attacker. Further, the count/ordering of the polynucleotide introduced in forming the final genome/DNA sequence comprising of distinct or the same type base pairs is also unknown to an attacker i.e., how may chemical elements (A, T, C, G, S) are present in the DNA sequence is unknown to an adversary.

(d) *DNA signature strength*: For an attacker, gauging the final signature length is extremely difficult. An IP designer may select DNA signature of any particular length to derive secret security constraints for embedding into the design. Thus, the unknown parameter of truncation length, further enhances the security strength of encrypted genome/DNA signature.

(e) *Encoding rules*: Encoding rule-1, used for encoding DNA chemical elements, and encoding rule-2, used for generating secret hardware security constraints corresponding to the encrypted DNA signature bits, are unknown to an adversary. Therefore, in case even if an adversary manages to compromise the encrypted DNA signature (which is anyhow extremely difficult), decoding the corresponding security constraints renders it further improbable.

(f) *Substitution* (S) *box*: Feistel cipher encryption process, during DNA signature encryption, uses S-box for performing substitution (confusion property). An IP designer may use same S-box for all 6-bit to 4-bit conversion or may use different S-boxes for each 6-bit to 4-bit conversion. The type of S-box used during performing substitution/confusion process in encryption is also unknown to an attacker.

Encrypted DNA signature expresses the aforesaid security features against brute-force attack, tampering attack, and fake assertion of IP ownership right. This is because without finding the exact encrypted DNA signature, the attacker cannot tamper with the security constraints. Further, the attacker will become incapable to regenerate the DNA signature without knowledge of (a) to (f). Thus, from attacker's perspective, the probability of evading piracy detection and demonstrating a fake assertion of IP ownership right is almost zero (Sengupta and Chaurasia, 2022).

2.6.2 *Security analysis of structural obfuscation*

Structural obfuscation enables the coprocessor design security by obscuring the design architecture (in terms of changing components interconnectivity) but without compromising actual functionality. This, therefore, hinders an adversary to reverse engineering the design by identifying its design functionality and hardware architectural details successfully. Therefore, the strength of obfuscation should be higher for enabling robust security against reverse engineering. The obfuscation in the design during high level synthesis leads to obscurity in RT level design and further it leads to obscuring gate-level netlist followed by RT level synthesis. Therefore, the strength of obfuscation is assessed in terms of the number of gates modified. The more the number of affected gates, more is the strength of obfuscation and impossible it is for an attacker to successfully modify the RT level

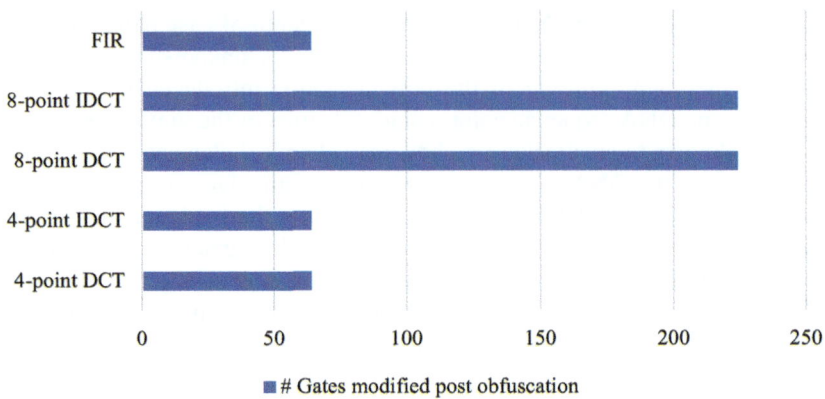

Figure 2.20   Obfuscation strength in terms of modified gates

description of the design. Figure 2.20 shows the number of modified gates corresponding to different coprocessor designs (Sengupta and Chaurasia, 2022).

2.6.3   Security analysis of encrypted DNA signature

Encrypted DNA biometric signature post embedding into the coprocessor design (as secret hardware security constraints) ensures the security in terms of enabling detective control against piracy and nullifying fraudulent claim of ownership. Security against fake assertion of IP ownership right is evaluated in terms of definitive proof of ownership, that is analyzed using the probability of coincidence (P_c) metric. The P_c is computed as follows (Sengupta and Rathor, 2019):

$$Pc = \left(1 - \frac{1}{h}\right)^g \tag{2.4}$$

where "h" signifies the count of registers required in the scheduled DFG design pre-embedding DNA secret security constraints and "g" signifies the number of encrypted DNA biometric security constraints inserted into the scheduled DFG design. The "P_c" signifies the probability of detecting encrypted DNA security constraints in an unsecured design by coincidence. Therefore, from an IP designer's perspective, it is necessary and desirable to have lesser P_c value to prove IP ownership and nullify fraudulent claims. The "P_c" comparison for different coprocessor designs based on varying sizes of encrypted DNA signature for different numbers of base pairs (AT/GC) in genome and different numbers of polynucleotide (leading/lagging strand in DNA) is shown in Table 2.6 (Sengupta and Chaurasia, 2022). As evident from Table 2.6, the higher strength of encrypted DNA signature results in lower P_c value (as desirable). Further, P_c comparison of different approaches is shown in Figure 2.21. As evident from figure, lower P_c is achieved through encrypted DNA signature

Table 2.6 P_c *value for different coprocessor designs (after implanting encrypted DNA constraints)*

# DNA base pairs (AT/GC)	#Polynucleostride (leading/lagging strand in DNA)	Encrypted DNA signature size	FIR	DFT	4-point DCT	8-point DCT
2	4	32	1.39E−2	9.3E−2	1.3E−2	1.2E−1
4	9	64	1.4E−3	8.7E−3	1.4E−3	1.6E−2
6	17	128	1.4E−3	7.59E−5	1.4E−3	2.5E−4

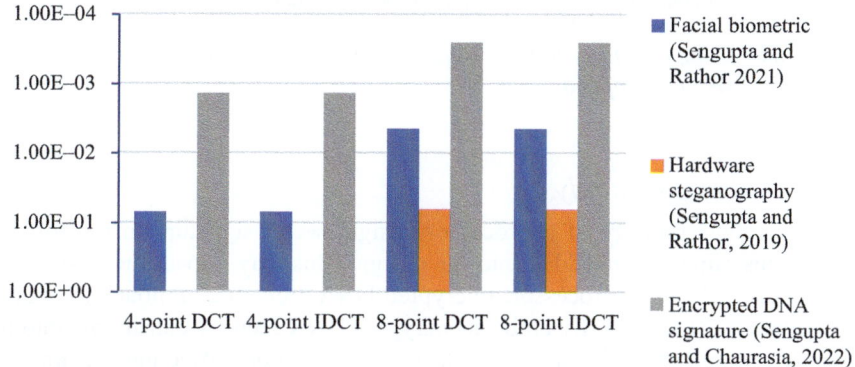

Figure 2.21 P_c *comparison of the different hardware security approaches*

based approach as compared to other contemporary approaches such as facial biometric and hardware steganography. This is because the number of generated security constraints for embedding into the design in approach (Sengupta and Chaurasia, 2022) is larger than the other approaches.

For a design, security against tampering attack is assessed through its ability of tolerance against tampering. The larger is the key-space, the harder it is for an attacker to guess the exact encrypted DNA signature to tamper. The tamper tolerance (TT) is measured as follows (Sengupta and Rathor, 2021):

$$TT = (v)^g \tag{2.5}$$

where "v" denotes the type of signature variables used for embedding into the coprocessor design (Sengupta and Chaurasia, 2022; Sengupta and Rathor, 2021). A comparison of tamper tolerance is shown in Figure 2.22. Since the number of generated security constraints through encrypted DNA biometric approach is significantly higher, thus the tamper tolerance ability of encrypted DNA biometric is substantially stronger than (Sengupta and Rathor, 2021).

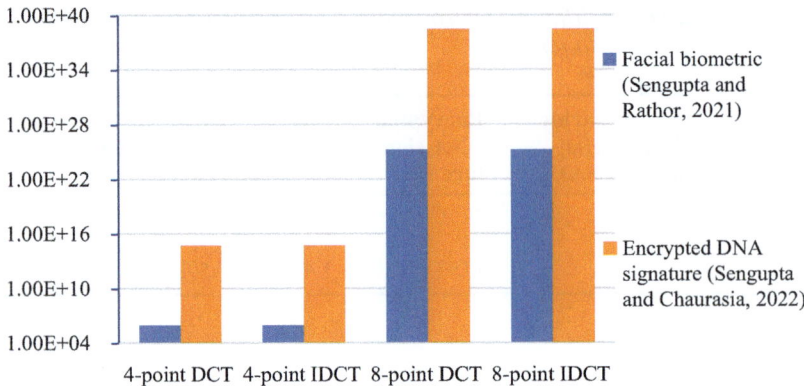

*Figure 2.22   Comparison of tamper tolerance ability for different hardware
security approaches*

2.6.4   Design cost analysis

To enable robust security of coprocessor design, secret signature in the form of security constraints is embedded into the design. This may impact the design cost overhead of hardware coprocessor. Encrypted DNA biometric approach (Sengupta and Chaurasia, 2022) covertly embeds encrypted DNA security constraints into the design during register allocation phase of HLS without affecting design cost overhead. The resulting design cost of embedding encrypted DNA signature is evaluated using the following design metric:

$$C_f(Z_i) = \omega_1 \frac{T_d}{T_{\max}} + \omega_2 \frac{V_d}{V_{\max}} \tag{2.6}$$

where "Z_i" denotes the IP designer-specified architectural constraints, V_d and T_d are area of the design and the latency, respectively, $V_{\max}$ and $T_{\max}$ denote the maximum area of the design and latency, ω_1 and ω_2 are the normalizing weight factors corresponding to latency and area in the cost function. The design cost comparison corresponding to pre and post embedding encrypted DNA signature is shown in Figure 2.23. As apparent from the figure that the approach (Sengupta and Chaurasia, 2022) results in zero design cost overhead. This is because no extra hardware resources (registers) were required during the embedding of encrypted DNA signature into the design.

The implementation run time of the integrated defense methodology using structural (architectural) obfuscation and encrypted DNA signature for ensuring the security for different DSP coprocessor benchmarks is shown in Table 2.7 (Sengupta and Chaurasia, 2022). As apparent from the table, it is capable of inserting robust encrypted DNA impression into different hardware coprocessor designs at very less implementation time (in terms of embedding time).

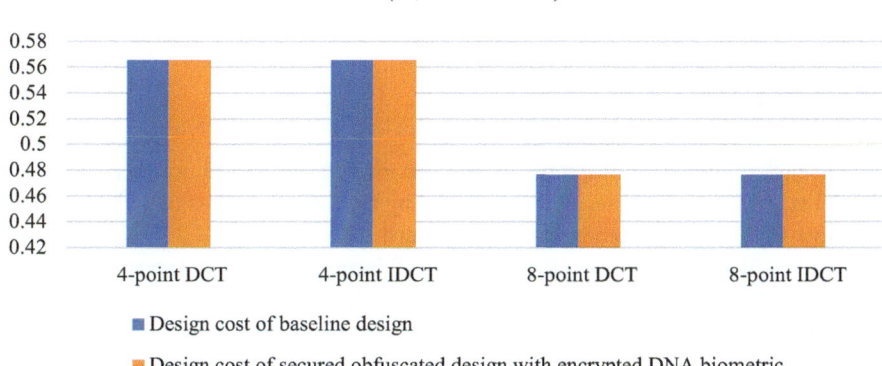

Design cost comparison for embedded DNA signature strength
(32,64 and 128 bits)

■ Design cost of baseline design

■ Design cost of secured obfuscated design with encrypted DNA biometric

Figure 2.23   Design cost (before and after embedding encrypted DNA security constraints into structurally obfuscated design)

Table 2.7   Implementation run time for different coprocessor designs

Coprocessors	Implementation time (sec)
DCT 4-point	2.323
IDCT 4-point	2.323
DCT 8-point	2.491
IDCT 8-point	2.491
FIR filter	2.904

2.7   Conclusion

Hardware coprocessors are the indispensable part of consumer electronics and computing systems for performing numerous applications and computationally intensive tasks. Therefore, ensuring the security of underlying hardware coprocessors designs is crucial from the perspective of safety and integrity of the end consumer. This chapter discusses integrated defense mechanism using multi-level structural obfuscation and embedding encrypted DNA biometric signature during HLS. This therefore enables the robust security of coprocessor designs against the threats of reverse engineering and piracy. Encrypted DNA signature results in the generation of larger security constraints, therefore, resulting into lesser P_c value (strength of ownership proof) and enabling higher tamper tolerance ability. Further, the embedding of encrypted DNA signature into structurally obfuscated design during HLS results into almost zero design cost overhead. This, therefore, exhibits sturdy security against IP piracy and fake assertion of ownership right at insignificant design cost. Thus, structural obfuscation and encrypted DNA biometric-based security methodology achieves robust security against reverse engineering attack and pirated designs, thereby ensuring the integration of only authentic coprocessor designs into CE and computing systems.

2.8   Questions and exercise

1. Discuss the need of biometric-based security for hardware coprocessors.
2. What are the possible security threats during the design process of a coprocessor?
3. How DNA-based biometric is more robust than hash-based digital signature?
4. Discuss the applications of different DSP hardware coprocessors?
5. How DNA-based biometric is more robust than hardware watermarking?
6. What security threat is handled by structural obfuscation and how it safeguards the design?
7. Why is the need of securing hardware coprocessors?
8. Explain the process of generating binarized DNA signature.
9. Discuss Feistel cipher encryption process and its security strength.
10. Discuss the key generation process?
11. Explain the process of encoding encrypted DNA signature for generating the secret hardware security constraints.
12. Explain the process of generating encrypted DNA signature embedded modified register allocation information.
13. Discuss THT-based structural obfuscation and its impact on the design in terms of security?
14. Explain piracy detection process?
15. Explain the impact of DNA signature strength on probability of coincidence and tamper tolerance.
16. Discuss the limitations of the DNA biometric-based hardware security approach. If any?
17. Explain the process of the generating scheduled coprocessor design.
18. Explain the security property of encrypted DNA signature.
19. Explain the security property of permutation and substitution process in Feistel cipher process.
20. Explain the different phases of HLS process to generate secure coprocessor design with structural obfuscation and encrypted DNA signature.

References

15 nm open cell library. Available: https://si2.org/open-cell-library/, last accessed on January 2020.

Bhardwaj, A. and S. Akhter (2021), "Multi feedback LFSR based watermarking of FSM," in: *2021 7th International Conference on Signal Processing and Communication (ICSC)*, 2021, pp. 357–361, doi:10.1109/ICSC53193.2021.9673360.

Chaurasia, R. and A. Sengupta (2022), "Crypto-genome signature for securing hardware accelerators," in: *2022 IEEE 19th India Council International Conference (INDICON)*, Kochi, India, 2022, pp. 1–6, doi:10.1109/ INDICON56171.2022.10039955

Chun, K., C. Sun, and N. Yang (2017), "An improved LOMB algorithm for HRV analysis on a PPG sensor for low-cost DSP processor," in: *2017 IEEE International Conference on Consumer Electronics – Taiwan (ICCE-TW)*, 2017, pp. 405–406, doi:10.1109/ICCE-China.2017.7991167.

Colombier, B., U. Mureddu, M. Laban, O. Petura, L. Bossuet, and V. Fischer (2017), "Complete activation scheme for FPGA-oriented IP cores design protection," in: *2017 27th International Conference on Field Programmable Logic and Applications (FPL)*, 2017, pp. 1–1, doi:10.23919/FPL.2017.8056772.

Cruz, J., P. Gaikwad, A. Nair, P. Chakraborty, and S. Bhunia (2022), "Automatic hardware trojan insertion using machine learning," *arXiv:2204.08580.*

Elnaggar, R. and K. Chakrabarty (2018), "Machine learning for hardware security: opportunities and risks," *Journal of Electronic Testing*, vol. 34, no. 2, p. 183201, doi: 10.1007/s10836-018-5726-9.

He, G., C. Dong, Y. Liu, and X. Fan (2020), "IPlock: an effective hybrid encryption for neuromorphic systems IP core protection," in: *2020 IEEE 4th Information Technology, Networking, Electronic and Automation Control Conference (ITNEC)*, 2020, pp. 612–616, doi: 10.1109/ITNEC48623.2020.9085144.

Huang, Z., Q. Wang, Y. Chen, and X. Jiang (2020), "A survey on machine learning against hardware Trojan attacks: recent advances and challenges," *IEEE Access*, vol. 8, pp. 10796–10826, doi:10.1109/ACCESS.2020.2965016.

Koushanfar, F., S. Fazzari, C. McCants, W. Bryson, P. Song, and M. Potkonjak (2012), "Can EDA combat the rise of electronic counterfeiting?," in: *Proceedings of DAC*, San Francisco, CA, pp. 133–138, doi: 10.1145/2228360.2228386.

Lao, Y., B. Yuan, C. H. Kim, and K. K. Parhi (2017), "Reliable PUF-based local authentication with self-correction," *IEEE Transactions on Computer-Aided Design of Integrated Circuits and Systems*, vol. 36, no. 2, pp. 201–213, doi:10.1109/TCAD.2016.2569581.

Liakos, K. G., G. K. Georgakilas, S. Moustakidis, P. Karlsson, and F. C. Plessas (2019), "Machine learning for hardware Trojan detection: a review," in: *2019 Panhellenic Conference on Electronics & Telecommunications (PACET)*, pp. 1–6, doi: 10.1109/PACET48583.2019.8956251.

Pekez, N., N. Kaprocki, and J. Kovačević (2018), "Firmware update procedure for audio systems based on CS4953xx DSP family," in: *2018 International Conference on Smart Systems and Technologies (SST)*, 2018, pp. 29–34, doi:10.1109/SST.2018.8564683.

Pilato, C., S. Garg, K. Wu, R. Karri, and F. Regazzoni (2018), "Securing hardware accelerators: a new challenge for high-level synthesis," *IEEE Embedded Systems Letters,* vol. 10, no. 3, pp. 77–80, doi: 10.1109/LES.2017.2774800.

Plaza, S. M. and I. L. Markov (2015), "Solving the third-shift problem in IC piracy with test-aware logic locking," *IEEE Transactions on Computer-Aided Design*

of Integrated Circuits and Systems, vol. 34, no. 6, pp. 961–971, doi: 10.1109/
TCAD.2015.2404876.

Rathor, M. and A. Sengupta (2020), "IP core steganography using switch based
key-driven hash-chaining and encoding for securing DSP kernels used in CE
systems," *IEEE Transactions on Consumer Electronics*, vol. 66, pp. 251–260,
doi: 10.1109/TCE.2020.3006050.

Sarihi, A., A. Patooghy, P. Jamieson, and A.-H. A. Badawy (2022), "Hardware
trojan insertion using reinforcement learning," arXiv:2204.04350.

Sengupta, A. and S. Bhadauria (2016), "Exploring low cost optimal watermark for
reusable IP cores during high level synthesis," *IEEE Access*, vol. 4, pp. 2198–
2215, doi:10.1109/ACCESS.2016.2552058.

Sengupta, A. and R. Chaurasia (2022), "Securing IP cores for DSP applications
using structural obfuscation and chromosomal DNA impression," *IEEE
Access*, vol. 10, pp. 50903–50913, doi:10.1109/ACCESS.2022.3174349.

Sengupta, A. and M. Rathor (2019a), "Crypto based dual phase hardware steganography for securing IP cores", *IEEE Letters of the Computer Society
(LOCS)*, vol. 2, no. 4, pp. 32–35.

Sengupta, A. and M. Rathor (2019b), "IP core steganography for protecting DSP
kernels used in CE systems," *IEEE Transactions on Consumer Electronics*,
vol. 65, no. 4, pp. 506–515, doi: 10.1109/TCE.2019.2944882.

Sengupta, A. and M. Rathor (2021), "Facial biometric for securing hardware
accelerators," *IEEE Transactions on VLSI Systems*, vol. 29, no. 1, pp. 112–
123, doi: 10.1109/TVLSI.2020.3029245.

Sengupta, A., D. Roy, and S. P. Mohanty (2018), "Triple-phase watermarking for
reusable IP core protection during architecture synthesis," *IEEE Transactions
on Computer-Aided Design of Integrated Circuits and Systems*, vol. 37, no. 4,
pp. 742–755, doi: 10.1109/TCAD.2017.2729341.

Wang, X., Y. Zheng, A. Basak, and S. Bhunia (2015), "IIPS: infrastructure IP for
secure SoC design," *IEEE Transactions on Computers*, vol. 64, no. 8,
pp. 2226–2238, doi:10.1109/TC.2014.2360535.

Xie, Y., C. Bao, and A. Srivastava (2017), "Security-aware 2.5D integrated circuit
design flow against hardware IP piracy," *Computer*, vol. 50, no. 5, pp. 62–71,
doi:10.1109/MC.2017.121.

Xue, M., C. Gu, W. Liu, S. Yu, and M. O'Neill (2020), "Ten years of hardware
Trojans: a survey from the attacker's perspective," *IET Computers & Digital
Techniques*, vol. 14, no. 6, pp. 231–246, https://doi.org/10.1049/iet-
cdt.2020.0041.

Zalivaka, S. S., A. A. Ivaniuk, and C.-H. Chang (2019), "Reliable and modeling
attack resistant authentication of arbiter PUF in FPGA implementation with
trinary quadruple response," *IEEE Transactions on Information Forensics and
Security*, vol. 14, no. 4, pp. 1109–1123, doi:10.1109/TIFS.2018.2870835.

Chapter 3

Facial signature-based biometrics for hardware security and IP core protection

Anirban Sengupta[1] and Rahul Chaurasia[1]

The chapter describes a robust physiological biometrics-based methodology using facial signature biometric for hardware security and intellectual property (IP) core protection (Sengupta and Rathor, 2021). In this methodology, first, the facial biometric signature of IP vendor is extracted and subsequently transformed into encoded constraints for hardware security using multi-level encoding specified by IP vendor. Next, these generated secret constraints for hardware security are inserted inside the design at the register allocation stage of high-level synthesis (HLS) process. These embedded facial security constraints act as digital evidence to enable security of IP core design against several security threats. The hardware security methodology based on facial biometric enables the robust and seamless detective control against the threat of IP piracy. Additionally, it also protects the ownership rights of authentic IP vendor, in case if a rogue system on chip (SoC) integrator or an adversary at foundry, falsely claims the IP ownership.

The rest of the chapter has been organized as follows: Section 3.1 provides the introduction of the chapter; Section 3.2 highlights the importance of HLS for designing digital signal processing (DSP) coprocessors; Section 3.3 presents the discussion and analysis of alternative techniques used for Intellectual property protection (IPP) of DSP coprocessors; Section 3.4 explains the features of facial biometric for IPP and its advantage over fingerprint biometrics; Section 3.5 shows the summary of facial biometric methodology for IPP; Section 3.6 explains the details of facial biometric methodology for IPP; Section 3.7 presents a case study of designing safeguarding N-point discrete Fourier transform (DFT) design using facial biometrics; Section 3.8 discusses security properties of facial biometric methodology for hardware security; Section 3.9 provides analysis and discussion and Section 3.10 concludes the chapter.

3.1 Introduction

A hardware IP is a reusable unit (block of data/logic) of computational function, Boolean logic, a cell, register transfer level (RTL) or a gate structure, is also

[1]Department of Computer Science and Engineering, Indian Institute of Technology Indore, India

known as the intellectual property of an IP designer. IP cores are envisioned for performing the execution of computationally and data-intensive tasks. Therefore, it is realistic to design them as dedicated reusable IP cores to accelerate the device performance with higher efficacy. IP core(s) is/are integrated in several consumer electronics (CE) or computing systems to perform computational and data-intensive tasks with higher efficacy. Therefore, IP cores are emerged as an integral part of several electronic items and devices such as cell phones, tablets, hearing machines, and digital imaging systems. Additionally, they are also used in several fields like Internet of Things (IoT), healthcare and robotics, etc. However, based on the computational capability and design size, they are categorized into two different types: micro-IPs and macro-IPs (are essentially bigger logic). Logic gates, combinational and sequential circuits (register and memory) are some of the examples of micro-IPs. On the other hand, digital signal processors (DSPs), central processing units (CPUs), and application-specific cores such as joint photographer expert group (JPEG) engines, moving picture expert group (MPEG) engines, digital filters like finite impulse response (FIR) filter and infinite impulse response (IIR) filter, falls under the category of macro-IPs. These DSP cores facilitate several applications like image compression–decompression, digital data filtration, audio processing, etc., which are computationally intensive in nature.

Further, based on the reconfigurability, IP cores can be classified as soft IPs and hard IPs. IP cores which are generally available as synthesizable register transfer level code in the form of either schematic design (.bdf file) or hardware description language (.vhd/.vhdl file), are called as soft IP. Further, soft IP cores are also availed in the format of gate level netlist (usually expressed in Verilog code with '.V' extension) by an IP designer. It describes the design interconnectivity in terms of various cells that are present within it and the output of synthesis process at logic level. RTL and gate-level netlist both allow post synthesis processing steps such as placement, routing, and downloading into reconfigurable platforms such as (field programmable gate arrays) FPGAs. On the other hand, hard IP cores are generally available as a layout format (fixed masked layout) of chip designs in the graphic data system (GDS) or layout editor documentation (LEF) format. Unlike soft IP cores, hard IPs cannot be modified by chip designers or system integrators. Further, demerit of a hard IP design is that it does not allow to be used in another foundry (for fabrication) for which it is not targeted to. This is because design at layout level comprises of process foundries and a design rule, which incapacitates the use of layout in another foundry except to whom it was targeted. Therefore, due to more flexibility (in terms of modifying functionality) and greater portability (can be reused), soft IPs are preferred over Hard IPs. However, soft IP cores are exposed to greater IP protection risk than hard IPs as they can be modified by system integrators.

So far, we discussed the IP classification and their importance in CE systems. Now, we discuss the need of IP core protection against the threat of piracy and fraudulent IP ownership claim. An IP core can be completely owned by an IP designer/vendor or may be provided to buyers on a licensed basis. In general, a SoC of consumer electronics/computing systems consists of many reusable IP cores. Further, design complexity, productivity gap, and time to market are some of the major factors that enforce the chip designers to design these reusable IP cores from third-party design houses rather than designing them all under a single integrated device manufacturer (IDM). This phenomenon leads to the crucial security threats to the design chain of electronics systems (Pilato *et al.*, 2018; Mahdiany *et al.*, 2001; Schneiderman, 2010; Plaza and Markov, 2015; Castillo *et al.*, 2007; Colombier and Bossuet, 2015; Newbould *et al.*, 2002; Roy *et al.*, 2008; Koushanfar *et al.*, 2012). Therefore, along with design, the security aspect of such IP cores is also crucial. Thus, from the perspective of SoC integrator, the reusability of IP cores leads to several concerns to the designer such as:

(a) Reliability of the IP vendor (supplier).
(b) Security against pirated versions.
(c) Risk associated in case if IP is faulty.
(d) Whether the IP is leveraging of the lesser design area, delay and power.
(e) Detection of any malicious logic inside the design, etc.

A system integrator needs to ensure the authenticity of the reusable IP core design before integrating into SoC design. This is because the pirated design may contain the malicious logic that may lead to safety and reliability consequences to end consumer. Further, from the perspective of IP designer, a rogue SoC integrator may falsely claim the ownership rights upon receiving the IP core from an IP designer/vendor. Therefore, the robust security of IP cores is crucial against the threats of IP piracy and false claim of IP right. There are several security mechanisms for safeguarding the IP cores against the threats of IP piracy and nullifying false claim of IP ownership.

Now, we look into the significance of physical biometrics and their importance in IPP. As there are two major categories of biometrics namely, behavioral biometrics and physical biometrics. Behavioral biometric exploits uniquely identifiable and measurable patterns in human activities. However, they are not as reliable as physical biometrics from the aspect of security and unique identification and authentication. Conversely, physical biometric approaches such as facial biometric, fingerprint biometric, palmprint biometric, and retina biometric exploit naturally unique physiological characteristics of human body. Therefore, are capable to associate unique physiological features of IP vendor with the design to secure it against piracy and false IP ownership claims. Therefore, in case of fraudulent claim of IP ownership, original IP vendor may easily prove his/her authorship by generating the exact security constraints as that of embedded constraints into an authentic design. On the other hand, non-biometric approaches are based on watermarking and steganography. However, in case of non-biometric-based security methodology, an adversary may easily replicate or decode the auxiliary secret signature. Further, it may not be possible for an authentic IP vendor to prove the authorship easily in case of non-biometric approaches. This is because these security approaches do not associate unique identity of IP vendor into the design (Sengupta and Rathor, 2021).

This chapter mainly focuses on safeguarding the hardware IP cores against the threats of false claim of IP ownership and IP piracy, using contact-less facial biometric-based hardware security methodology (Sengupta and Rathor, 2021). To ensure robust security against both these threats, first the unique facial signature of corresponding to facial image of an IP vendor is generated. Subsequently, the facial signature is encoded using IP vendor-specified encoding rules, to generate secret hardware security constraints. These hardware security constraints are then covertly inserted into the design during HLS phase of very large-scale integration (VLSI) design process. The embedded facial constraints act as robust digital evidence for enabling detective control against pirated designs during the piracy detection process. Additionally, uniqueness of facial security constraints renders security against replication of secret signature by an adversary for proving false IP ownership claim successfully. Thus, the facial biometric signature ensures detective control against pirated IP cores before being integrated into the system and nullify false IP ownership claim from an adversary (Sengupta and Rathor, 2021).

3.2   Importance of HLS for designing DSP co-processors

So far, we discussed the need of hardware IP cores and the necessity of their protection against crucial security threats. Now, we discuss the importance of HLS to design DSP co-processors efficiently. Hardware co-processor is generally employed to carry out computationally intensive portion of a complex algorithm or application. For example, in consumer electronic systems, several DSP IP cores are used to perform the applications based on image and audio/video processing. DSP hardware IP cores are envisioned to execute complex DSP algorithms such as DFT (discrete Fourier transform), DCT (discrete cosine transform), and FIR (finite impulse response) filter, which are large in size and require massive computations. Therefore, owing to high complexity, it is not preferable to design such IP cores from lower abstraction level of very large-scale integration (VLSI) design process. Hence to design, DSP co-processors (IP cores) are meant to be synthesized from higher abstraction level into a hardware form using the HLS framework of VLSI design process. This is because designing at lower level of abstraction such as RT-level or gate level design involves complex design structure and huge design time, which does not remain pragmatic from a designer's perspective. The RT-level design may include several functional unit resources such as multiplexers, demultiplexers as the interconnect hardware for adder, subtractor, multiplier operations of the design along with registers and latches as storage hardware. On the other hand, gate level design includes thousands of gates even for a smaller design. Therefore, from a designer's perspective, it is only realistic to synthesize from higher abstraction level of the VLSI design process. HLS is a process of converting algorithmic description of an application (corresponding to hardware IP) into its equivalent register transfer level counterpart, which consists of several phases such as compilation, transformation, scheduling, allocation, binding, datapath synthesis, and controller synthesis, as shown in Figure 3.1. Now, we discuss the details of different phases of HLS and their importance in terms of their functionality and flexibility offered to generate low-cost design and possible integration of robust security at minimal design cost overhead.

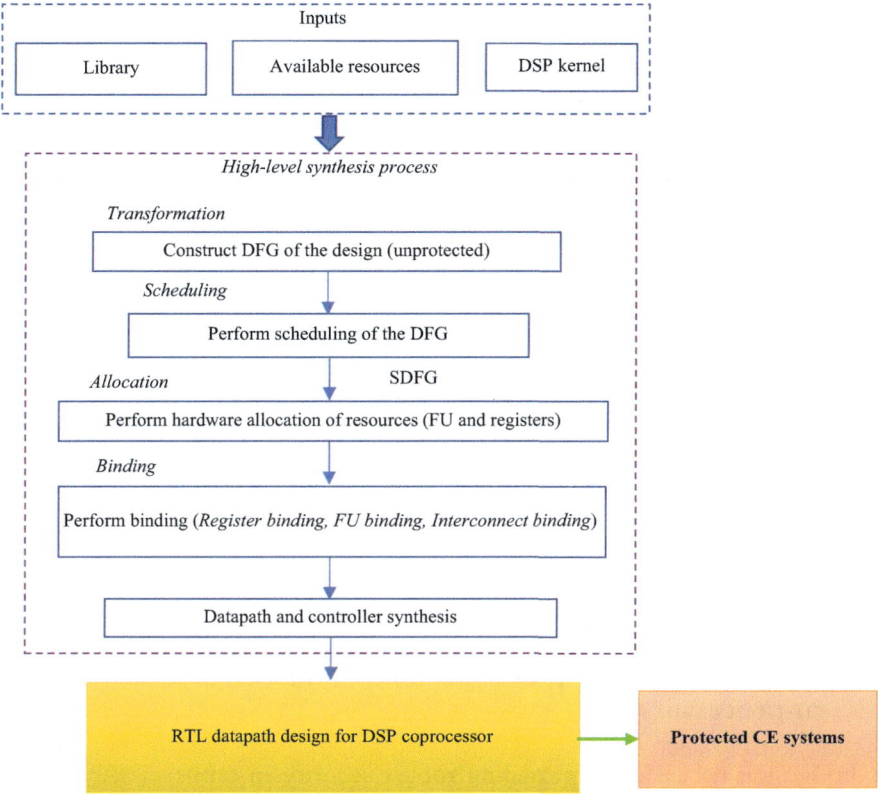

Figure 3.1   HLS design flow for generating RTL data path design

Transformation phase: this phase of HLS (as shown in Figure 3.1) is responsible to transform the sample DSP application into its respective data flow graph (DFG). However, from the perspective of integrating security during this phase, the structural transformation can be performed, which obscures the design and thereby makes it unobvious for an adversary to identify design functionality and hardware architectural details. Therefore, thwarting reverse engineering and RTL design alteration.

Scheduling phase: this phase is responsible for generating the scheduled DFG of the design based on IP designer selected resource constraints. Scheduling phase offers flexibility to integrate design space exploration mechanism to explore and obtain low-cost solution while satisfying design constraints such as area, time, and power. It thus helps an IP designer to obtain low-cost resource constraints for scheduling the DFG of the design. This may result in an optimal design area and latency. Further, this phase can also be exploited for enabling the security of the design in terms of embedding secret digital evidence.

Allocation phase: in this phase, functional units (FUs) like adders, multipliers, subtractors, etc. are allocated to respective operations such as addition, multiplication, and subtraction, respectively. Further, storage elements like registers are also allocated to the respective storage variables (carrying input, intermediate, and output values) of the design. This phase also offers the flexibility to enable security in terms of embedding hardware security constraints into the design during register allocation. Following hardware security methodologies such as facial biometric (Sengupta and Rathor, 2021), fingerprint biometric (Sengupta and Rathor, 2020), and watermarking (Sengupta and Bhadauria, 2016) implant covert constraints for hardware security during the register allocation stage of HLS.

Binding phase: this phase is responsible to perform binding of allocated hardware and resources. In this phase, register binding, functional binding, and interconnect binding (binding of multiplexers and demultiplexers) are performed. An IP designer may also exploit the binding phase to modify the design to enable security mechanism without a change in functionality.

Finally, by performing the datapath and controller synthesis, corresponding RTL design can be generated. Thus, HLS offers flexibility in terms of generating low-cost design and incorporating robust security into the design with lesser design cost overhead (Coussy *et al.*, 2009; McFarland *et al.*, 1988).

3.3   Alternative techniques used for IPP of DSP co-processors

To bridge the necessity of providing robust security in terms of IPP of DSP coprocessors, several powerful techniques have been presented in the literature. Some of the important hardware security techniques against IP piracy include fingerprint biometric, secret watermarking, hash-based digital signature, and IP steganography. Each hardware security approach has their own distinctive characteristics and security features. These security approaches can be classified into signature-based and non-signature-based approaches. However, signature-based approaches can be further classified into biometric and non-biometric approaches. IP watermarking and hash-based digital signature approaches employ the signature-based non-biometric approach. Further, fingerprint biometric approach employs signature-based security mechanism for hardware security. On the other hand, hardware steganography is non-signature-based approach that employs the embedding of secret stego constraints into the design for enabling hardware security against IP piracy. Now, let us discuss each of these techniques in terms of their security mechanism and offered security strength.

3.3.1   Fingerprint biometric

Biometric fingerprinting methodology (Sengupta and Rathor, 2020) enables the protection of DSP co-processors against IP piracy by exploiting the fingerprint information of an IP vendor. To do so, biometric fingerprint of an IP vendor is first transformed into a biometric signature (digital template), followed by inserting into the IP design (as encoded biometric constraints) during the register allocation phase of HLS. This embedded fingerprint signature acts as security evidence to detect piracy. In the fingerprint biometric approach, following sub-steps are performed to generate secret fingerprint signature: (a) capture the fingerprint of IP vendor by performing the scanning using optical scanner device; (b) subsequently, pre-process the captured image that includes three sub-processes: image enhancement, binarization, and thinning. Image enhancement employs fast Fourier transform (FFT), to magnify and reconnect the broken ridges. On the other hand, binarization transforms the image into two intensity levels ("0" for low and "255" for high) and thinning reduces the thickness of the ridge lines to one pixel width. Post pre-processing, thinned fingerprint image is exploited to extract its minutiae points (points where ridge lines are bifurcated into branches or ended abruptly) which leverages the unique features an IP vendor (c) next, minutiae points are represented in its corresponding binary form. Subsequently, a biometric template is formulated by concatenating the signatures of each minutiae point. Next, digital template is converted into covert constraints for hardware security depending on the encoding rule defined by the IP vendor. Subsequently, these secret fingerprint constraints are inserted into the IP design. These embedded secret hardware security constraints act as security evidence to detect IP piracy. However, higher implementation complexity of fingerprint biometric and its dependence on optical scanner during capturing are a bottleneck. Additionally, external elements such as dirt and lubricant may adversely impact the capturing of accurate minutiae points and thereby affecting the accurate generation of fingerprint digital template. On the other hand, facial biometric approach (Sengupta and Rathor, 2021) is contact-less and does not depend on optical scanner during signature generation. Therefore, facial biometric approach offers more flexibility and accuracy during signature generation process.

3.3.2   Hardware watermarking and steganography

Cui and Chang (2007) presented watermarking-based security scheme during logic synthesis level. Sengupta and Bhadauria (2016) presented watermarking-based security which comprises of four different watermarking variables encoding for embedding a dynamic watermark in the IP core design during architectural synthesis. Koushanfar *et al.* (2005) presented watermarking approach based on binary encoding scheme to implant watermarking constraints for intellectual property protection. Further, Le Gal and Bossuet (2012) presented an IP watermarking in HLS. In this approach, watermark is generated based on a set of mathematical relations between the IP input data, the initial values of the internal computation, and the output of IP. These watermarking-based security approaches are dependent on auxiliary security features such as types of signature variables and their combination, mapping mechanism, and signature strength. Therefore, this dependence of watermarking security methodologies on auxiliary parameters renders the watermark susceptible to typical security threats. This is because an attacker upon compromising the involved auxiliary security parameters may replicate the watermark of the design and embed into pirated versions. This may lead to evasion during IP piracy detection process. Therefore, in case of watermarking-based security, an adversary may successfully evade the piracy detection process. Further, watermarking approaches do not uniquely associate the identity of original IP vendor with the design. Therefore, it may lead several challenges for a genuine IP vendor to prove ownership, in case if an adversary falsely claims the IP rights. Ni and Gao (2005) presented detector-based watermarking approach that emphasizes itself on the detection of the watermarking signatures, using detached detector to distill watermarking seeds that contain unique traits of the original design and generate meaningful watermarking messages. However, as the watermark does not uniquely associate the identity of authentic IP vendor, therefore it may not guarantee awarding of IP ownership to genuine IP vendor always.

Furthermore, Sengupta and Rathor (2019a) presented hardware steganography methodology for detecting piracy of DSP kernel by implanting the concealed stego mark into the DSP design without using any external signature. The amount of concealed evidence which is meant for embedding is fully under the control of designer through a "thresholding" parameter. The stego mark depends upon specific design data, stego-encoder, entropy threshold, and encoding rules. Sengupta and Rathor (2019b) presented crypto-based steganography for hardware protection, which implants the stego constraints during multiple phases of behavioral synthesis process such as during register allocation and hardware allocation stage. Rathor and Sengupta (2020) presented hardware steganography using key-driven hash-chaining for securing such IP cores integrated into CE systems. In this technique, secret imperceptible stego-marks are generated by performing hash-chaining process that incorporates switches, strong large stego-keys, mapping rules, and hash blocks. This intricated methodology for stego-mark generation using steganography approach makes it sturdier than watermarking. However, if the specific design data, stego-encoding process, and threshold value are leaked to an attacker, then the resolution of steganography is defeated. Additionally, higher complexity is involved during stego-constraint generation that results into a cumbersome piracy detection process. Further, in case of IP ownership, it does not render as much robust security as the facial biometric-based hardware security methodology. On the contrary, facial biometric methodology offers robust and seamless detection of pirated IP cores through embedded secret facial security constraints corresponding to the original IP vendor. This is because facial biometric-based security methodology is capable of generating a large number of unique facial security constraints than the watermarking and steganography-based approaches. These embedded facial security constraints of IP vendor facial image act as invisible digital evidence to ensure the protection of IP rights of the original IP vendor. Thus, facial biometric methodology ensures both seamless detection of pirated versions of IP core as well as nullifies false IP ownership claim.

3.3.3   Hash-based digital signature

Sengupta *et al.* (2019) presented a non-biometric-based hardware security approach using encrypted-hashing-based digital signature. This security methodology first encodes the DFG of target DSP design into bitstream. Subsequently, it generates bitstream digest by using secure hash algorithm. Further, RSA encryption is used as an encryption technique using IP vendor's private key. This results in producing the digitally signed hash value. Next, the generated digital signature is coded into secret security constraints using IP vendor-specified encoding rule. These digital signature-based security constraints are then inserted into the target design during architectural synthesis. Thereby generating digital signature-secured design. However, an adversary upon decoding the bitstream encoding mechanism and RSA encryption key may compromise the security provided by the encrypted hash-based digital signature approach. Therefore, an attacker can regenerate the digital signature to evade the piracy detection process. The major limitations of digital signature approach are:

1.   Highly dependent on auxiliary security factors such as encoding rule, hash algorithm, and encryption key.
2.   Incurs higher implementation complexity for generating digital signature during verification and authentication.
3.   Vulnerable, as the uniqueness of digital signature is not always guaranteed. Therefore, an attacker can reuse it for claiming the IP authorship.

Whereas in the facial biometric methodology, facial security constraints are produced based on naturally unique facial features of a genuine IP vendor. Therefore, it is highly unlikely for an adversary to successfully claim the IP ownership by replicating the embedded facial security constraints corresponding to facial features of the original IP vendor. Further, facial signature is non-vulnerable to theft and therefore it cannot be reused by an adversary to evade piracy detection. Further, it does not depend on any external key, thereby is robust against key-based attacks. Additionally, facial biometric approach does not involve complex process to generate facial nodal points. Thus, hardware security approach based on facial biometric offers robust security than digital signature, against the threats of IP piracy and false IP authorship claim (by an adversary).

3.4   Features of facial biometrics for IPP and its advantages over fingerprint biometrics

So far, in a nutshell, we have discussed the contemporary hardware security techniques such as watermarking, hardware steganography, digital signature, and fingerprint biometric for securing an IP core design. Facial biometric renders stronger and more robust security of hardware IP cores than non-biometric approaches such as watermarking, hardware steganography, and digital signature in terms of definite proof of ownership, seamless piracy detection, and tamper tolerance ability. Further, as compared to fingerprint biometric approach (Sengupta and Rathor, 2020), facial biometric incurs lesser implementation complexity and ensures accurate feature extraction for performing piracy detection and authorship verification. This makes facial biometric approach more feasible for ensuring robust security of IP cores than fingerprint biometric.

Features:
First, we discuss the features of facial biometric for IPP:

1. Facial biometric leverages naturally unique facial feature of IP vendor to ensure IPP for hardware security.
2. It renders contact-less biometric process for protecting hardware co-processors using captured facial image, where encoded biometric signature is produced and secretly implanted within the design using HLS framework.
3. Is proficient of offering lesser complexity in terms of implementation, zero dependence on external factors as it is contact-less.
4. Enables seamless and robust detection of pirated versions of IP cores before their integration into SoCs of CE systems.
5. It associates unique facial identity of IP vendor within design in terms of embedded secret facial constraints and therefore it ensures to nullify false IP ownership claim from an adversary and protects ownership rights of a genuine IP vendor.
6. Incurs almost zero design cost overhead for embedding the facial security constraints during HLS to ensure robust security corresponding to different DSP applications.
7. Does not rely on any external key, therefore, it is not sensitive to key-based attacks.
8. Renders non-replicability of facial signature due to uniqueness of facial features.

All these features of facial biometric security methodology make it aptly suitable for hardware security.

Advantages:
Now, we discuss the advantages facial biometric security methodology (Sengupta and Rathor, 2021) over fingerprint biometric (Sengupta and Rathor, 2020). The facial biometric technique provides many advantages as compared to fingerprint biometric, such as:

1. Obtaining the facial biometric does not rely on optical scanning device during capturing and validation. Therefore, in case of facial biometric, it exhibits almost zero probability of inadvertently losing any biometric information, as it never requires recapturing of facial image of an IP vendor. This is because the accurate fingerprint minutiae feature point extraction requires optical scanning device with good resolution and similar surface area. Thus, for fingerprint biometric, if verification is performed by recapturing the fingerprint again, then there is every possibility of losing any biometric information. This may hinder the verification process.

2. Facial biometric nodal feature points generation process has no impact of external factors like grease and dirt unlike fingerprint biometric.

3. Even if an adversary manages to compromise/forge the stored facial image, it is not possible for him/her to falsely authenticate his/her ownership as the authentication process requires crucial details of several security parameters such as (a) specific grid magnitude and spacing for subjecting facial image; (b) location of facial nodal feature points; (c) type and number of facial features; (d) specific assembling order of features for producing facial template; (e) employed encoding rules. For an adversary, it is very challenging to decode or regenerate exact facial biometric information to falsely validate IP ownership. On the other hand, saved fingerprint image can be forged and reused by an adversary during false ownership validation, as fingerprint approach does not involve so many security variables.

4. In the fingerprint approach, the overall quality of captured fingerprint image relies on the quality of scanning device, position of fingerprint on scanning device surface area and pressure applied, and so on. Further, it requires pre-processing of captured fingerprint image using following sub-processes: (a) enhancement using FFT; (b) binarization; and (c) thinning. Unlike fingerprint biometric, facial biometric methodology does not require such preprocessing steps for facial image and subsequent generation of signature.

The summary of comparative analysis between facial and fingerprint biometric is shown in Table 3.1. As evident from the table, facial biometric methodology

Table 3.1 Facial biometric vs fingerprint biometric for hardware security

S. no.	Characteristics/parameters	Facial biometric approach	Biometric finger-printing approach
1.	Dependence on external factors	Independent	Grease, dirt may affect the exact generation of fingerprint
2.	Mechanism	Contact-less	Requires optical scanner
3.	Device required	Camera	Scanner
4.	Implementation complexity	LESS	HIGH
5.	Robustness during IP piracy detection and ownership verification process	Strong	Strong
6.	Dependence on optical scanner during authentication	Not dependent	Dependent
7.	Injury prone	No	Yes
8.	Probability of coincidence (secret digital evidence) and tamer tolerance ability (defence against brute force attack)	Robust	Robust
9.	Accuracy	More	Relatively less
10.	Security	Includes more parameters	Includes less parameters
11.	Pre-processing	Not required	Image enhancement using FFT

ensures lesser implementation complexity and seamless signature regeneration than fingerprint biometric approach.

3.5   Summary of facial biometric methodology for IPP

This chapter discusses facial biometric-driven methodology for hardware security by implanting IP vendor's unique facial biometric evidence within the design (Sengupta and Rathor, 2021). Facial biometric is a contact-less and robust hardware security approach that exploits naturally unique facial features from the facial image of IP designer/vendor to ensure piracy detection and nullifying false IP ownership claim. Figure 3.2 shows the overview of facial biometric-driven hardware security. As apparent from figure, facial biometric methodology first generates the facial signature corresponding to facial image of an IP vendor. Subsequently, the produced facial signature is transformed into secret information for inserting into the target DSP IP core design. The embedding process includes the following subprocesses:

1. Transforming the algorithmic description (behavioral description/transfer function) of the chosen DSP application into DFG.
2. Subsequently, generating scheduled DFG of the design based on resource constraints chosen by IP designer.
3. Generating secret hardware security constraints (secret information) based on IP vendor-specified encoding rule, for embedding inside the design during the stage of register allocation (of HLS).

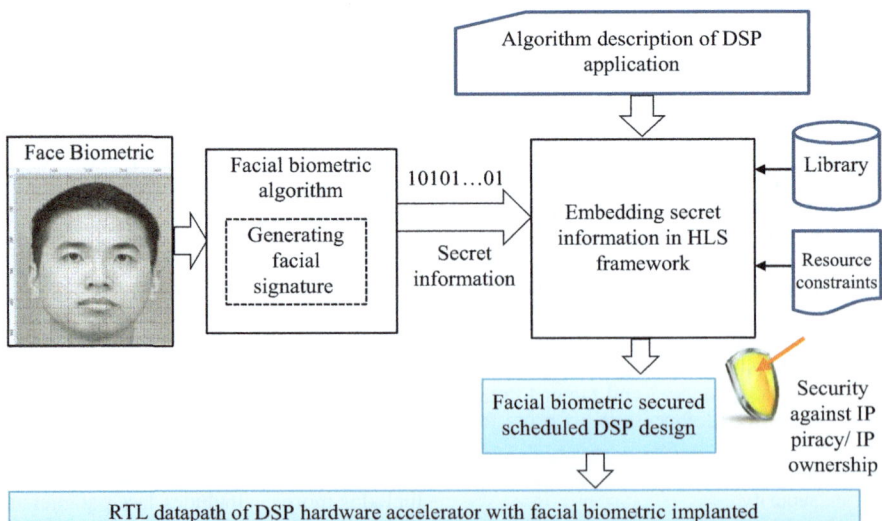

Figure 3.2   Overview of facial biometric methodology for hardware security

4. Performing the embedding of IP vendor selected facial security constraints into the scheduled DSP design using HLS. Thus, the facial biometric secured hardware IP design is obtained.

Next, we discuss the summary of major steps involved in facial biometric-based hardware security methodology. The process of securing target hardware IP core using facial biometric includes various steps, as shown in Figure 3.3:

(a) Capturing the face using high-resolution digital camera and placing it into a specific grid size/spacing (selected by true IP vendor for secret signature generation).
(b) Deriving the nodal points on captured facial image corresponding to the chosen facial features set by IP vendor.

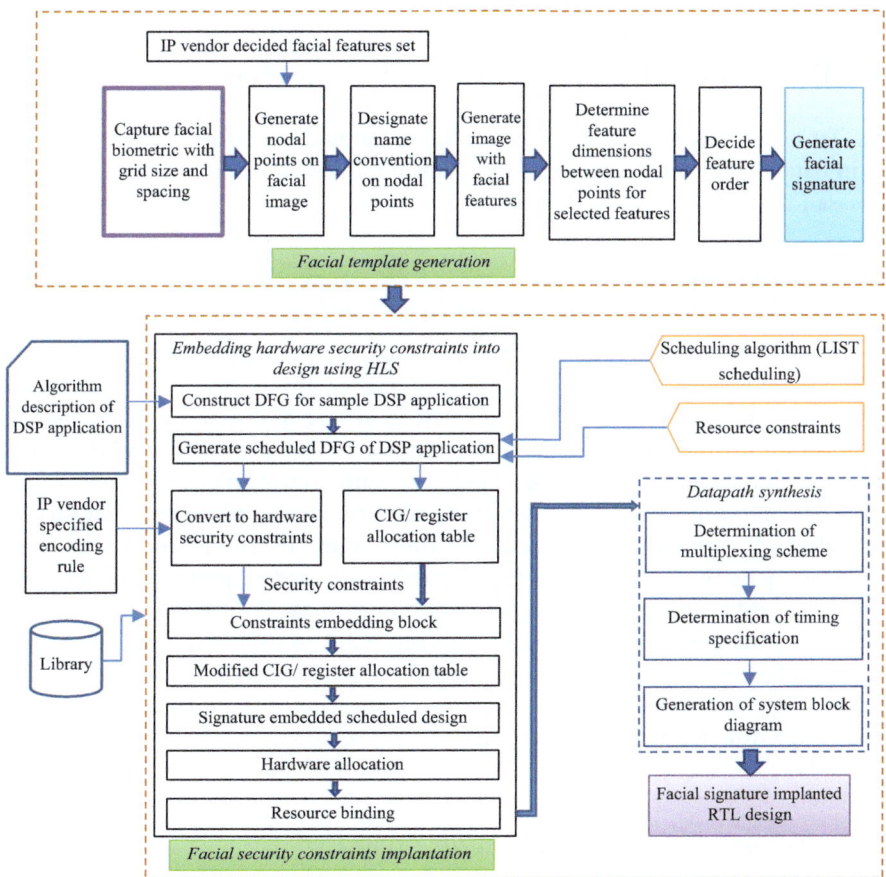

Figure 3.3   Process flow of generating secured IP cores using facial biometric approach

(c) Assigning the naming resolution to nodal points and thereby generating the image with facial features.

(d) Calculating feature dimensions (using Manhattan distance) for the IP vendor chosen facial features for generating facial signature. Subsequently, converting the dimensional information of each feature into equivalent binary.

(e) Subsequently, facial signature as a digital template is generated after concatenating the features based on chosen feature's order by the IP vendor.

(f) Next, each bit of generated signature in converted into covert constraints for hardware security by operating the encoding rules decided by the IP vendor.

(g) Thereafter, the implantation of derived security constraints inside the design, during register allocation stage of HLS is performed.

(h) Thus, post embedding the facial biometric information corresponding to facial image of IP vendor, secured RTL design is obtained using HLS, followed by datapath and controller synthesis.

3.6   Details of facial biometric methodology for IPP

Now, let us discuss the facial biometric-based hardware security methodology for IPP in detail. The process of securing hardware IP cores using facial biometric is discussed through the following steps.

3.6.1   Capturing facial biometric of IP vendor and subjecting to a specific grid size and spacing

To produce the facial biometric information, first, the facial image of an IP vendor is taken using a high-resolution digital imaging device. The captured facial image is then subjected to a specific grid size and spacing. This enables the generation of accurate facial nodal feature points and subsequently determining facial feature dimensions. This also helps during facial biometric signature regeneration process for hardware security (while proving IP ownership and detecting pirated versions), where the feature locations and dimensions would easily be revived from the facial image with grid magnitude and spacing. A sample captured facial image corresponding to an IP vendor with specific grid magnitude and spacing is shown in Figure 3.4. As shown in the figure, the grid size chosen by IP vendor for placing the facial image is [450 pixels × 550 pixels]. This, therefore, helps in generating the accurate nodal feature points and feature dimensions for generating facial biometric information (Sengupta and Rathor, 2021).

3.6.2   Generate facial nodal feature points

Post obtaining the captured facial image with a specific grid size, facial nodal feature points are generated based on IP vendor-decided facial features. IP vendor elects what facial features are to be transformed into facial signature (in the form of

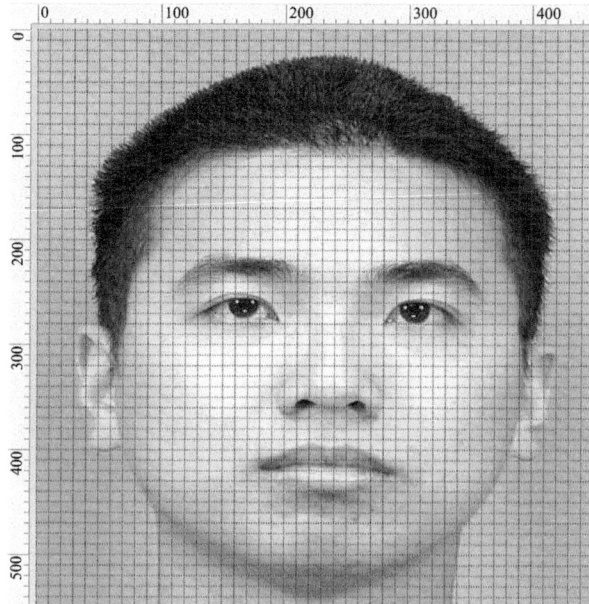

Figure 3.4   Sample-captured facial image of IP vendor with specific grid size and spacing

digital template) and implanted into the IP design. However, IP vendor may vary (by keeping or discarding) the strength of facial features to obtain digital template of desired strength (size). Let us say, following facial features are selected by an IP vendor for the purpose of facial signature generation:

1. HFH = Height of Forehead
2. IPD= Inter Pupillary Distance
3. BOB = Bio-Ocular Breadth
4. IOB = Inter-Ocular Breadth
5. OB = Ocular Breadth
6. WNR = Width of Nasal Ridge
7. WF = Width of face
8. HF = Height of Face
9. WNB = Width of Nasal Base
10. NB = Nasal Breadth
11. OCW = Oral Commissure Width

Subsequently, as per the IP vendor-chosen set of facial features, nodal points are located on the facial image. Since each facial feature is identified as the measure of distance between two conforming nodal points. Therefore, the nodal points for the selected features are located by marking their end points. This is performed by scanning the facial image from left to right and top to bottom. A sample facial

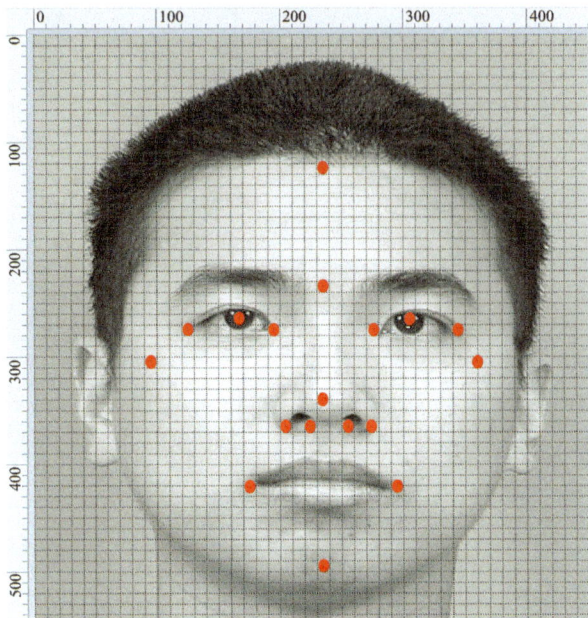

Figure 3.5   Sample facial image of IP vendor with designated nodal feature point

image of IP vendor with designated nodal points is shown in Figure 3.5, where each nodal point is marked with red color (red dots). As shown in the figure, there are a total of 18 facial nodal points corresponding to IP vendor-selected 11 facial features. These nodal points are used for computing (measuring) the dimensions of facial features chosen by the IP vendor.

3.6.3   Assign naming convention on facial nodal feature points

Once nodal points are located on the captured facial image (based on the facial features chosen by IP vendor), their naming resolution is made. The facial image with a naming resolution of facial nodal points is shown in Figure 3.6, where corresponding nodal points are designated using P1, P2, ..., P18. Further, the naming resolution of nodal point pairs for each facial feature is shown in Table 3.2, where each facial feature is represented using nodal points pair. For example, facial feature "HFH (Height of Forehead)" is represented between nodal points P1 and P2. Similarly, facial feature "IPD (Inter Pupillary Distance)" is represented between nodal points P3 and P4. Thus, the naming resolution of all the nodal point pairs with respect to facial features is performed. Subsequently, IP vendor-chosen features are located on the facial image by connecting the respective nodal points. The facial image with IP vendor chosen eleven features is shown in Figure 3.7, where each facial feature is represented using a different color. Furthermore,

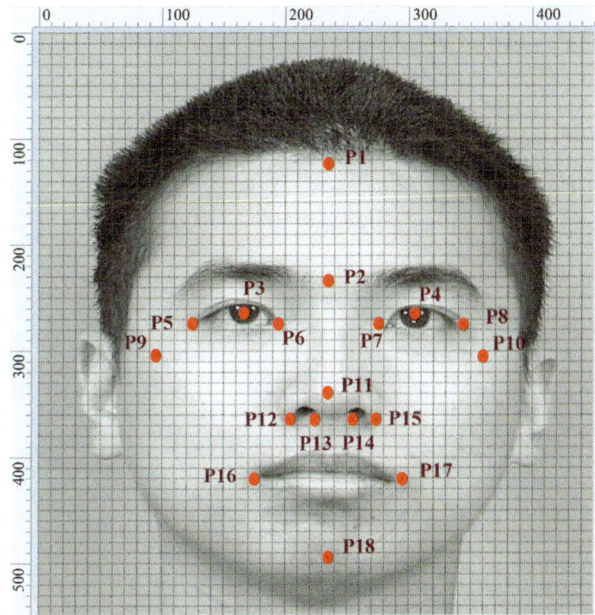

Figure 3.6   Facial image of IP vendor with nodal points naming conventions

Table 3.2   Co-ordinates of each facial feature, calculated using nodal points marked on the facial image

S. no.	Facial features	Naming convention of points	Co-ordinates (a1, b1)–(a2, b2)
1.	HFH	(P1)–(P2)	(235, 125)–(235, 235)
2.	IPD	(P3)–(P4)	(165, 265)–(305, 265)
3.	BOB	(P5)–(P8)	(125, 275)–(345, 275)
4.	IOB	(P6)–(P7)	(195, 275)–(275, 275)
5.	OB	(P5)–(P6)	(125, 275)–(195, 275)
6.	WNR	(P2)–(P11)	(235, 235)–(235, 340)
7.	WF	(P9)–(P10)	(95, 305)–(360,305)
8.	HF	(P1)–(P18)	(235, 125)–(235,495)
9.	WNB	(P13)–(P14)	(225, 365)–(255, 365)
10.	NB	(P12)–(P15)	(205, 365)–(275, 365)
11.	OCW	(P16)–(P17)	(175, 420)–(295, 420)

Figure 3.8 depicts the skeleton of 11 facial features on the facial image selected by the IP vendor.

3.6.4   Determining feature dimensions

Post obtaining facial image with designated nodal points respective to IP vendor-selected facial features, determination of dimension of each feature is performed.

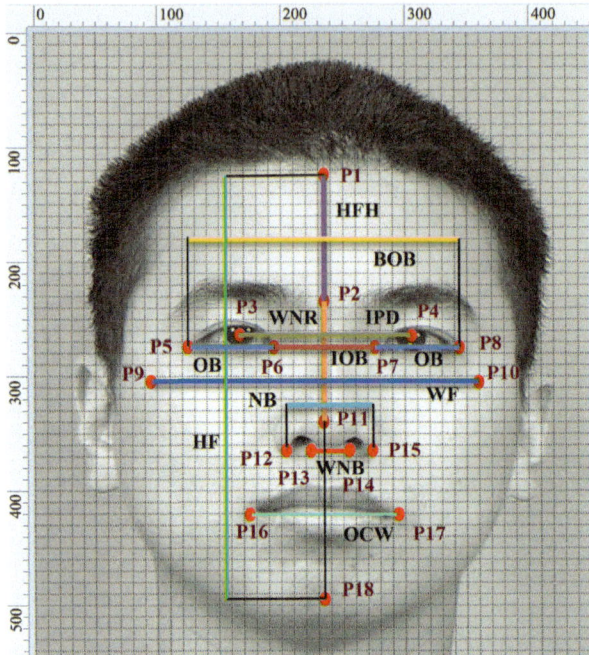

Figure 3.7   IP vendor facial image with selected facial features

For this purpose, first, the co-ordinates (p, q) of each nodal point are located. The respective co-ordinates are obtained using the location of nodal point on the facial image with a specific grid magnitude. For all the nodal points respective to each facial feature, co-ordinates are shown in Table 3.2, where (a1, b1) and (a2, b2) indicate the location of the first and second nodal points, respectively. Next, based on the co-ordinate information of all the nodal points pair, feature dimension is determined using Manhattan distance (for each facial feature). Each feature is represented as the measure of Manhattan distance between respective nodal points. The computed feature dimension (in decimal format) corresponding to each IP vendor-selected facial feature is shown in Table 3.3. Subsequently, the feature dimension of each facial feature is transformed to their equivalent binarized form. Thus, the binarized feature dimension corresponding to IP vendor-selected facial feature is obtained.

3.6.5   Generating facial signature for IP vendor-defined feature order

Once the dimensions of all facial features (F) chosen by IP vendor are determined using Manhattan distance, IP vendor specifies the order of feature concatenation for producing the facial biometric signature. The facial signature is produced by performing the concatenation of each feature dimension in the form of their binary equivalent as per the order chosen by the vendor. There can be multiple possible ordering of features for generating the facial template. For example, corresponding to

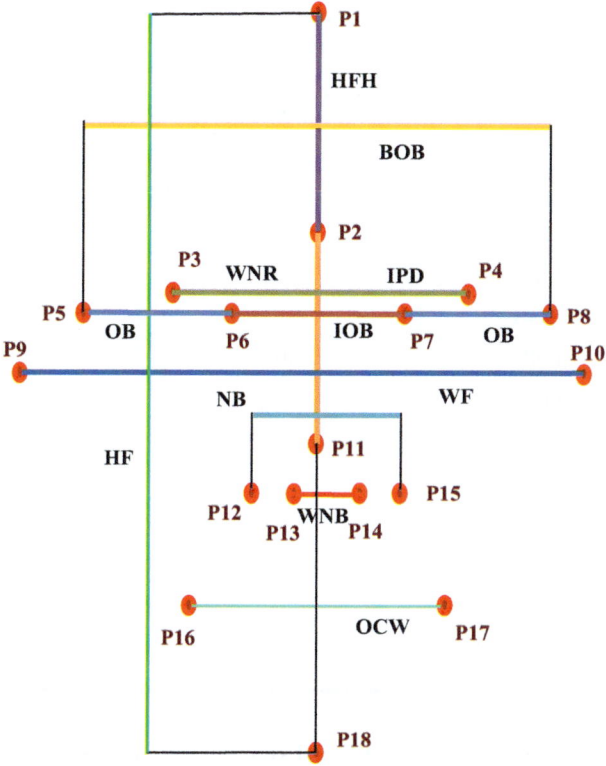

Figure 3.8   Skeleton of IP vendor-selected set of facial features

11 facial features, there can be "2^11= 2,048" different ways to generate facial signature. Let us say if an IP vendor chosen feature concatenation order is as follows:

(IPD) & (HFH) & (IOB) & (BOB) & (WNR) & (OB) & (HF) & (WF) & (NB) & (WNB) & (OCW).

Therefore, by concatenating the binarized information of each facial feature as per the IP vendor-decided concatenation, the generated facial signature is:

"10001100110111010100001101110011010011000110110110010
10000100110001101111011111000".   (3.1)

The strength of generated facial signature is 81 bits, consisting of 40, 0s, and 41, 1s. However, an IP vendor may vary the signature strength by varying the strength of selected facial features for signature generation. This therefore offers flexibility to an IP vendor to generate facial signature of desired strength. Thus, to protect larger designs, facial signature of larger strength can be produced by picking a greater number of facial features. On the other hand, for securing the designs of medium size, a smaller (appropriate) size facial signature can be obtained by picking lesser number of facial features.

Table 3.3   Determining feature dimension of IP vendor-selected facial features and their binarized representation

S.no.	Facial features	Feature dimension (Manhattan distance) = |a2–a1|+|b2–b1|	In binary
1.	HFH	110	1101110
2.	IPD	140	10001100
3.	BOB	220	11011100
4.	IOB	80	1010000
5.	OB	70	1000110
6.	WNR	105	1101001
7.	WF	265	100001001
8.	HF	370	101110010
9.	WNB	30	11110
10.	NB	70	1000110
11.	OCW	120	1111000

However, exact regeneration of facial signature is impossible for an adversary as it depends on several intricate security parameters such as (a) grid size/spacing used in determining the precise coordinates of nodal points; (b) type of chosen feature set by genuine IP vendor (among the exhaustive features for producing digital template); and their ordering (c) position of signature bits (0s,1s) in digital template, all are unknown to an adversary.

Case study of designing secured N-point DFT using facial biometrics

So far, we discussed the process of generating facial biometric signature corresponding to captured facial image of an IP vendor. Now, let us discuss the methodology Bf designing secured N-point DIT-FFT IP core using facial biometric-based hardware security, as a case study. First, we discuss the mathematical function (behavioral description) of DIT-FFT application. The DIT-FFT is used to convert time-domain signal into frequency-domain signals. The mathematical description of the DIT-FFT is shown below (Salivahanan and Vallavaraj, 2001; Sengupta and Mohanty, 2019):

$$Y(k) = \sum_{n=0}^{N-1} x(n) B_N^{nk}, \quad 0 \le k \le N-1 \tag{3.2}$$

where $Y(k)$ represents the DFT of a sequence $x(n)$, $x(n)$ corresponds to a sequence of N values and B_N indicates the complex-valued phase factor. Now, to decimate N-valued sequence (as shown in (3.2)), the decimated *x(n)* can be obtained by decimating the sequence into two N/2 point sequences (even and odd values of *x(n)*):

$$Y(k) = \sum_{n=0,n \ even}^{N-1} x(n)B_N^{nk} + \sum_{n=0,n \ odd}^{N-1} x(n)B_N^{nk} \tag{3.3}$$

Now, by replacing $n = 2r$ if n is even and $n = 2r+1$ if n is odd:

$$\begin{aligned} Y(k) &= \sum_{r=0}^{N/2-1} x(2r)B_N^{2rk} + \sum_{r=0}^{N/2-1} x(2r+1)B_{N/2}^{(2r+1)k} \\ &= \sum_{r=0}^{N/2-1} x(2r)B_{N/2}^{rk} + B_N^k \sum_{r=0}^{N/2-1} x(2r+1)B_{N/2}^{rk} \end{aligned} \tag{3.4}$$

Here, $B_N^2 = B_{N/2}$. Now the above equation can be written as

$$Y(k) = A(k) + B_N^k C(k), \quad k = 0, 1, \dots \frac{N}{2} - 1 \tag{3.5}$$

In the above (3.5), $A(k)$ and $C(k)$ are the two $N/2$ point DFTs of odd and even sequences respectively and "k" value for both $A(k)$ and $C(k)$ belongs between 0 and $(N/2)-1$ as shown in (3.5). Now based on (3.5), the derived signal flow graph for $N = 8$ (for the sake of brevity) will have following representation as shown below:

$$Y(0) = A(0) + B_8^0 * C(0) \tag{3.5.1}$$

$$Y(1) = A(1) + B_8^1 * C(1) \tag{3.5.2}$$

$$Y(2) = A(2) + B_8^2 * C(2) \tag{3.5.3}$$

$$Y(3) = A(3) + B_8^3 * C(3) \tag{3.5.4}$$

$$Y(4) = A(0) + B_8^4 * C(0) \tag{3.5.5}$$

$$Y(5) = A(1) + B_8^5 * C(1) \tag{3.5.6}$$

$$Y(6) = A(2) + B_8^6 * C(2) \tag{3.5.7}$$

$$Y(7) = A(3) + B_8^7 * C(3) \tag{3.5.8}$$

Thus, the above equations ((3.5.1)–(3.5.8)) represent the outputs from $Y(0)$ to Y (7). Similarly, each of the generated $N/2$-point DFTs ($A(x)$ and $C(x)$) can also be decimated further into two $N/4$-point DFTs. Note: *in a N-point DFT there will be (Log 2^N) stages. For example, in case of 8-point DFT there will be three stages e.g., stage-1 with four 2-point DFTs, stage-2 with two four N/4-point DFTs and stage-3 will have one 8-point DFT.*

Next, we convert the above equations corresponding to DIT-FFT into the control DFG, the DFG for DIT-FFT application (for $N=8$) is shown in Figure 3.9.

Subsequently, DFG is scheduled based on the available dependency information of storage variables, IP vendor-specified resource constraints and scheduling algorithm. To schedule the design, let us consider an IP vendor-specified resource, constraints are two adder and two multipliers and LIST scheduling algorithm is used for performing the scheduling. The scheduled DFG design corresponding to DIT-FFT is shown in Figure 3.10, where J0–J39 are the storage variables and required registers (K^1, K^2, ..., K^{24}) are designated using different colors corresponding to DIT-FFT design. As evident, six control steps (CS_0–CS_5) were required to schedule the design and obtain the final output. Next, the register

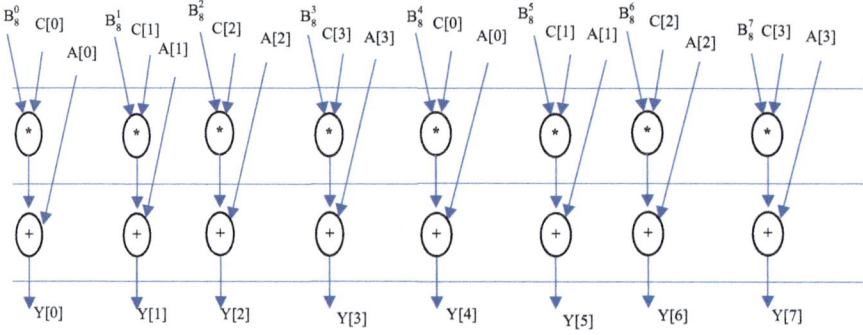

Figure 3.9   DFG of N-point DIT-FFT application for N = 8

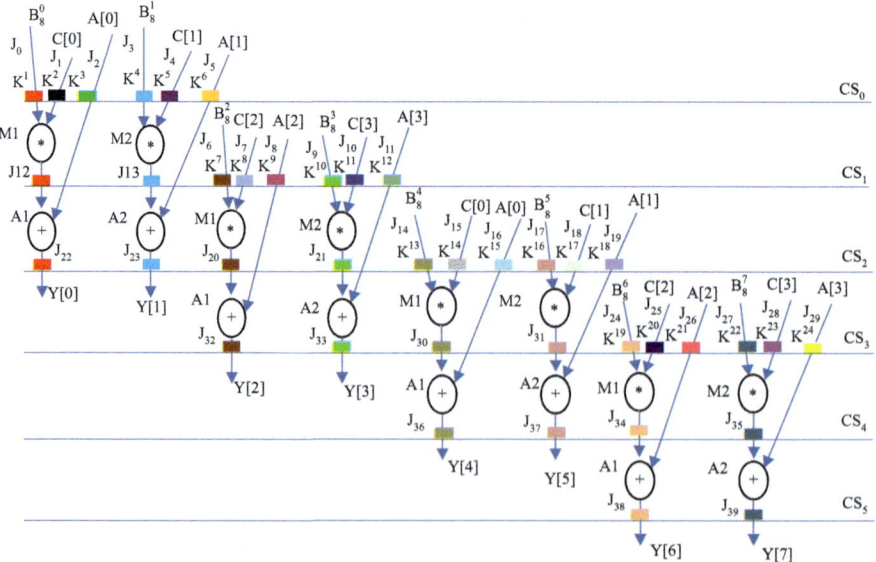

Figure 3.10   Scheduled DFG of N-point DIT-FFT application for N = 8

allocation table corresponding to scheduled DFG can be obtained. Furthermore, scheduled DFG is subjugated to generate covert constraints for hardware security. The process of generating the hardware security constraints accepts the following inputs: (a) target DSP application as a scheduled DFG/transfer function; (b) vendor specified secret encoding rule; and (c) facial biometric digital signature, and generates the secret constraints for hardware security corresponding to generated signature (from facial biometric). The scheduled DFG is used to extract the storage variables required, which are used to form the security constraints pair subsequently. Let us consider if an IP vendor-specified encoding rule to generate covert constraints for hardware security corresponding to facial biometric signature is as follows:

- Signature bit "1" signifies the embedding of security constraints between odd–odd storage variable pair $<Jx, Jy>$ of the scheduled DFG, where x and y represent the specific storage variable used for pairing.
- Signature bit "0" signifies the embedding of security constraints between even–even storage variable pair $<Jx, Jy>$ of the scheduled DFG. The encoding is shown in Figure 3.11.

Then the resulting secret constraints for hardware security corresponding to facial signature bit "0" and "1" are shown in Tables 3.4 and 3.5, respectively.

Next, we perform the embedding of generated secret constraints into the design during HLS. The constraints embedding block accept the following two inputs such as (a) generated security constraints corresponding to target scheduled design and facial biometric signature and (b) register allocation information. The register allocation table is used for embedding the hardware security constraints by locally altering (modifying) the register assignments using the following constraint rule such that "two storage variables in a pair cannot be assigned to the same register." Finally, the facial signature implanted modified register allocation table corresponding to DIT-FFT is obtained. The pre-embedding (without facial signature) and post embedding (with facial signature) register allocation table is shown in Tables 3.6 and 3.7, respectively, where the columns represent the number of registers needed to accommodate the storage variables of the design. However, a register can be used by multiple

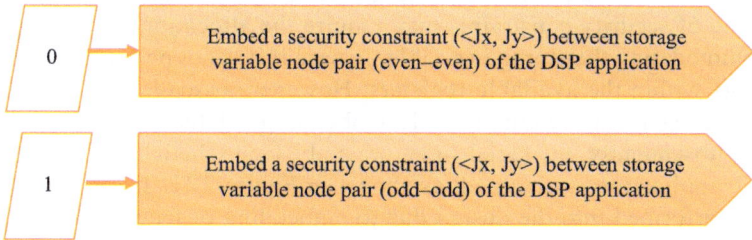

*Figure 3.11 Encoding rule for hardware security constraints generation
(Sengupta and Rathor, 2021)*

Table 3.4 Facial biometric signature-based secret hardware security constrains corresponding to signature bit "0"

$\langle J_0, J_2 \rangle$	$\langle J_0, J_{26} \rangle$	$\langle J_2, J_{14} \rangle$	$\langle J_2, J_{38} \rangle$	–
$\langle J_0, J_4 \rangle$	$\langle J_0, J_{28} \rangle$	$\langle J_2, J_{16} \rangle$	$\langle J_4, J_{18} \rangle$	–
$\langle J_0, J_6 \rangle$	$\langle J_0, J_{30} \rangle$	$\langle J_2, J_{18} \rangle$	$\langle J_4, J_{20} \rangle$	–
$\langle J_0, J_8 \rangle$	$\langle J_0, J_{32} \rangle$	$\langle J_2, J_{20} \rangle$	$\langle J_4, J_{22} \rangle$	–
$\langle J_0, J_{10} \rangle$	$\langle J_0, J_{34} \rangle$	$\langle J_2, J_{22} \rangle$	–	–
$\langle J_0, J_{12} \rangle$	$\langle J_0, J_{36} \rangle$	$\langle J_2, J_{24} \rangle$	–	–
$\langle J_0, J_{14} \rangle$	$\langle J_0, J_{38} \rangle$	$\langle J_2, J_{26} \rangle$	–	–
$\langle J_0, J_{16} \rangle$	$\langle J_2, J_4 \rangle$	$\langle J_2, J_{28} \rangle$	–	–
$\langle J_0, J_{18} \rangle$	$\langle J_2, J_6 \rangle$	$\langle J_2, J_{30} \rangle$	–	–
$\langle J_0, J_{20} \rangle$	$\langle J_2, J_8 \rangle$	$\langle J_2, J_{32} \rangle$	–	–
$\langle J_0, J_{22} \rangle$	$\langle J_2, J_{10} \rangle$	$\langle J_2, J_{34} \rangle$	–	–
$\langle J_0, J_{24} \rangle$	$\langle J_2, J_{12} \rangle$	$\langle J_2, J_{36} \rangle$	–	–

Table 3.5 Facial biometric signature-based secret hardware security constrains corresponding to signature bit "1"

$\langle J_1, J_3 \rangle$	$\langle J_1, J_{27} \rangle$	$\langle J_3, J_{15} \rangle$	$\langle J_3, J_{39} \rangle$	–
$\langle J_1, J_5 \rangle$	$\langle J_1, J_{29} \rangle$	$\langle J_3, J_{17} \rangle$	$\langle J_5, J_7 \rangle$	–
$\langle J_1, J_7 \rangle$	$\langle J_1, J_{31} \rangle$	$\langle J_3, J_{19} \rangle$	$\langle J_5, J_9 \rangle$	–
$\langle J_1, J_9 \rangle$	$\langle J_1, J_{33} \rangle$	$\langle J_3, J_{21} \rangle$	$\langle J_5, J_{11} \rangle$	–
$\langle J_1, J_{11} \rangle$	$\langle J_1, J_{35} \rangle$	$\langle J_3, J_{23} \rangle$	$\langle J_5, J_{13} \rangle$	–
$\langle J_1, J_{13} \rangle$	$\langle J_1, J_{37} \rangle$	$\langle J_3, J_{25} \rangle$	–	–
$\langle J_1, J_{15} \rangle$	$\langle J_1, J_{39} \rangle$	$\langle J_3, J_{27} \rangle$	–	–
$\langle J_1, J_{17} \rangle$	$\langle J_3, J_5 \rangle$	$\langle J_3, J_{29} \rangle$	–	–
$\langle J_1, J_{19} \rangle$	$\langle J_3, J_7 \rangle$	$\langle J_3, J_{31} \rangle$	–	–
$\langle J_1, J_{21} \rangle$	$\langle J_3, J_9 \rangle$	$\langle J_3, J_{33} \rangle$	–	–
$\langle J_1, J_{23} \rangle$	$\langle J_3, J_{11} \rangle$	$\langle J_3, J_{35} \rangle$	–	–
$\langle J_1, J_{25} \rangle$	$\langle J_3, J_{13} \rangle$	$\langle J_3, J_{37} \rangle$	–	–

storage variables but not at the same time. In Table 3.7, the storage variables marked in red color are indicating the local alterations, post embedding the secret security constraints into the design (covertly). Moreover, sometimes the embedding of security constraints may require the allocation of new register(s) for accommodating the storage variable, in case if it is not possible to satisfy the constraints within the available registers. However, as evident from Table 3.7, no extra register was required while embedding all the facial security constraints. Subsequently, the facial signature embedded scheduled DFG of N-point DIT-FFT application is obtained, as shown in Figure 3.12. It reflects the modification performed into the design due to embedding of facial signature. Thus, post embedding the facial biometric information corresponding to facial image of IP vendor, secured RTL design is obtained using HLS, followed by datapath and controller synthesis.

Table 3.6 *Register allocation information corresponding to scheduled DIT-FFT design (before embedding facial biometric signature)*

CS	K^1	K^2	K^3	K^4	K^5	K^6	K^7	K^8	K^9	K^{10}	K^{11}	K^{12}	K^{13}	K^{14}	K^{15}	K^{16}	K^{17}	K^{18}	K^{19}	K^{20}	K^{21}	K^{22}	K^{23}	K^{24}
CS_0	J_0	J_1	J_2	J_3	J_4	J_5	–	–	–	–	–	–	–	–	–	–	–	–	–	–	–	–	–	–
CS_1	J_{12}	–	–	J_{13}	–	–	J_6	J_7	J_8	J_9	J_{10}	J_{11}	–	–	–	–	–	–	–	–	–	–	–	–
CS_2	J_{22}	–	–	J_{23}	–	–	J_{20}	–	–	J_{21}	–	–	J_{14}	J_{15}	J_{16}	J_{17}	J_{18}	J_{19}	–	–	–	–	–	–
CS_3	–	–	–	–	–	–	J_{32}	–	–	J_{33}	–	–	J_{30}	–	–	J_{31}	–	–	J_{24}	J_{25}	J_{26}	J_{27}	J_{28}	J_{29}
CS_4	–	–	–	–	–	–	–	–	–	–	–	–	J_{36}	–	–	J_{37}	–	–	J_{34}	–	–	J_{35}	–	–
CS_5	–	–	–	–	–	–	–	–	–	–	–	–	–	–	–	–	–	–	J_{38}	–	–	J_{39}	–	–

Table 3.7 *Modified register allocation information post embedding facial security constraints corresponding to scheduled DIT-FFT design*

CS	K^1	K^2	K^3	K^4	K^5	K^6	K^7	K^8	K^9	K^{10}	K^{11}	K^{12}	K^{13}	K^{14}	K^{15}	K^{16}	K^{17}	K^{18}	K^{19}	K^{20}	K^{21}	K^{22}	K^{23}	K^{24}
CS_0	J_0	J_1	J_2	J_3	J_4	J_5	–	–	–	–	–	–	–	–	–	–	–	–	–	–	–	–	–	–
CS_1	–	J_{12}	–	–	J_{13}	–	J_6	J_7	J_8	J_9	J_{10}	J_{11}	–	–	–	–	–	–	–	–	–	–	–	–
CS_2	–	J_{22}	–	–	J_{23}	–	J_{20}	–	–	J_{21}	–	–	J_{14}	J_{15}	J_{16}	J_{17}	J_{18}	J_{19}	–	–	–	–	–	–
CS_3	–	–	–	–	–	–	J_{32}	–	–	J_{33}	–	–	J_{30}	–	–	J_{31}	–	–	J_{24}	J_{25}	J_{26}	J_{27}	J_{28}	J_{29}
CS_4	–	–	–	–	–	–	–	–	–	–	–	–	J_{36}	–	–	J_{37}	–	–	J_{34}	–	–	J_{35}	–	–
CS_5	–	–	–	–	–	–	–	–	–	–	–	–	–	–	–	–	–	–	J_{38}	–	–	J_{39}	–	–

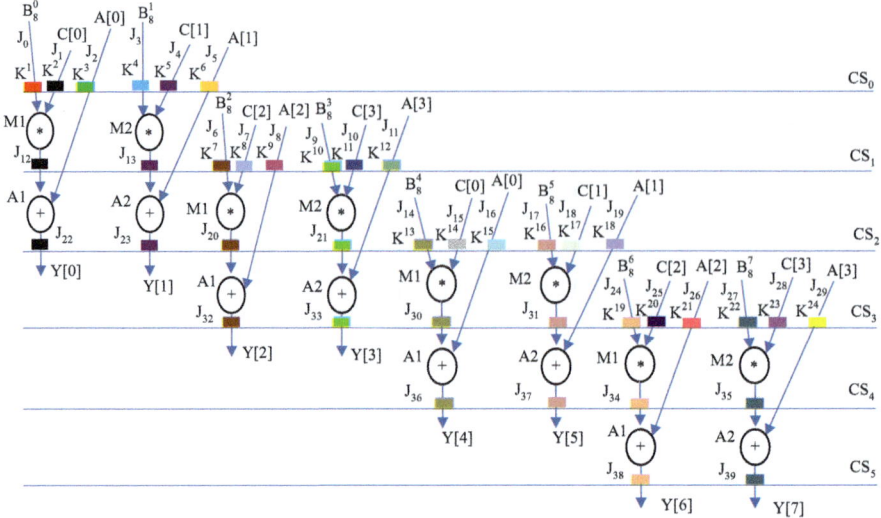

Figure 3.12   Facial signature embedded scheduled DFG of N-point DIT-FFT application for N = 8

3.7   Security properties of facial biometric methodology for hardware security

Facial biometric-based hardware security methodology renders the following security properties (Sengupta and Rathor, 2021):

1. *Zero probability of proving false IP ownership (for an adversary)*: Facial biometric methodology renders robust security against fraudulent claim of IP ownership. It is highly unlikely for an adversary to successfully prove the false claim of IP ownership, because the IP ownership verification process requires the exact matching of digital template of embedded signature (to be extracted from design under test) to the genuine vendor's regenerated facial signature. In other words, positions of "0" and "1" bits and also the strength of 0s and 1s in the digital template of facial biometric are matched. Hence, during the matching process, only the digital evidence of the original IP owner would exactly coincide with the inserted facial signature. This is because of the uniqueness of facial features of an individual; therefore, the facial template of an adversary cannot be the same as that of an original IP vendor, even in case if adversary is a twin or look alike.

2. *Zero probability of evading piracy detection (for an adversary)*: It is highly unlikely for an adversary to evade piracy detection process to successfully infuse the pirated designs into SoCs of CE systems. This is because it is highly challenging (practically impossible) for an adversary to regenerate the embedded facial security constraints corresponding to an original IP vendor

and infuse them into pirated designs. The facial security constraints re-generation process requires decoding of the following crucial security parameters:

(a) IP vendor-specified grid size, used for subjecting the captured facial image.

(b) Exact locations of the nodal points which are not known to the adversary.

(c) Type and number of facial features selected among the features set for generating facial biometric signature (digital template) which are unknown to an adversary.

(d) Concatenation order of the binarized facial features (vendor selected) for generating digital evidence which is also not known to the adversary.

(e) Encoding rules specified by an IP vendor to encode the binarized facial biometric signature to generate secret facial security constraints, which are also not known to an adversary. Thus, for an adversary, it is highly challenging to produce the same digital evidence as that of the original IP vendor and thereby cannot regenerate exact facial security constraints to evade piracy detection.

3. *Security against brute force attack*: Employing brute force attack for finding the exact facial signature combination amongst the huge signature space (2^l) is practically impossible. This disables an attacker in evading piracy detection and proving false IP ownership.

4. *Tamper tolerance*: Facial biometric-based hardware security methodology results in larger number of unique security constraints than IP watermarking, steganography and digital signature. Therefore, the exact regeneration of facial biometric-based secret hardware security constraints is practically impossible. Further, tampering of facial security constraints by finding the exact facial signature is also practically impossible for an adversary.

5. *Security against key extraction*: Since the facial biometric-based hardware security methodology does not employ any key for enabling the security. Therefore, it is not susceptible to key mining attacks in contrast to crypto-based dual-phase hardware steganography and steganography exploiting switch-based key-driven hash-chaining and encoding based approaches (Sengupta and Rathor, 2019; Rathor and Sengupta, 2020).

6. The pre-stored facial biometric image of IP vendor with specific grid size is used for piracy detection. However, in case if the stored image is leaked to an adversary, the exact regeneration of facial signature from compromised/leaked facial biometric image is not possible. This is because the security parameters discussed earlier (from point 2. (a) to (d)) are all unknown to an adversary. In the facial biometric approach, IP vendor does not store his/her facial signature. However, in case if an adversary even manages to derive the exact facial signature, the generation of secret hardware security constraints is not possible because of following details unknown to an adversary:

(a) Final facial signature strength used by IP vendor for embedding.

(b) Secret encoding rule specified by IP vendor for facial security constraints generation.

(c) Final order of storage variables during security constraints generation (storage variables can be either sorted in ascending, descending, or alternate order, etc.).

3.8   Analysis and discussion

This section analyzes the hardware security methodology using facial biometric approach (Sengupta and Rathor, 2021). The following subsections discuss the security strength of facial biometric technique in terms of (1) analyzing the impact of varying facial biometric signature on security strength; (2) analyzing the strength of embedded facial biometric signature in terms of probability of coincidence and tolerance against tampering attack.

3.8.1   Analyzing the security strength based on varying facial signature

An IP vendor may vary the facial signature by adding or deleting his/her facial features from the captured facial image. The more the number of IP vendor-selected facial features, the more is the size of the facial signature. This therefore enables a greater number of secret facial security constraints to be implanted into the design for enabling stronger digital evidence and robust security. Further, the type and strength of facial features elected by the IP vendor also impacts the number of generated encoded constraints for hardware security. The resultant of varying facial signature on the strength of hardware security constraints is assessed using different facial images chosen from the dataset (Multimedia Laboratory Dataset, 2020). The impact of distinct facial signatures, for the similar strength of facial features, on constraints for hardware security, is mentioned in Figure 3.13. As mentioned, the size of facial signature respective to the facial image_1 is 81 bits, where strength of 0s and 1s is 40 and 41 respectively, which represents the security constraints to be implanted for hardware security. Similarly, the biometric template for the facial image_2 is 82 bits, where strength of 0s and 1s is 39 and 43 respectively, which represents the security constraints to be implanted for hardware security. Further, the impact of varying facial features on constraints for hardware security (corresponding to Image_1) is shown in Table 3.8. As evident from the table that the strength of hardware security constraints to be inserted respective to a facial image can be varied by choosing larger/lesser number of facial features. On the contrary, in other techniques (Sengupta and Bhadauria, 2016; Koushanfar et al., 2005; Hong and Potkonjak, 1999; Le Gal and Bossuet, 2012) for reducing or augmenting the strength of security constraints, the vendor needs to vary the signature combination/size. In these techniques, it is difficult to perform a prior estimation that how different signature combinations and strength would impact the strength of constraints that are to be inserted into the target design.

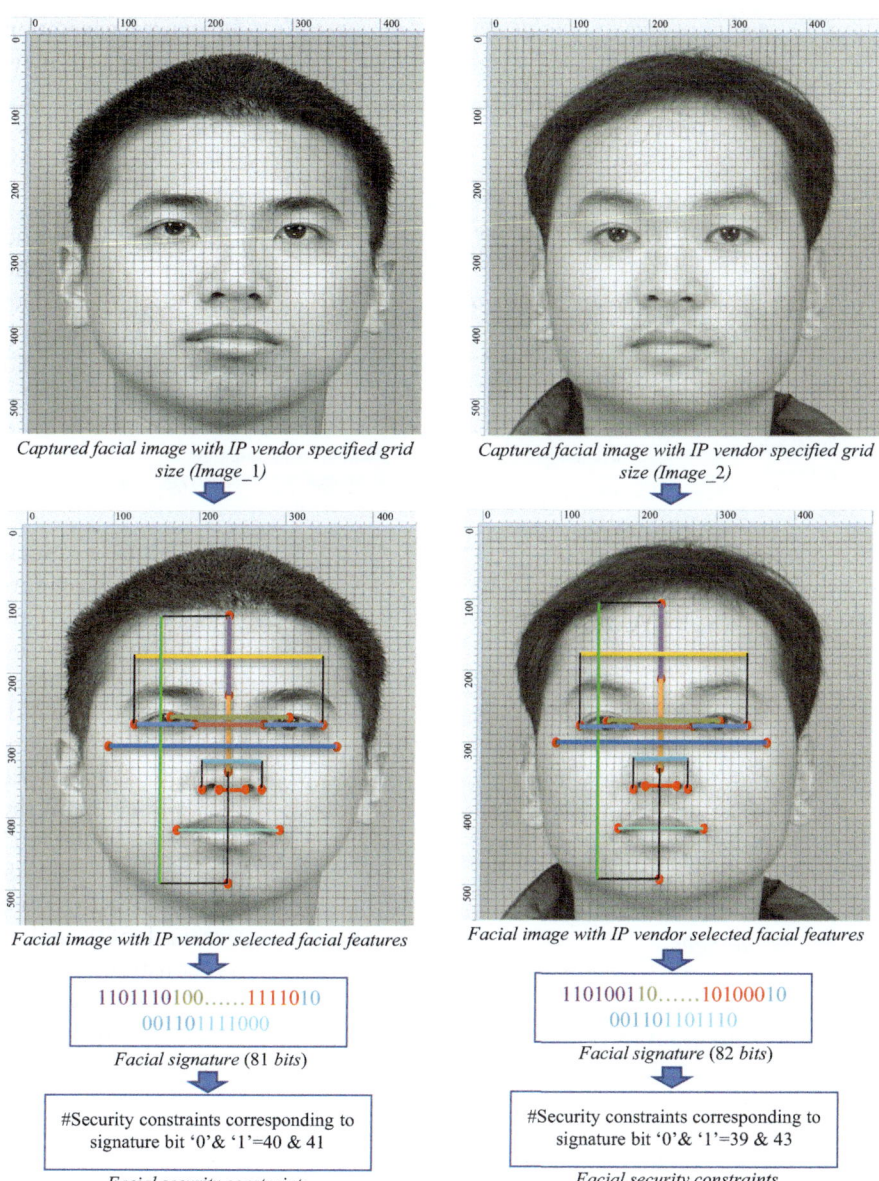

Figure 3.13   Facial biometric signature and its respective covert constraints for hardware security for different facial images (Image_1 and Image_2)

3.8.2   Security analysis

Security against fraudulent claim of IP ownership is analyzed using probability of coincidence (Xp) metric. The value of Xp indicates the dissimilarity index of detecting genuine facial security constraints in an unsecured design; hence for

Table 3.8   Impact of varying facial features on constraints for hardware security corresponding to Image_1

IP vendor-selected facial features	#Security constraints
(IPD) & (HFH) & (IOB) & (BOB) & (WNR) & (OB) & (HF) & (WF)	62
(IPD) & (HFH) & (IOB) & (BOB) & (WNR) & (OB) & (HF) & (WF) & (NB)	69
(IPD) & (HFH) & (IOB) & (BOB) & (WNR) & (OB) & (HF) & (WF) & (NB) & (WNB)	74
(IPD) & (HFH) & (IOB) & (BOB) & (WNR) & (OB) & (HF) & (WF) & (NB) & (WNB) & (OCW)	81

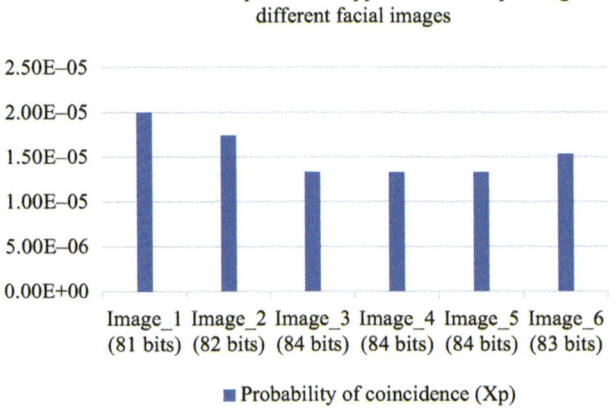

Figure 3.14   Variation in Xp for 8-point DCT application for different facial images

asserting definite proof of IP ownership, value of Xp must be lesser. The value of Xp is computed as follows:

$$Xp = \left(1 - \frac{1}{K^n}\right)^l \tag{3.6}$$

where "K^n" denotes the number of registers/colors required for accommodating the storage variables of the design during register allocation, before implanting secret constraints and "l" denotes the strength of secret constraints (storage variable pairs) for hardware security. The "Xp" value achieved using facial biometric approach for 8-point DCT and DIT-FFT design is shown in Figures 3.14 and 3.15 respectively, for different size facial signatures of different facial images. *Note*: the facial images (Image_1 to Image_6) are adopted from (Multimedia Laboratory Dataset, 2020). As apparent from the figures, a very low value of "Xp" is attained for all the facial

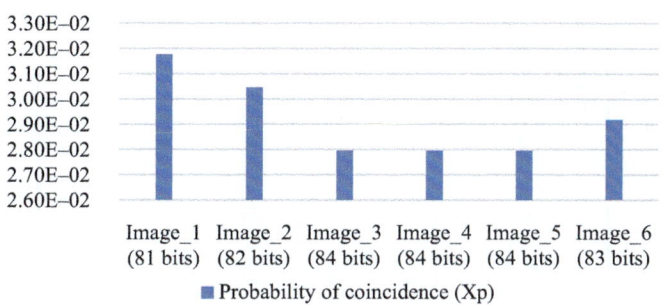

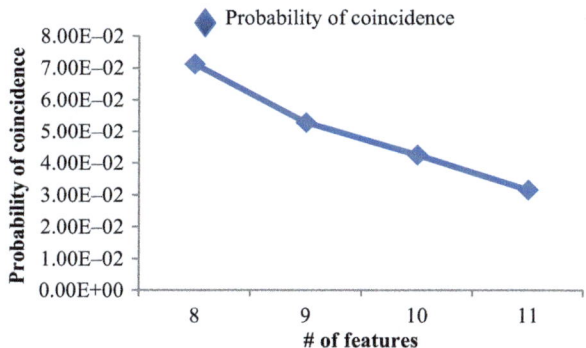

Figure 3.15   Variation in Xp for DIT-FFT application for different facial images

Figure 3.16   Variation in Xp for different number of IP vendor-selected facial features (for facial Image_1)

images for the same strength of facial features (F=11, as discussed earlier in Section 3.6.5). Further, the impact of varying facial features on Xp is shown in Figure 3.16. As apparent, the "Xp" value decreases with the increase in number of facial features. This is because a greater number of facial features results in a greater number of corresponding constraints for hardware security in terms of embedding into the design. Therefore, for ensuring definitive proof of ownership (i.e., lower "Xp"), strength of facial features should be higher for facial signature generation.

The facial biometric for hardware security (Sengupta and Rathor, 2021) is also compared with contemporary approaches such as hardware steganography (Sengupta and Rathor, 2019) and fingerprint biometric (Sengupta and Rathor, 2020). The comparison of "Xp" of facial biometric with hardware steganography is shown in Table 3.9. As mentioned in the table, the "Xp" of facial biometric is analyzed for different facial images with respect to different DSP applications. As

Table 3.9 *Comparison of Xp for facial biometric and hardware steganography corresponding to different facial images*

Bench-marks	# colors (registers)	Facial biometric						Hardware steganography	
		Image_1		Image_2		Image_3, 4 and 5		# stego-constraints	Xp
		(# biometric constraints)	Xp	(# biometric constraints)	Xp	(# biometric constraints)	Xp		
8-point DCT	8	81	2.01E−5	82	1.75E−5	84	1.34E−5	13	1.8E−1
								24	4.1E−2
								43	3.2E−3
8-point IDCT	8	81	2.01E−5	82	1.75E−5	84	1.34E−5	13	1.8E−1
								24	4.1E−2
								43	3.2E−3
FIR	8	81	2.01E−5	82	1.75E−5	84	1.34E−5	20	6.9E−2
								57	4.9E−4
MPEG	14	81	2.47E−3	82	2.29E−3	84	1.98E−3	21	2.1E−1
								52	2.1E−2
								59	1.3E−2
FFT	24	81	3.18E−2	82	3.05E−2	84	2.80E−2	21	4.09E−1
								52	1.09E−1
								59	8.11E−2

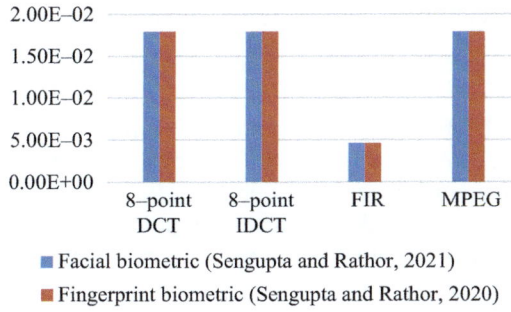

Figure 3.17   Comparison Xp for facial biometric and fingerprint biometric approaches

apparent, the facial biometric approach attains much lower "Xp" (supporting definitive proof of ownership) unlike hardware steganography. The reason being is that the strength of encoded constraints generated using steganography are significantly lesser than the facial biometric methodology. Additionally, the comparison of "Xp" of facial biometric methodology with fingerprint biometric-based hardware security is shown in Figure 3.17. As apparent from the figure, the "Xp" value is the same for facial biometric approach and fingerprint biometric (due to enabling the embedding of almost equal constraints into the hardware). The constraints from fingerprint biometric has been obtained by employing the following encoding rule: bit "0" is encoded as a constraint using storage variable pair (Vx and Vy), where x is "0" and y can be of any integer value and bit "1" is encoded as a constraint using storage variable pair (Vx and Vy), where x is "1" and y can be of any prime integer value.

This encoding rule is different than the encoding rule used for facial biometric based hardware security.

3.9   Conclusion

This chapter discussed a facial biometric-based hardware security methodology (Sengupta and Rathor, 2021) which enables robust security of DSP applications against the threat of IP piracy and fraudulent IP ownership claim. The robustness of facial biometric methodology lies in the fact that the naturally unique facial features of an IP vendor are exploited to generate a unique facial signature, which in turn results in robust digital evidence for inserting into the target design. Unlike non-biometric approaches, an adversary cannot replicate the facial signature of genuine IP vendor to prove false IP ownership. Further, the robust encoding and IP vendor-specified security parameters make it practically impossible for an adversary to evade piracy detection process. Further, the process of verification of IP ownership is seamless and highly robust using facial biometric methodology.

At the end of this chapter, a reader gains the knowledge about the following:

- The significance of robust hardware security against the threats IP piracy and false claim of IP rights.
- Importance of HLS for designing secure DSP coprocessors.
- Contact-free facial biometric methodology for hardware security, in terms of enabling detective control against IP piracy.
- A facial biometric secured design flow of N-point DIT-FFT.
- Comparative study of different hardware security approaches.

3.10   Questions and exercise

1. What is an intellectual property (IP) core?
2. What are the possible security threats during the design process of a hardware IP?
3. Discuss the significance of physical biometrics and their importance in IPP.
4. Discuss the importance of HLS for designing DSP coprocessors.
5. Discuss the different phases of HLS process for generating RTL datapath design.
6. How facial biometric-based hardware security methodology is more robust than hash-based digital signature?
7. Discuss the comparative analysis between different hardware security methodologies.
8. How hardware security methodology using facial biometric is more robust than hardware watermarking?
9. How hardware security methodology using facial biometric is more robust than hardware steganography?
10. Discuss the features of facial biometrics for IPP and its advantage over fingerprint biometrics.
11. Explain the process of generating binarized facial biometric signature from a sample facial image.
12. What are facial nodal points?
13. Explain the process of generating scheduled N-point DIT-FFT design.
14. Discuss the process of facial security constraints generation with a sample facial image.
15. Explain the impact of differing facial features on facial signature and probability of coincidence.
16. Discuss the limitations of the methodology for hardware security using facial biometric. If any?
17. Explain the security property of facial biometric methodology for hardware security.
18. Discuss the security parameters responsible for enabling the robustness of facial biometric methodology.
19. Discuss different facial features.
20. Discuss the process of determining facial feature dimensions.

21. Analyze the case study of securing N-point DIT-FFT design with facial biometric approach.
22. Discuss the process of inserting the facial security constraints during register allocation phase.
23. How the different phases of HLS can be exploited for hardware security?

References

15 nm open cell library. Available: https://si2.org/open-cell-library/, last accessed on January 2020.

Castillo, E., U. Meyer-Baese, A. García, L. Parrilla, and A. Lloris (2007), "IPP@HDL: efficient intellectual property protection scheme for IP cores," *IEEE Transactions on Very Large Scale Integration (VLSI) Systems*, vol. 15, no. 5, pp. 578–591.

Colombier, B. and L. Bossuet (2015), "Survey of hardware protection of design data for integrated circuits and intellectual properties," *IET Computers & Digital Techniques*, vol. 8, no. 6, pp. 274–287.

Coussy, P., D. D. Gajski, M. Meredith and A. Takach (2009), "An introduction to high-level synthesis," *IEEE Design and Test of Computers*, vol. 26, no. 4, pp. 8–17, doi:10.1109/MDT.2009.69.

Cui, A. and C. Chang (2007), "Watermarking for IP protection through template substitution at logic synthesis level," in: *Proc. ISCAS*, New Orleans, LA, 2007, pp. 3687–3690.

Hong, I. and M. Potkonjak (1999), "Behavioral synthesis techniques for intellectual property security," in: *Proc. Design Autom. Conf.*, June 1999, pp. 849–854.

Koushanfar, F., S. Fazzari, C. McCants, *et al.* (2012), "Can EDA combat the rise of electronic counterfeiting?," in: *DAC Design Automation Conference*, San Francisco, CA, pp. 133–138.

Koushanfar, F., I. Hong and M. Potkonjak (2005), "Behavioral synthesis techniques for intellectual property protection," *ACM Transactions on Design Automation of Electronic Systems*, vol. 10, no. 3, pp. 523–545.

Le Gal, B. and L. Bossuet (2012), "Automatic low-cost IP watermarking technique based on output mark insertions," *Design Automation for Embedded System*, vol. 16, no. 2, pp. 71–92.

Mahdiany, H. R., A. Hormati and S. M. Fakhraie (2001), "A hardware accelerator for DSP system design," in: *Proc. ICM*, pp. 141–144.

McFarland, M. C., Alice C. Parker, and Raul Camposano (1988), "Tutorial on high-level synthesis," in: *Proceedings of the 25th ACM/IEEE Design Automation Conference (DAC '88)*, Washington, DC: IEEE Computer Society Press, pp. 330–336.

Multimedia Laboratory Datasets, Available: http://mmlab.ie.cuhk.edu.hk/datasets.html, last accessed on June 2020.

Newbould, R. D., J. D. Carothers, and J. J. Rodriguez (2002), "Watermarking ICs for IP protection," *Electronics Letters*, vol. 38, no. 6, pp. 272–274.

Ni, M. and Z. Gao (2005), "Detector-based watermarking technique for soft IP core protection in high synthesis design level," *in Proc. CCS*, Hong Kong, pp. 1348–1352.

Pilato, C., S. Garg, K. Wu, R. Karri and F. Regazzoni (2018), "Securing hardware accelerators: a new challenge for high-level synthesis," *IEEE Embedded Systems Letters*, vol. 10, no. 3, pp. 77–80.

Plaza, S. M. and I. L. Markov (2015), "Solving the third-shift problem in IC piracy with test-aware logic locking," *IEEE Transactions on Computer-Aided Design of Integrated Circuits and Systems*, vol. 34, no. 6, pp. 961–971.

Rathor, M. and A. Sengupta (2020), "IP core steganography using switch based key-driven hash-chaining and encoding for securing DSP kernels used in CE systems," *IEEE Transactions on Consumer Electronics*, vol. 66, pp. 251–260, doi: 10.1109/TCE.2020.3006050.

Roy, J. A., F. Koushanfar, and I. L. Markov (2008), "EPIC: ending piracy of integrated circuits," in: *2008 Design, Automation and Test in Europe*, pp. 1069–1074.

Salivahanan, S. and A. Vallavaraj (2001), *Digital Signal Processing*, McGraw-Hill Education (India) Pvt. Limited, ISBN: 9780074639962.

Schneiderman, R. (2010), "DSPs evolving in consumer electronics applications," *IEEE Signal Processing Magazine*, vol. 27, no. 3, pp. 6–10.

Sengupta, A. and S. Bhadauria (2016), "Exploring low cost optimal watermark for reusable IP cores during high level synthesis," *IEEE Access*, vol. 4, pp. 2198–2215.

Sengupta, A. and S. P. Mohanty (2019), *IP Core Protection and Hardware-Assisted Security for Consumer Electronics*, The Institute of Engineering and Technology (IET), ISBN: 978-1-78561-799-7, e-ISBN: 978-1-78561-800-0.

Sengupta, A. and M. Rathor (2019a), "IP core steganography for protecting DSP kernels used in CE systems," *IEEE Transactions on Consumer Electronics*, vol. 65, no. 4, pp. 506–515.

Sengupta, A. and M. Rathor (2019b), "Crypto-based dual-phase hardware steganography for securing IP cores," *IEEE Letters of the Computer Society*, vol. 2, no. 4, pp. 32–35.

Sengupta, A. and M. Rathor (2020), "Securing hardware accelerators for CE systems using biometric fingerprinting," *IEEE Transactions on VLSI Systems*, vol. 28, pp. 1979–1992, doi: 10.1109/TVLSI.2020.2999514.

Sengupta, A. and M. Rathor (2021), "Facial biometric for securing hardware accelerators," *IEEE Transactions on VLSI Systems*, vol. 29, no. 1, pp. 112–123, doi: 10.1109/TVLSI.202 0.3029245.

Sengupta, A., E. R. Kumar and N. P. Chandra (2019), "Embedding digital signature using encrypted-hashing for protection of DSP cores in CE," *IEEE Transactions on Consumer Electronics*, vol. 65, no. 3, pp. 398–407.

Chapter 4

Secured convolutional layer hardware co-processor in convolutional neural network (CNN) using facial biometric

Anirban Sengupta[1] and Rahul Chaurasia[1]

The chapter describes a methodology for designing a secured custom reusable hardware co-processor intellectual property (IP) core for convolutional neural network (CNN)-based convolutional layer [1]. In this methodology, first the convolutional layer hardware IP core is designed using high-level synthesis (HLS) process. HLS transforms the behavioral description/transfer function corresponding to the convolution operation of CNN into scheduled data flow graph (DFG) design to realize the behavior of convolutional layer. Since the reusable IP cores may be vulnerable to the hardware threat of IP piracy, therefore, facial biometric of IP vendor is integrated within the design. To do so, first, the facial signature corresponding to facial biometric image of an IP vendor is generated. Subsequently, facial biometric signature in the form of encoded hardware security constraints (digital evidence) is integrated into the IP design during the register allocation module of HLS. HLS-based design methodology results in a custom reusable convolutional layer hardware co-processor IP core with lesser implementation complexity and robust security. The embedded facial biometric-driven digital evidence enables the robust detective control against IP piracy. Therefore, it ensures the integration of genuine reusable hardware co-processor IP cores in the system-on-chip (SoCs) of consumer electronics (CE) systems, thereby also ensuring the integrity and safety of end consumers.

The organization of the chapter is as follows: Section 4.1 summarizes the introduction of the chapter; Section 4.2 discusses the motivation for designing CNN-based custom convolutional layer reusable IP core; Section 4.3 presents the benefits of the methodology to the end consumer; Section 4.4 presents the discussion on similar existing works; Section 4.5 provides background on CNN framework; Section 4.6 presents an overview of the approach; Section 4.7 provides the details of the approach; Section 4.8 provides analysis and discussion; and Section 4.9 concludes the chapter.

[1]Department of Computer Science and Engineering, Indian Institute of Technology Indore, India

4.1   Introduction

Interpretation of images is not only important for human brain but also equally important in the context of machines. A human brain can analyze and understand the image context easily, however, to program a machine to do the things with similar efficacy is highly challenging. To interpret images, computing machines treat them as an array of pixels. The scenario where the task is to interpret an input image and produce a classification output like a human brain does is the initial point of evolution of convolutional neural network (CNN). The initial work on CNN is said to be done by Kunihiko Fukushima and Yann LeCun in 1980. Further in 2012, Alex Krizhevsky revolutionized this field by using the artificial neural network to substantially bring down the image classification error. With this innovation, several tech giants such as Google, Facebook, Pinterest, and Instagram, started using CNNs to provide several different applications to the end users. As a result, a lot of applications such as face recognition, medical imaging, autonomous driving, and biometric authentication have emerged that people encounter in their daily life. In addition to that CNN is widely used to facilitate the following applications such as object/curve detection, image classification, sentiment detection, image segmentation, voice analyzing, automatic tagging, and video surveillance. Furthermore, CNN frameworks are also employed in the domain of computer vision and natural language processing as they are capable of achieving higher accuracy in feature extraction. CNN frameworks offer higher accuracy but on the other hand are also highly computationally intensive, especially the convolutional layer. Therefore, owing to the high computational intensiveness of the convolutional layer, their realization as hardware co-processors is very crucial. The designed convolutional layer reusable co-processor IP core can be used for all the aforementioned applications. Further, it can be employed in graphics processor, Internet of Things (IoT) devices, and in several portable or wearable electronics items (Sengupta and Chaurasia, 2022).

The main highlights of the methodology for designing secured reusable IP core for the convolutional layer of CNN are the following (Sengupta and Chaurasia, 2022):

1. It discusses the methodology for generating a secured CNN convolutional layer reusable hardware co-processor IP core for enabling the accelerated performance of CNN.
2. Ensures hardware-based parallel computation of pixels corresponding to each kernel, thereby accelerating the computation process or feature map generation.
3. The designed IP core is safeguarded against IP piracy using facial biometric of IP vendor. The embedded facial biometric signature acts as digital evidence for discerning between the pirated and authentic IP versions.
4. Provides robust security by employing multi-level encoding of facial biometric, thereby thwarting an adversary to evade the piracy detection process.

4.2   Why to design secured CNN convolutional layer co-processor IP core?

To design secured computing and consumer electronics systems, the integration of only secured reusable co-processor IP core into SoCs must be ensured. This is because an adversary may induce external threats of IP piracy. Pirated designs may contain secret malicious logic (known as hardware Trojan) that may be intentionally implanted by an adversary, may suffer from reliability hazards, or may not be satisfactory in terms of design constraints. These may remain undetected/unchecked as the pirated IP cores may not be rigorously tested before their integration into their system on chip. In the context of hardware Trojan, for example an implanted malicious logic may cause excessive heat dissipation, performance degradation, aging of components, and malfunctioning of the device. A pirated version of an IP core may result into misclassification of the images, inaccurate object detection and producing erroneous results in medical imaging. Therefore, for ensuring the correct output functionality in electronics devices as well as integrity and safety of the end consumer, the detection followed by isolation of pirated versions before their integration into SoCs is crucial. Further, for protecting the brand value of system designer integration of only secured IP core versions into the SoC is also important (Sengupta and Chaurasia, 2022).

4.3   Benefits of the approach

The secured convolutional layer IP core design offers several benefits from the perspective of the end consumer and the system designer (Sengupta and Chaurasia, 2022):

1. The designed convolutional layer reusable hardware co-processor IP core is capable of achieving parallelism during feature extraction (output feature map generation) by performing pixel computation parallelly, corresponding to each feature kernel.
2. To ensure robust security of convolutional IP core design, it integrates facial biometric-based naturally unique hardware security constraints into the design using HLS. Therefore, it is capable of providing detective control against pirated versions of IP core before integration into systems. This also provides the safety of the end consumer against fake components causing malfunctioning or degrading performance of the application/device.
3. Further, it ensures the correctness during object/curve detection without compromising spatial information (at corner/boundary pixels of the image).
4. Secured convolutional co-processor IP core is capable of performing hardware-based computation of six-pixels in parallel (two output pixels corresponding to each kernel).
5. Moreover, the generated secured convolutional layer IP core design incurs zero design overhead while offering unique security against pirated/fake IP versions.

4.4 Summary of existing approaches in the literature

In the literature, a few approaches targeting field programmable gate array (FPGA)-based hardware accelerators for CNN have been presented.

FPGA-based solutions for CNN framework: Kyriakos *et al.* (2019) have presented a hardware accelerator for CNN using FPGA framework. In this approach, parallel computation has been achieved at convolutional layer and fully connected layers of CNN. However, at the output layer it follows pipelined behavior. Tsiktsiris *et al.* (2018) presented an FPGA-based hardware accelerator for performing applications based on graphics processing. In this approach, they presented a portable USB accelerator for some basic algorithms of digital image processing such as Sobel edge detection and grayscale imaging. It therefore eliminates the requirement of heavy workstations. Liu *et al.* (2017) have presented an FPGA-based scalable framework that exploits parallelism in hardware acceleration corresponding to CNN. It employs parallelism at the task level, layer level, loop level, and operator level. Further, in this approach, an optimal design space has been explored to attain maximum throughput within FPGA constraints. The considered FPGA constraints are chip space, computational resources, clock frequency, and bandwidth of external memory. Shen *et al.* (2019) presented an end-to-end framework for achieving the accelerated performance of CNN on FPGAs. It also targets minimum user effort for the same and is centered around a CNN accelerator generator. The accelerator generator accepts FPGA specifications and a CNN model as its input for generating optimized CNN accelerator register transfer level (RTL) design correspondingly. Additionally, it employs CNN inference microservice and device driver, which provides an easy access for end user applications to CNN accelerator based on FPGA. All these approaches have presented FPGA-based solution for accelerating the performance of CNN framework through mapping CNN onto FPGA. However, these approaches have not targeted designing custom IP core for CNN especially for convolutional layer which is highly computationally intensive in nature. Further, they have not incorporated any hardware security mechanism to ensure reliable performance of the framework and protection against IP piracy. Unlike the aforementioned approaches, Sengupta and Chaurasia (2022) have presented a secured design of IP core corresponding to CNN convolutional layer. This design methodology is also capable of accelerating the performance of CNN through custom reusable IP core for convolutional layer as well as integrates facial biometric-based hardware security mechanism to provide IP piracy detective control.

Solutions for convolutional computation of CNN: Srivastava and Sarawadekar (2020) have presented a methodology by using FPGA-based accelerators for accelerating the performance of CNN interface by employing optimization at algorithmic level using depth wise separable convolution rather than standard convolution. In this approach, for a single layer of CNN, a pipeline-driven architecture for depth wise separable convolution, activation, and pooling operations has been presented. This methodology mainly focusses on FPGA-based acceleration of convolution process but not on implementing the entire CNN. Bai *et al.* (2018) have presented a scalable depth wise separable convolution instead of standard convolution. This therefore results into lesser number of operations than standard convolution. This therefore enables the possibility of integration of CNNs into small size embedded systems and portable devices like mobile phones. The presented accelerator can be employed for different sizes of FPGA while considering hardware resources and processing speed in a balanced way. However, both these approaches do not preserve accuracy while performing feature extraction. Chang *et al.* (2017) have presented a methodology for designing hardware accelerator for convolution operation by employing efficient matrix multiplication. In this approach, operation of CNN has been simulated using RISC-V rocket processor for performing image classification. Ma *et al.* (2018) presented a methodology for accelerating deep neural network on FPGA by enabling the optimization of convolution operation based on looping. Typically, convolution process involves multiply and accumulate operations with four levels of loops. This may result in the requirement of a large design space. However, in this approach, such limitations have been overcome by analyzing and optimizing the design objectives of CNN accelerator. It therefore has presented a specific data flow and architecture corresponding to CNN hardware accelerator to maximize the resource utilization and minimize the memory access, to achieve high performance. Guo *et al.* (2016) have demonstrated design flow for performing CNN mapping onto embedded FPGA. The presented CNN accelerator architecture is programmable and flexible which includes data quantization strategy and compilation tool. Data quantization strategy is employed for reducing the bit-width down to 8-bit, whereas a compilation tool is used for mapping a certain CNN model onto hardware. The presented hardware is faster and power efficient than its contemporary accelerators on FPGA. However, it does not preserve accuracy. Kim *et al.* (2019) have presented a hierarchical convolution computing-based algorithm for CNNs. In this algorithm, to accelerate the convolution, it does not mandate to retain parameters while reducing multiplier–accumulator operations. It therefore achieves faster convolutional computation but achieves efficiency for smaller feature maps only. It also fails to preserve accuracy against feature extraction.

All these approaches have targeted FPGA-based solution to accelerate convolutional computation of CNN framework. However, these approaches do not ensure efficient and reliable performance corresponding to feature map generation. Further, these approaches also do not provide secure custom solution for convolutional computation. On the contrary, Sengupta and Chaurasia (2022) have presented a secure hardware design of RTL IP core corresponding to CNN convolutional layer. This custom RTL design corresponding to CNN convolutional co-processor IP core offers the following features:

1. Secured HLS-based design methodology.
2. Security against IP piracy threat by integrating facial biometric of IP vendor as digital evidence.
3. Accelerated hardware-based computation of output feature map in terms of performing parallel pixel computation while preserving accuracy.

4.5 Background on CNN framework

As we have seen several applications of CNN earlier, CNN is a kind of algorithm that is specifically useful for the applications that exploit image-related data or involve the processing of pixel data. Usually, a CNN framework accepts image data (pixel matrix) as input and exploit it for performing tasks such as feature extraction, object detection, image classification, etc. To do so, the working of CNN involves several functional layers such as (a) convolutional layer, (b) pooling layer, (c) flattening layer, and (d) fully connected (FC) layer. The layered architecture of CNN is shown in Figure 4.1. Each of these functional layers is responsible for the execution of some specific tasks and the output of one layer (post processing) is fed as input to the next subsequent layer for further processing. For example, the first layer (convolutional layer) is responsible for performing convolution (dot product). To do so, it convolves kernels/filters over input image matrix, specifically on the receptive field of image. The segment of input image (of the same dimension as the kernel) over which kernel operates is called as receptive field. Thus, it generates feature map corresponding to each kernel. For example, if "N" kernels are convolved over input image, then "N" feature maps are generated. Kernels/filters are exploited for performing the extraction of features from the input image, and

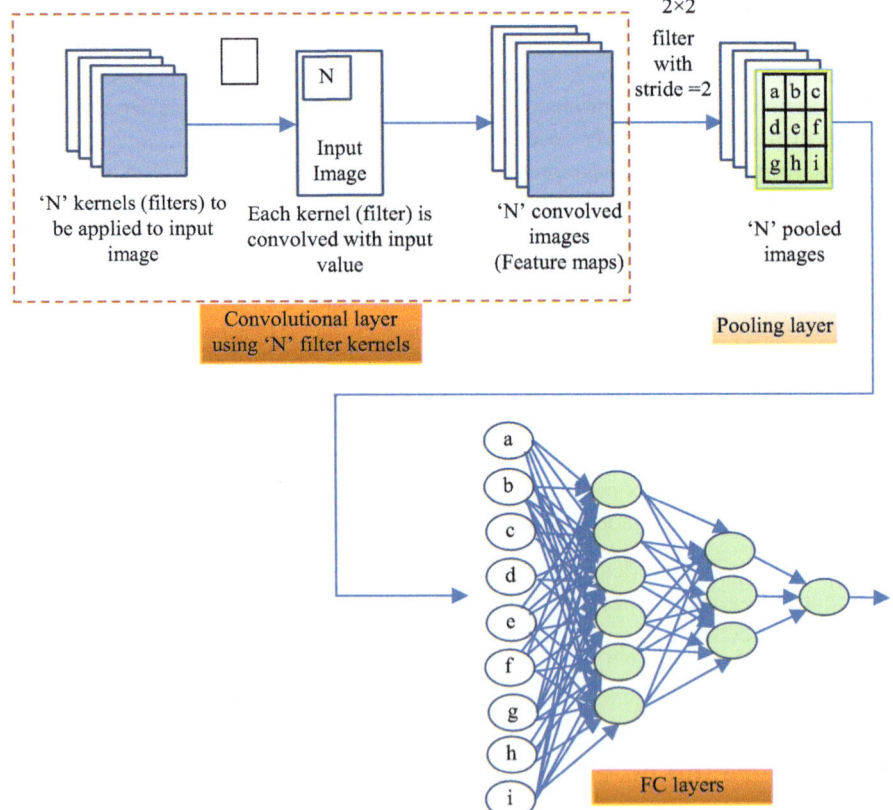

Figure 4.1   Layered architecture of CNN

therefore are also acknowledged as a feature extractor. Kernel/filters (considered 2D for the sake of brevity) are vectors consisting of weights that resemble the curve (in case of filter as a curve detector) which convolves over the input image based on stride/shift value. Further, a feature map contains the extracted/detected information from input image corresponding to applied filter. Moreover, each kernel during convolution convolves over input image independently for producing the output feature map. Therefore, more the kernels/filters, the better the performance in terms of accurate detection of patterns and visual features. Further, depending on their weights, each conv 2D kernel is capable of identifying features from an image. Thus, feature maps (convolved images) are generated as the output of convolutional layer. Subsequently, post convolutional layer, the generated feature maps are processed through the next subsequent layer i.e., the pooling layer. In this layer, pooling (down sampling) operation (Max pooling is preferred) corresponding to each feature map is performed independently. Pooling down samples (reduces) the spatial size of corresponding feature map. Therefore, this results into lesser number of parameters and computations. Pooling down samples the image by maintaining the original shape from the feature map. Thus, pooling layer provides assistance to fully connected layer in terms of minimizing the computation. Post processing through convolutional layer and pooling layer, output feature map is converted into vectors (1D array) called the feature vector/flattened layer. The output of the flattened layer is then processed through the FC layer. In the FC layer, all the features are collectively transferred into this network. Subsequently, outputs are generated as probabilistic values (corresponding to particular feature or class) by employing either soft max or logistic function. For example, if the program is predicting some object of the image, it will have high values in the activation maps that represent the features corresponding to that object. Basically, a FC layer looks at features that strongly correlate to a particular object or class. Thus, enabling accurate object detection (Albawi *et al.*, 2017; Gu *et al.*, 2018; Zeiler and Fergus, 2013).

4.6   Overview of the approach for designing a secured CNN convolutional co-processor IP core using facial biometric

So far, we have discussed the basic functionality of CNN framework through its layered architecture. Now we discuss the overview of the methodology for designing secured CNN convolutional co-processor IP core (Sengupta and Chaurasia, 2022). In this methodology, HLS-based design technique has been followed for designing secured reusable co-processor IP core for CNN convolutional process. Since CNN framework is computationally intensive, specifically its convolutional layer, therefore realizing a secured reusable co-processor IP core corresponding to convolutional process is relevant and could be used in smaller embedded and portable computing system designs. To do so, first the transfer function of convolution is transformed into DFG and is unrolled twice. Unrolling twice enables parallel output pixel computation during feature map generation corresponding to each kernel (*Note: multiple filter kernels are used for convolution with the input image matrix for faster and accurate detection/classification*). Subsequently, the DFG is scheduled using resource constraints supplied by IP designer. The scheduled design is then exploited for extracting its corresponding register allocation information, which is further used to embed security mark (as a digital evidence) in the form of encoded hardware security constraints. The embedded security therefore enables detective control against pirated versions of the design. To embed security against piracy of the design, facial biometric information of IP vendor has been exploited. To do so, first facial biometric of IP vendor is converted into digital signature based on IP vendor-opted facial features set and their ordering. Subsequently, facial biometric signature is encoded into secret hardware security constraints. Facial biometric-based naturally unique encoded hardware security constraints are then implanted into the design during higher abstraction phase of HLS. These facial security constraints post embedding into the IP core design are responsible for enabling detection and isolation of pirated versions before their integration into computing systems.

4.7   Details of the approach

Now, let us discuss the approach in detail (Sengupta and Chaurasia, 2022).

4.7.1   HLS flow of the approach for designing secured convolutional hardware IP core in CNN

HLS-based design flow for generating secured IP core design corresponding to the CNN convolutional layer is shown in Figure 4.2. To generate secured IP core design, following steps are followed:

1. First, the transfer function of convolution process (capable of computing two output pixels corresponding to each kernel in single execution using parallel sliding window) is transformed into DFG. To construct the DFG, mathematical description corresponding to process of feature map generation (by convolving kernel matrices over input image matrix) is obtained. Subsequently, it is transferred into its corresponding DFG.
2. Next, DFG design is scheduled during the scheduling block execution. The scheduling block accepts following inputs: DFG of the design, resource constraints, scheduling algorithm, and dependency information of operational nodes to generate scheduled design flow graph design.
3. Next, hardware resources are allocated to operation of the scheduled design and their binding is performed. Subsequently, register allocation information corresponding to scheduled design is extracted. The generated register allocation information is used as a foundation for performing the embedding of security (digital evidence).
4. Next, security in the form of encoded digital evidence is implanted into the design. To do so, security embedding block accepts following inputs: (a) register allocation information corresponding to scheduled design and (b) facial

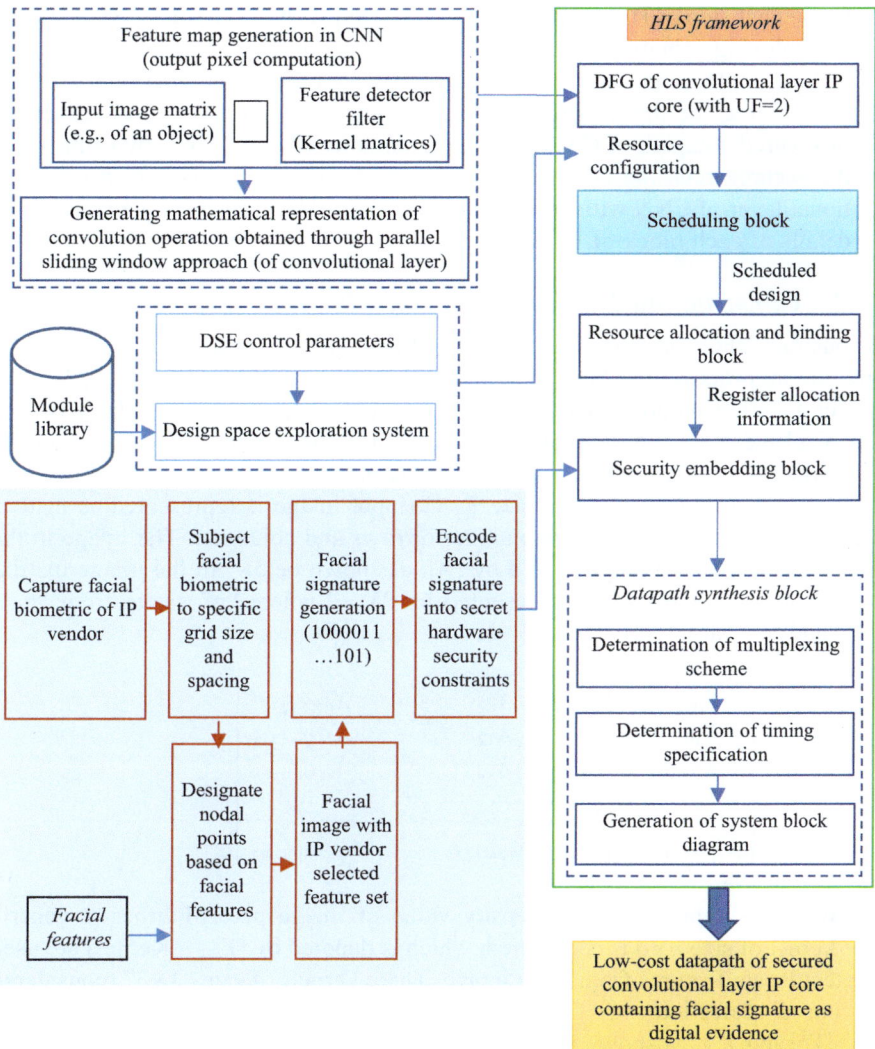

Figure 4.2   Design flow for generating secured convolutional layer hardware co-processor IP core using HLS-based approach

biometric-driven secret hardware security constraints. To derive facial biometric-based secret hardware security constraints, first, captured facial biometric of IP vendor is transformed into digital signature based on chosen facial feature set. Subsequently, facial biometric-based generated digital signature is encoded into secret hardware security constraints. These hardware security constraints are covertly inserted into the design during register allocation phase of HLS. Embedded security constraints offer robust security against piracy in terms of enabling detective control against pirated versions.

5. In the next step of HLS, datapath is synthesized with the aid of ascertaining the multiplexing scheme corresponding to each of the functional resources (e.g., adders, multipliers, and registers). Synthesis process is then followed by determining the timing specifications (controller synthesis) and development of secured datapath of CNN convolutional layer IP core. Thus, by using HLS, the secured reusable co-processor IP core design corresponding to convolutional layer of CNN with embedded facial biometric security is generated. The details of each block of HLS framework are discussed in next subsections.

4.7.2   Constructing DFG of CNN convolutional IP core

To construct DFG corresponding to convolutional layer, the transfer function of the convolution process has been exploited (Sengupta and Chaurasia, 2022). The transfer function incorporates the details of input image and kernel (in terms of their pixel indices to get convoluted, as an input) and generates the output pixels corresponding to output feature map (as an output).

Consider an input image of size $F \times G$ (input image is represented as matrix, where "F" and "G" represent the number of rows and columns). The image in the form of pixel matrix is represented by $[X]$, as shown below. In the image matrix, each image pixel/element is represented by "X_{uv}," where "u" varies from "0 to $F-1$" and "v" varies from "0 to $G-1$."

$$[X] = \begin{bmatrix} X_{00} & X_{01} & X_{02} & \cdots & X_{0(G-1)} \\ X_{10} & X_{11} & X_{12} & \cdots & X_{1(G-1)} \\ X_{20} & X_{21} & X_{22} & \cdots & X_{2(G-1)} \\ \vdots & \vdots & \vdots & \ddots & \vdots \\ X_{(F-1)0} & X_{(F-1)1} & X_{(F-1)2} & \cdots & X_{(F-1)(G-1)} \end{bmatrix}_{F \times G}$$

Here, "X" represents the intensity value of image pixel. Further, a generic filter/kernel of size $p \times q$ is considered, which is denoted by $[Y]_{pxq}$. Kernels are used to detect/locate features from input image. Three kernels of size "3×3" (considered for curve detection in the discussed methodology) are represented as $[Y^1]$, $[Y^2]$ and $[Y^3]$, as shown below:

$$[Y^1] = \begin{bmatrix} y_{00}^1 & y_{01}^1 & y_{02}^1 \\ y_{10}^1 & y_{11}^1 & y_{12}^1 \\ y_{20}^1 & y_{21}^1 & y_{22}^1 \end{bmatrix}_{3 \times 3} \quad [Y^2] = \begin{bmatrix} y_{00}^2 & y_{01}^2 & y_{02}^2 \\ y_{10}^2 & y_{11}^2 & y_{12}^2 \\ y_{20}^2 & y_{21}^2 & y_{22}^2 \end{bmatrix}_{3 \times 3}$$

$$[Y^3] = \begin{bmatrix} y_{00}^3 & y_{01}^3 & y_{02}^3 \\ y_{10}^3 & y_{11}^3 & y_{12}^3 \\ y_{20}^3 & y_{21}^3 & y_{22}^3 \end{bmatrix}_{3 \times 3}$$

In the kernel matrices, pixel values are represented as "y_{ab}^t," where "a" and "b" varies from "0 to 2" and "t" represents kernel/filter number.

Since, in the convolutional IP core design methodology, "same convolution" is performed, therefore, the size of input image matrix is increased by augmenting zero-rows and zero-columns based on the following function:

$$A = \frac{(L-1)}{2} \tag{4.1}$$

where "A" corresponds to the number of zero rows/columns to be added on each side of input image matrix (left, right, top and bottom) and "L" corresponds to the size of kernel, i.e., $L = 3$ for "3 × 3" matrix. Therefore, input image matrix size is increased by 2 (by adding a zero row/column on each side of image matrix). The image matrix post performing zero padding $[X']$ is shown below:

$$[X'] = \begin{bmatrix} 0 & 0 & 0 & 0 & 0 & 0 & 0 \\ 0 & X_{00} & X_{01} & X_{02} & \cdots & X_{0(G-1)} & 0 \\ 0 & X_{10} & X_{11} & X_{12} & \cdots & X_{1(G-1)} & 0 \\ 0 & X_{20} & X_{21} & X_{22} & \cdots & X_{2(G-1)} & 0 \\ 0 & \vdots & \vdots & \vdots & \ddots & \vdots & 0 \\ 0 & X_{(F-1)0} & X_{(F-1)1} & X_{(F-1)2} & \cdots & X_{(F-1)(G-1)} & 0 \\ 0 & 0 & 0 & 0 & 0 & 0 & 0 \end{bmatrix}_{R \times S}$$

where "R × S" corresponds to the size of padded input image matrix. The image matrix size is (R × S) = (F + 2)×(G + 2). Further, a generic representation corresponding to matrix $[X']$ is shown below:

$$[X'] = \begin{bmatrix} Z_{00} & Z_{01} & Z_{02} & \cdots & Z_{0(S-1)} \\ Z_{10} & Z_{11} & Z_{12} & \cdots & Z_{1(S-1)} \\ Z_{20} & Z_{21} & Z_{22} & \cdots & Z_{2(S-1)} \\ \vdots & \vdots & \vdots & \ddots & \vdots \\ Z_{(R-1)0} & Z_{(R-1)1} & Z_{(R-1)2} & \cdots & Z_{(R-1)(S-1)} \end{bmatrix}_{R \times S} \tag{4.2}$$

Pixel values from (4.2) are denoted by Z_{uv}, where "u" varies from 0 to "R−1" and "v" varies from "0 to S−1."

Therefore, the size of the feature map corresponding to an input matrix (augmented) of size "R × S" and "N" filters of size "p × q," can be computed using the following equation:

$$[(R-p+1) \times (S-q+1)] \times N \tag{4.3}$$

Further, output matrix corresponding to the "same convolution" between kernel matrix and input matrix is denoted by $[O_v]$. However, dimensions of $[O_v]$ are the same as that of input image matrix pre-padding (i.e., R × S). The values corresponding to output matrix are denoted by O_i^j, where "i" varies in the range from "0 to $[(R-p+1) \times (S-q+1) - 1]$" and "j" indicates the kernel number.

Now, we discuss the computation of pixel value or matrix element corresponding to output feature map. The output pixel(s) value is/are computed as follows:

$$O_v = \sum_{R,p\,=\,\text{lower value}}^{R,p\,=\,\text{upper value}} \left(\sum_{S,q\,=\,\text{lower value}}^{S,q\,=\,\text{upper value}} X'_{R\times S} \times Y_{p\times q} \right) \tag{4.4}$$

where O_v represents the computed output pixel matrix corresponding to feature map, Y_{pxq} represents the size of kernel matrix, and $X'_{R\times S}$ represents the size of image matrix.

For the sake of demonstration, convolution of kernel matrix (kernel matrix $[Y^1]$, with twice unrolled) over image matrix is shown below:

Considering the input image as shown in (4.2):

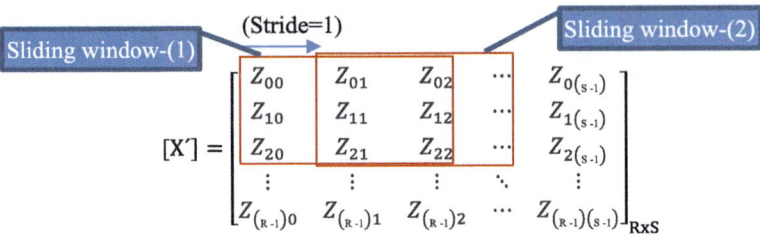

And the kernel matrix $[Y^1]$.

Since, in the methodology for generating secured convolutional IP core (Sengupta and Chaurasia, 2022), two pixels are computed simultaneously corresponding to a specific kernel. Therefore, over an input image matrix, two sliding windows of a kernel convolves to compute two-pixel outputs simultaneously. For example, if the kernel size is "3 × 3" and is applied on the input image matrix (on receptive field of size "3 × 3") then, to compute first pixel output, convolve kernel $[Y^1]$ over first sliding window of input image:

$$[X'] = \begin{bmatrix} Z_{00} & Z_{01} & Z_{02} \\ Z_{10} & Z_{11} & Z_{12} \\ Z_{20} & Z_{21} & Z_{22} \end{bmatrix}_{3\times3} \quad [Y^1] = \begin{bmatrix} y^1_{00} & y^1_{01} & y^1_{02} \\ y^1_{10} & y^1_{11} & y^1_{12} \\ y^1_{20} & y^1_{21} & y^1_{22} \end{bmatrix}_{3\times3}$$

Similarly, to compute second pixel output, perform convolution of the kernel $[Y^1]$ over second sliding window of input image based on stride/shift value= "1":

$$[X'] = \begin{bmatrix} Z_{01} & Z_{02} & Z_{03} \\ Z_{11} & Z_{12} & Z_{13} \\ Z_{21} & Z_{22} & Z_{23} \end{bmatrix}_{3\times3} \quad [Y^1] = \begin{bmatrix} y^1_{00} & y^1_{01} & y^1_{02} \\ y^1_{10} & y^1_{11} & y^1_{12} \\ y^1_{20} & y^1_{21} & y^1_{22} \end{bmatrix}_{3\times3}$$

The mathematical representation of computing two-pixel outputs is as follows:

$$1^{st}\text{ output: } O_0 = \sum_{P\,=\,0\ P\,=\,2}^{R\,=\,0\ R\,=\,2} \left(\sum_{q\,=\,0\ q\,=\,2}^{S\,=\,0\ S\,=\,2} X'_{R\times S} \times Y_{p\times q} \right)$$

$$2^{nd}\text{ output: } O_1 = \sum_{\substack{R\,=\,0 \\ p\,=\,0}}^{\substack{R\,=\,2 \\ p\,=\,2}} \left(\sum_{\substack{S\,=\,1 \\ q\,=\,0}}^{\substack{S\,=\,3 \\ q\,=\,2}} X'_{R\times S} \times Y_{p\times q} \right) \tag{4.5}$$

Therefore, the output pixel computation corresponding to kernels ($N =$ "3") can be represented as follows (from (4.5)):

For kernel 1:

$$O_0^1 = \left[(Z_{00} \times y_{00}^1) + (Z_{01} \times y_{01}^1) + (Z_{02} \times y_{02}^1)\right]$$
$$+ \left[(Z_{10} \times y_{10}^1) + (Z_{11} \times y_{11}^1) + (Z_{12} \times y_{12}^1)\right]$$
$$+ \left[(Z_{20} \times y_{20}^1) + (Z_{21} \times y_{21}^1) + (Z_{22} \times y_{22}^1)\right]$$

$$O_1^1 = \left[(Z_{01} \times y_{00}^1) + (Z_{02} \times y_{01}^1) + (Z_{03} \times y_{02}^1)\right]$$
$$+ \left[(Z_{11} \times y_{10}^1) + (Z_{12} \times y_{11}^1) + (Z_{13} \times y_{12}^1)\right] \tag{4.6}$$
$$+ \left[(Z_{21} \times y_{20}^1) + (Z_{22} \times y_{21}^1) + (Z_{23} \times y_{22}^1)\right]$$

For kernel 2:

$$O_0^2 = \left[(Z_{00} \times y_{00}^2) + (Z_{01} \times y_{01}^2) + (Z_{02} \times y_{02}^2)\right]$$
$$+ \left[(Z_{10} \times y_{10}^2) + (Z_{11} \times y_{11}^2) + (Z_{12} \times y_{12}^2)\right]$$
$$+ \left[(Z_{20} \times y_{20}^2) + (Z_{21} \times y_{21}^2) + (Z_{22} \times y_{22}^2)\right]$$

$$O_1^2 = \left[(Z_{01} \times y_{00}^2) + (Z_{02} \times y_{01}^2) + (Z_{03} \times y_{02}^2)\right]$$
$$+ \left[(Z_{11} \times y_{10}^2) + (Z_{12} \times y_{11}^2) + (Z_{13} \times y_{12}^2)\right] \tag{4.7}$$
$$+ \left[(Z_{21} \times y_{20}^2) + (Z_{22} \times y_{21}^2) + (Z_{23} \times y_{22}^2)\right]$$

Similarly, the output pixel computation for third kernel can be represented as follows:

$$O_0^3 = \left[(Z_{00} \times y_{00}^3) + (Z_{01} \times y_{01}^3) + (Z_{02} \times y_{02}^3)\right]$$
$$+ \left[(Z_{10} \times y_{10}^3) + (Z_{11} \times y_{11}^3) + (Z_{12} \times y_{12}^3)\right]$$
$$+ \left[(Z_{20} \times y_{20}^3) + (Z_{21} \times y_{21}^3) + (Z_{22} \times y_{22}^3)\right]$$

$$O_1^3 = \left[(Z_{01} \times y_{00}^3) + (Z_{02} \times y_{01}^3) + (Z_{03} \times y_{02}^3)\right]$$
$$+ \left[(Z_{11} \times y_{10}^3) + (Z_{12} \times y_{11}^3) + (Z_{13} \times y_{12}^3)\right] \tag{4.8}$$
$$+ \left[(Z_{21} \times y_{20}^3) + (Z_{22} \times y_{21}^3) + (Z_{23} \times y_{22}^3)\right]$$

In (4.6), (4.7), and (4.8), each product term belongs to the elementary multiplication of $(Z_{uv} \times y_{ab}^t)$ from the matrices $[X'] \times [Y]_{pxq}$, where Z_{uv} and y_{ab}^t correspond to pixel/element values of the input matrix and the kernel matrix, respectively. Corresponding to computation of the first two-pixel values (O_0^1 and O_1^1) corresponding to kernel —"1" ($N = 1$), as shown in (4.6), the values of "u," "a," and "b" vary from "0 to 2." Further, "v" corresponding to Z_{uv} varies from "0 to 2" for pixel output "O_0^1" and varies from "1 to 3" for pixel output "O_1^1." Subsequently, in the remaining computations of the same row, a maximum value of "v" can reach up to "S−1." Further, the value of "u" varies from "0 to 2" in the initial row of feature matrix. However, for computing, the next row output value of the feature matrix, lower and upper values of "u" gets increased by "1." Subsequently, for the other remaining computations corresponding to output pixels, the maximum value of "u" can reach up to "R−1." Thus, by convolving a feature kernel over input image in sliding window manner, the output feature map is generated. Similarly, the remaining two kernels convolve over input image to generate their corresponding feature maps.

DFG construction: So far, we have discussed the mathematical representation of convolution process corresponding to convolving three kernels ($[Y^1]$, $[Y^2]$, and $[Y^3]$) in sliding window manner over input image. This therefore results into the detection of object/curve/feature that coincides with the kernel matrix weight pattern. This transfer function is then transformed into DFG corresponding to each kernel that computes two output values simultaneously (as shown in Figure 4.2). The DFG corresponding to kernel-"1" (computing two output pixel of feature map using sliding window) is shown in Figure 4.3, where $Z_{00}, Z_{01}, Z_{02}, Z_{10} \ldots \ldots Z_{22}$ and $Z_{01}, Z_{02}, Z_{03}, Z_{11} \ldots \ldots Z_{23}$ represent the pixel/element indices corresponding to receptive field of input image matrix ($[X']$) using sliding window. Further, $y_{00}^1, y_{01}^1, y_{02}^1, y_{10}^1 \ldots \ldots y_{22}^1$ represent pixel/element indices corresponding to kernel matrix-"1" ($[Y^1]$). The convolution process involves operations like multiplication and addition, which are denoted using "×" and "+," respectively, and (O_0^1 and O_1^1) corresponds to the output pixels corresponding to kernel-"1" ($[Y^1]$). Similarly, the DFG corresponding to other kernels

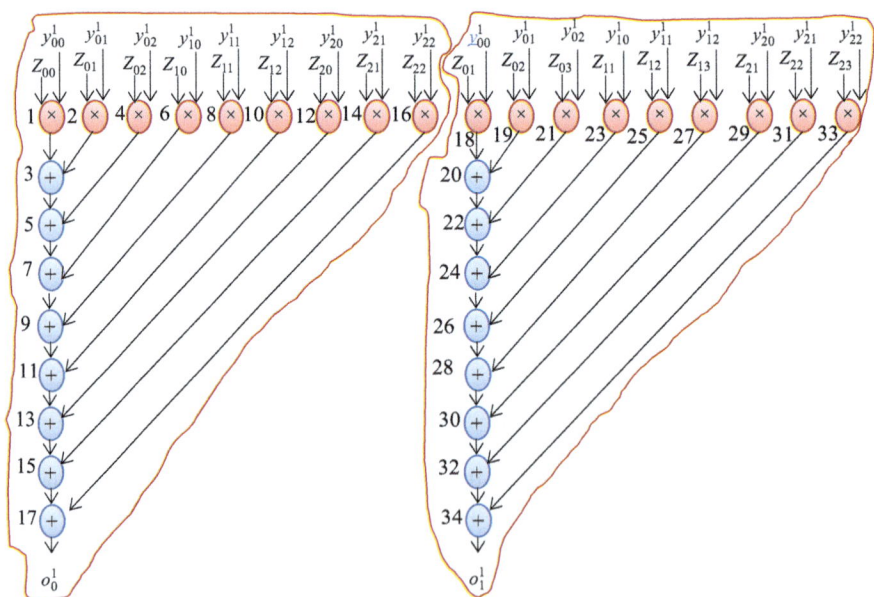

Figure 4.3   DFG of CNN convolutional layer IP core (corresponding to kernel-"1" ($[Y^1]$) of size "3×3") and twice unrolled using HLS-based approach

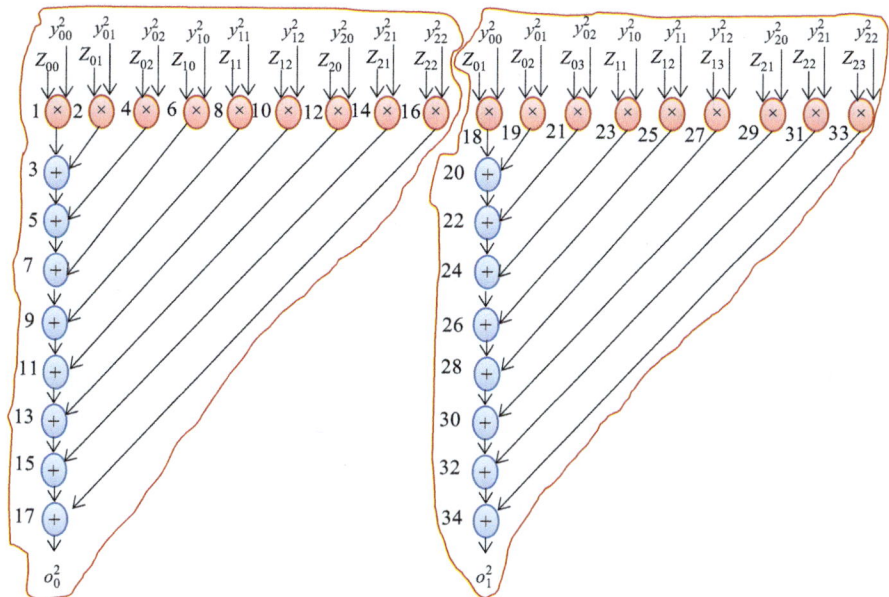

Figure 4.4 DFG of CNN convolutional layer IP core (corresponding to kernel-"2" ([Y^2]) of size "3×3") and twice unrolled using HLS-based approach

([Y^2] and [Y^3]) can be generated. The DFG corresponding to kernel [Y^2] and [Y^3] are shown in Figures 4.4 and 4.5, respectively, where (O_0^2 and O_1^2) and (O_0^3 and O_1^3) represent the output pixels (first two) of feature maps corresponding kernel-"2" ([Y^2]) and kernel-"3" ([Y^3]) respectively. Further, y_{00}^2, y_{01}^2, y_{02}^2, y_{10}^2......y_{22}^2 and y_{00}^3, y_{01}^3, y_{02}^3, y_{10}^3......y_{22}^3 represent pixel/element indices corresponding to kernel matrix-"2" ([Y^2]) and kernel-"3" ([Y^3]), respectively. Thus, convolutional IP core design computes total six pixels in one execution (through computing two pixels using each of the three kernels).

4.7.3 Scheduling the IP core design and generating register allocation information

Now, we discuss the details of scheduling block (as shown in Figure 4.2). Scheduling block is responsible for generating the scheduled design architecture. To do so, scheduling block accepts the following inputs:

- DFG of the design: DFG corresponding to convolutional process, as shown in Figures 4.3, 4.4, and 4.5 corresponding to each of the three kernels.
- Resource constraints: To generate a low-cost schedule, an optimal resource configuration is important. In this methodology, resource constraints have been obtained by performing design space exploration (Mishra and Sengupta, 2014). Design space exploration results into a low-cost resource configuration that

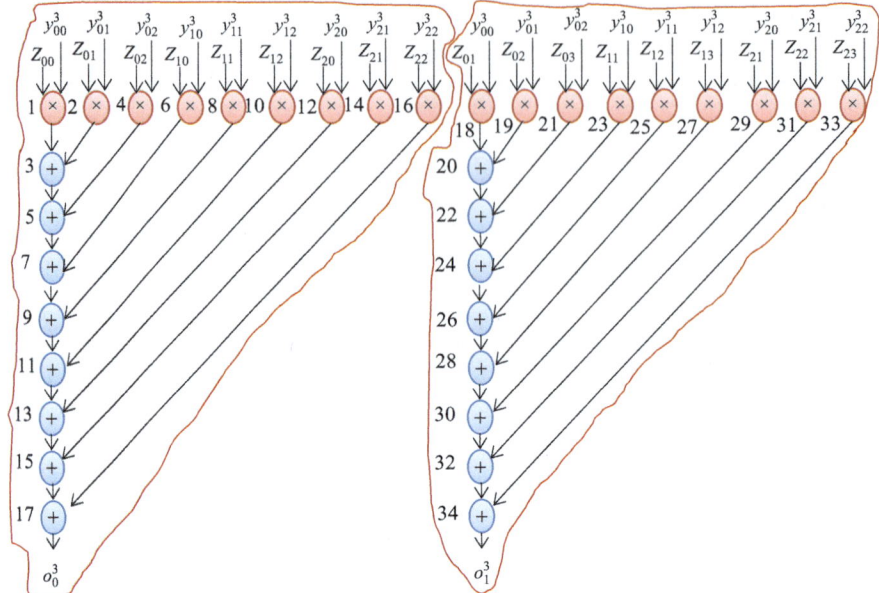

Figure 4.5   DFG of CNN convolutional layer IP core (corresponding to kernel-
"3" ([Y^3]) of size "3 × 3") and twice unrolled using HLS-based
approach

acts as the input resource constraints for the design process. This therefore
results into lesser design area and lower latency.

- Scheduling algorithm: LIST scheduling algorithm has been employed for
  generating the scheduled design based on resource constraints. It works by
  trying to schedule a maximum number of operations in a single control step
  subjected to resource constraints and data dependency. Dependency information of operational nodes is responsible to generate correct functional output
  (Sengupta and Roy, 2017; Wanhammar, 1999).

The scheduled DFG design of convolution process followed by hardware
allocation and their binding, corresponding to kernels [Y^1], [Y^2], and [Y^3], are
shown in Figures 4.6, 4.7, and 4.8, respectively. As evident, the DFG of convolution process corresponding to kernels is scheduled using the resources "1"
adder "A[1]" and 1 multiplier "M[1]." Further, during the hardware allocation, adder
(s) and multiplier (s) is/are assigned to operations like addition (+) and multiplication (×), respectively. Further, storage elements/registers (represented using
different colors) are assigned corresponding to storage variables (refer the input,
intermediate, and output values corresponding to design) during scheduling and
there binding is performed. As evident from Figure 4.6, the number of required
control steps for scheduled design to generate convolved output pixels (corresponding to feature kernel [Y^1]) using hardware resources (1 adder "A[1]" and

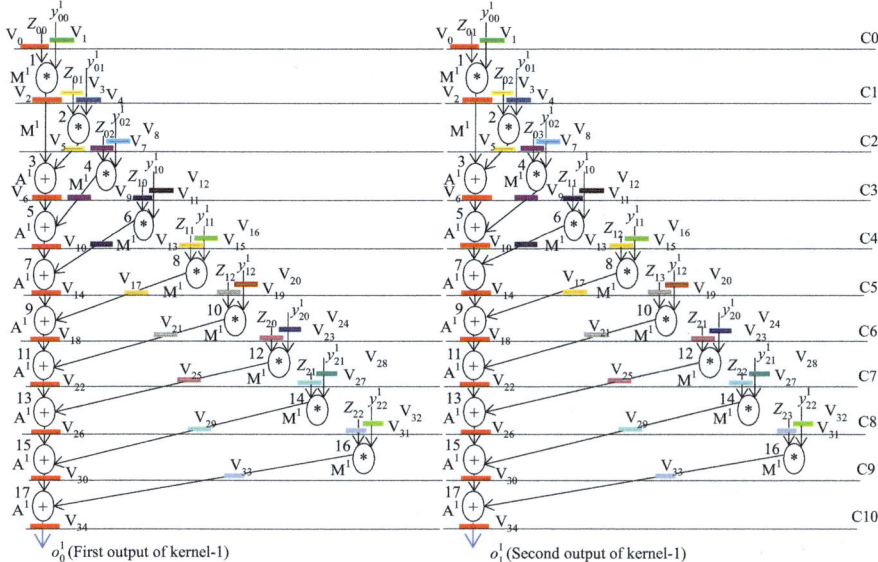

Figure 4.6   Scheduled data flow graph of CNN convolutional IP core (corresponding to kernel-"1" ([Y¹]) of size "3 × 3") and twice unrolled using HLS-based approach

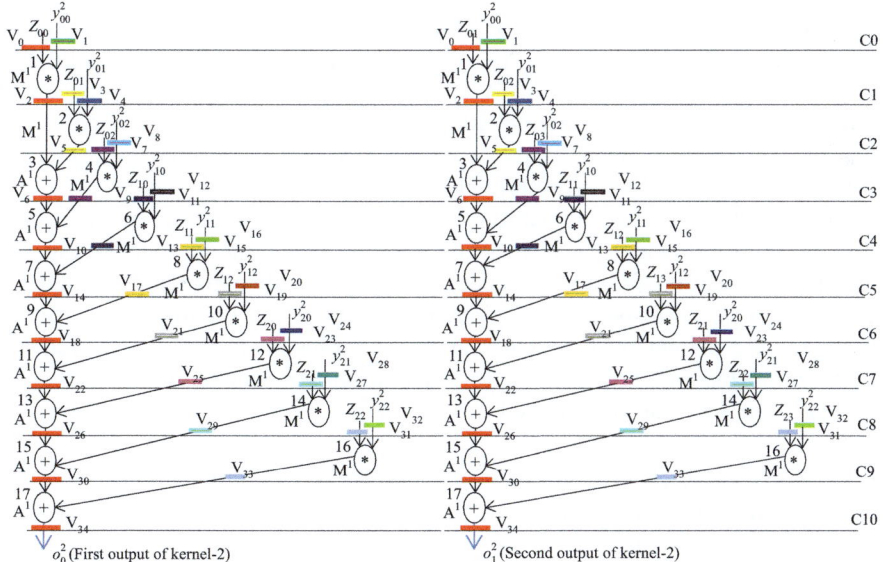

Figure 4.7   Scheduled data flow graph of CNN convolutional IP core (corresponding to kernel-"2" ([Y²]) of size "3 × 3") and twice unrolled using HLS-based approach

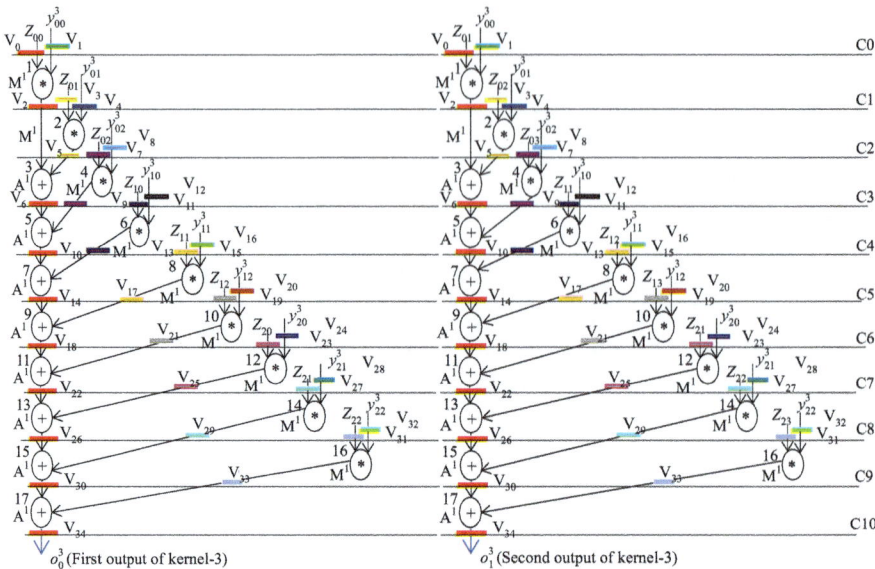

Figure 4.8   Scheduled data flow graph of CNN convolutional IP core (corresponding to kernel-"3" ([Y^3]) of size "3 × 3") and twice unrolled using HLS-based approach

1 multiplier "M^1") is ten (C0 to C10). Further, the required registers for accommodating the storage variables ($V_1, V_2, --, V_{34}$) corresponding to each unrolling of kernel [Y^1] are 18, represented as "$R^1, R^2, ---, R^{18}$." Similarly, the scheduled DFG design corresponding to kernel filter [Y^2] and [Y^3], as shown in Figures 4.7 and 4.8 requires ten control steps to generate convolved output pixels using resources (1 adder "A^1" and 1 multiplier "M^1").

Subsequently, register allocation information corresponding to scheduled design is extracted which is further used as a foundation for performing the embedding of security (digital evidence). The register allocation information for scheduled design corresponding to convolution process with respect to feature kernel [Y^1] are shown in Tables 4.1 and 4.2, respectively. Similarly, register allocation information corresponding to convolution process with respect to feature kernel [Y^2] and [Y^3] can be obtained.

4.7.4   Details of generating facial biometric signature

So far, we have discussed the details of generating scheduled design (and have extracted register allocation information) corresponding to convolution process for feature kernels. Now we discuss the process of generating biometric-based digital evidence (secret signature) by exploiting facial biometric of IP vendor. This facial biometric-based digital evidence is subsequently used for embedding into the design during register allocation phase for enabling detective control against IP

Table 4.1 Register allocation information of scheduled design corresponding to convolution process with respect to feature kernel $[Y^1]$ (for first pixel output computation)

Registers	R^1	R^2	R^3	R^4	R^5	R^6	R^7	R^8	R^9	R^{10}	R^{11}	R^{12}	R^{13}	R^{14}	R^{15}	R^{16}	R^{17}	R^{18}
C0	V_0	—	—	—	—	—	—	—	—	—	—	—	—	—	—	—	—	—
C1	V_2	V_1	V_3	—	—	—	—	—	—	—	—	—	—	—	—	—	—	—
C2	V_2	—	V_5	V_4	—	—	—	—	—	—	—	—	—	—	—	—	—	—
C3	V_6	—	—	—	V_7	V_8	—	—	—	—	—	—	—	—	—	—	—	—
C4	V_{10}	—	—	—	V_9	—	V_{11}	V_{12}	—	—	—	—	—	—	—	—	—	—
C5	V_{14}	—	—	—	—	—	V_{13}	—	V_{15}	V_{16}	—	—	—	—	—	—	—	—
C6	V_{18}	—	—	—	—	—	—	—	V_{17}	—	V_{19}	V_{20}	—	—	—	—	—	—
C7	V_{22}	—	—	—	—	—	—	—	—	—	V_{21}	—	V_{23}	V_{24}	—	—	—	—
C8	V_{26}	—	—	—	—	—	—	—	—	—	—	—	V_{25}	—	V_{27}	V_{28}	—	—
C9	V_{30}	—	—	—	—	—	—	—	—	—	—	—	—	—	V_{29}	—	V_{31}	V_{32}
C10	V_{34}	—	—	—	—	—	—	—	—	—	—	—	—	—	—	—	V_{33}	—

Table 4.2 Register allocation information of scheduled design corresponding to convolution process with respect to feature kernel $[Y^1]$ (for second pixel output computation).

Registers	R^1	R^2	R^3	R^4	R^5	R^6	R^7	R^8	R^9	R^{10}	R^{11}	R^{12}	R^{13}	R^{14}	R^{15}	R^{16}	R^{17}	R^{18}
C0	V_0	—	—	—	—	—	—	—	—	—	—	—	—	—	—	—	—	—
C1	V_2	V_1	V_3	—	—	—	—	—	—	—	—	—	—	—	—	—	—	—
C2	V_2	—	V_5	V_4	—	—	—	—	—	—	—	—	—	—	—	—	—	—
C3	V_6	—	—	—	V_7	V_8	—	—	—	—	—	—	—	—	—	—	—	—
C4	V_{10}	—	—	—	V_9	—	V_{11}	V_{12}	—	—	—	—	—	—	—	—	—	—
C5	V_{14}	—	—	—	—	—	V_{13}	—	V_{15}	V_{16}	—	—	—	—	—	—	—	—
C6	V_{18}	—	—	—	—	—	—	—	V_{17}	—	V_{19}	V_{20}	—	—	—	—	—	—
C7	V_{22}	—	—	—	—	—	—	—	—	—	V_{21}	—	V_{23}	V_{24}	—	—	—	—
C8	V_{26}	—	—	—	—	—	—	—	—	—	—	—	V_{25}	—	V_{27}	V_{28}	—	—
C9	V_{30}	—	—	—	—	—	—	—	—	—	—	—	—	—	V_{29}	—	V_{31}	V_{32}
C10	V_{34}	—	—	—	—	—	—	—	—	—	—	—	—	—	—	—	V_{33}	—

piracy. To generate secret signature corresponding to facial biometric of an IP vendor, following steps are followed (Sengupta and Chaurasia, 2022; Sengupta and Rathor, 2021; Chaurasia and Sengupta, 2022):

Step 1: Capture the facial biometric image of an IP vendor and subject it to a specific grid size for accurate and precise feature extraction.

Step 2: Mark nodal feature points on the facial image as per IP vendor-chosen facial features set for deriving facial biometric signature. Flexibility in the selection of facial features enables an IP vendor to vary the strength of signature by varying the number of chosen facial features.

Step 3: Generate facial image as per IP vendor-chosen final features set.

Step 4: Determine the dimensions of each facial feature and transform them into binarized form. To measure feature dimension, Manhattan distance is computed between two nodal points representing a feature.

Step 5: Concatenate binarized signature corresponding to facial feature from feature set. There are "2^T" possibilities for an IP vendor to perform concatenation of features to produce facial biometric signature, where "T" represents the numeral value of facial features.

To automate the facial biometric signature generation for protecting CNN convolutional IP core, facial biometric tool has been employed (CAD for Assurance, 2022). The screen shots of tool are shown in Figure 4.9(a) and (b). Figure 4.9(a) displays a sample facial image of an IP vendor and Figure 4.9(b) displays the generated biometric signature corresponding to facial biometric trait of an IP vendor. The generated facial biometric signature (digital evidence) is shown below:

'10000111101000001000001001011111101010111010011001011100
110011010101000101111111111101'.

$$(4.9)$$

The facial biometric-based digital evidence (secret signature) offers following benefits over other contemporary hardware security methodologies based on digital signature (Sengupta *et al.*, 2019) and IP steganography (Sengupta and Rathor, 2019).

1. Facial signature comprises of the naturally unique physical biometric information of IP vendor.
2. Does not require to store the generated facial biometric signature (facial biometric image with IP vendor-specified grid size is stored in encrypted form). Further, the discussed methodology does not require recapturing of facial biometric image of an IP vendor for detection/verification purpose.
3. Offers robust security. This is evident, as in case even if an adversary manages to compromise facial biometric image of an IP vendor, he/she cannot exactly regenerate the same encoded hardware security constraints corresponding to facial biometric. This is because to regenerate the facial security constraints, an adversary needs to exactly guess the details of several security parameters such as grid size (used for subjecting the facial image), feature set (comprising of the different facial features), feature order (for generating digital biometric template), signature encoding algorithm (to encode signature bit into secret

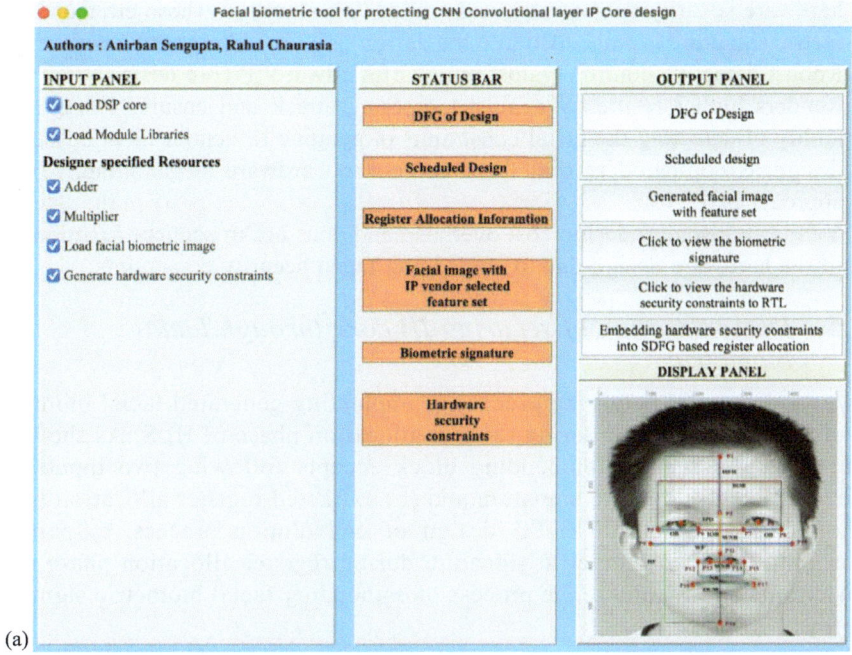

(a)

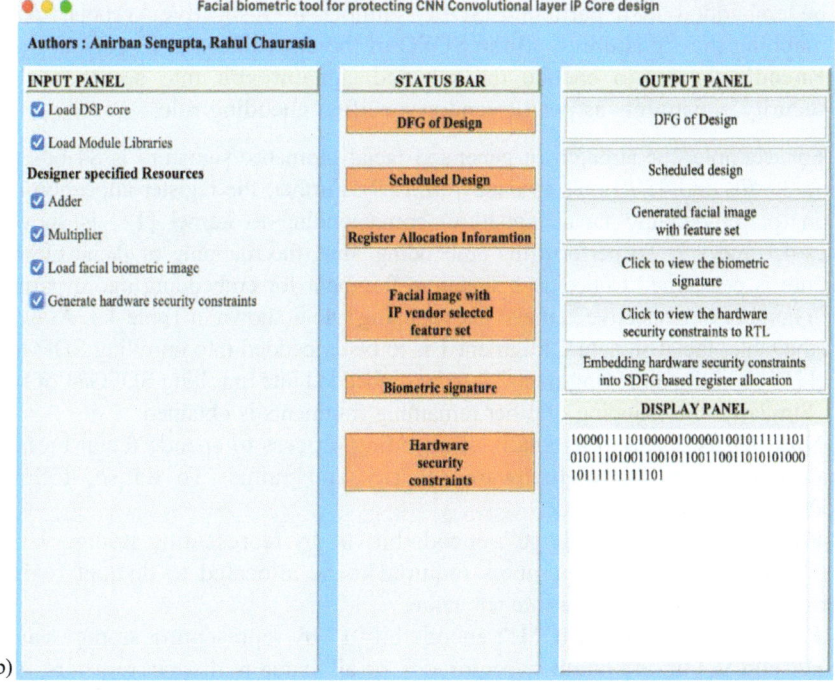

(b)

Figure 4.9   (a) Sample screenshot of facial biometric tool representing the generated facial image with IP vendor-chosen facial features. (b) Sample screenshot of facial biometric tool representing the generated signature (digital evidence) corresponding to facial biometric of an IP vendor.

hardware security constraints), and embedding algorithm (how the facial biometric signature is embedded into the design), all are unknown to an adversary.

4. Robust detective control against piracy of hardware IP core design.
5. Renders higher resiliency against tampering attack and ensures lesser prob-
    ability of detecting the facial constraints of genuine IP vendor in an unsecured
    design as compared to digital signature and hardware steganography-based
    approaches.
6. Yields almost zero design cost overhead and does not impact the functionality
    of the hardware design due to embedding facial security constraints.

4.7.5   Demonstration of securing IP core through facial biometric

Now, we discuss in detail the process of implanting generated facial biometric
signature into the design during register allocation phase of HLS. As shown in
Figure 4.2, the security embedding block accepts following two inputs: (a)
generated facial biometric signature and (b) extracted register allocation information corresponding to SDFG design of convolution process, to perform
embedding of facial biometric signature during register allocation phase (into
register allocation tables). The process of embedding facial biometric signature
employs two rules:

1. **Mapping rule**: for specifically selecting fragment/portion of the signature to
    be embedded in a particular SDFG number of respective kernels, thereby
    mapping the signature to all the SDFG of the convolutional IP core design.
2. **Encoding rule**: to encode the selected signature bit into secret hardware
    security constraints as per IP vendor-specified encoding rule.

For example, the strength of generated facial biometric signature is 84 bits (containing 39 #0s and 45 #1s) as evident from (4.9). Further, the register allocation information (pre-embedding facial signature) corresponding to kernel $[Y^1]$ is shown in
Tables 4.1 and 4.2. To perform the embedding, first, the mapping of facial biometric
signature is performed to decide a signature fragment for embedding into a particular
SDFG number of respective kernels. The mapping rule is shown in Table 4.3. As evident
from the table, facial signature fragment-1 is to be embedded into unrolling SDFG#2 of
kernel $[Y^1]$ and signature fragment-2 is to be embedded into unrolling SDFG#1 of kernel
$[Y^1]$. Similarly, the mapping of other remaining fragments is obtained.

Now, let us discuss the details of encoding process to encode facial biometric
signature bits into secret hardware security constraints. To do so, following
encoding rules are employed:

For facial signature bit "0": encode bit "0" by representing storage variable
pair (V_x and V_y) of even numbers required to be allocated to distinct registers,
where x and y can be of any integer value.

For facial signature bit "1": encode bit "1" by representing storage variable
pair (V_x and V_y) of odd numbers required to be allocated to distinct registers, where
x and y can be of any integer value.

For example, considering the encoding of facial biometric signature fragment—1 & 2. The signature fragment-1 is "10000111101000" (comprising of 8 #0s and 6 #1s) and signature fragment-2 is "00100000100101" (comprising of 10 #0s and 4 #1s). Therefore, the generated secret hardware security constraints corresponding to facial biometric signature fragments (1 & 2) based on the above encoding rules are shown in Tables 4.4 and 4.5, respectively. Similarly, for the other facial signature fragments (3, 4, 5, and 6), secret hardware security constraints are generated. Subsequently, the embedding of these encoded hardware security constraints is performed into corresponding SDFG design (during register allocation based on mapping rule as shown in Table 4.3). Facial biometric embedded SDFG design corresponding to kernel $[Y^1]$ is shown in Figure 4.10, where the modification in the design due to implanted facial signature has been represented using red color dotted boundary. Similarly, the

Table 4.3   IP vendor decision mechanism for selecting a specific 14-bit long facial signature fragment for embedding into a particular SDFG design (during register allocation) of Nth kernel using color mapping

Facial signature (14 bits each)	Signature fragment number based even–odd representation	Mapping rule: {design number of Nth kernel (N = 1,2,3) + (unrolling SDFG design #1, unrolling SDFG #2)}	SDFG number of respective kernels in which facial signature is implanted
10000111101000	Fragment (1 → odd)	{1+(1,2)} = {2,3}; 2 → even, 3 → odd	kernel $[Y^1]$ and unrolling SDFG #1
00100000100101	Fragment (2 → even)		Kernel $[Y^1]$ and unrolling SDFG #2
11111010101110	Fragment (3 → odd)	{2+(1,2)} = {3,4}; 3 → odd, 4 → even	Kernel $[Y^2]$ and unrolling SDFG #1
10011001011001	Fragment (4 → even)		Kernel $[Y^2]$ and unrolling SDFG #2
10011010101000	Fragment (5 → odd)	{3+(1,2)} = {4,5}; 4 → even, 5 → odd	Kernel $[Y^3]$ and unrolling SDFG #1
10111111111101	Fragment (6 → even)		Kernel $[Y^3]$ and unrolling SDFG #2

Table 4.4   Encoded storage variable pairs (hardware security constraints) corresponding to facial signature fragment-1

For bit "0"		For bit "1"	
$< V_0, V_2>$	$< V_0, V_{14}>$	$< V_1, V_3>$	–
$< V_0, V_4>$	$< V_0, V_{16}>$	$< V_1, V_5>$	–
$< V_0, V_6>$	–	$< V_1, V_7>$	–
$< V_0, V_8>$	–	$< V_1, V_9>$	–
$< V_0, V_{10}>$	–	$< V_1, V_{11}>$	–
$< V_0, V_{12}>$	–	$< V_1, V_{13}>$	–

signature embedded scheduled DFG design for kernels $[Y^2]$ and $[Y^3]$ can be generated. Further, post f embedding the facial biometric, register allocation information of scheduled design corresponding to convolution process with respect to feature kernel $[Y^1]$ is shown in Tables 4.6 and 4.7. Table 4.6 represents the modified/locally altered (post embedding facial security constraints) register allocation information corresponding to unrolling SDFG#1 of kernel $[Y^1]$. On the other hand, Table 4.7 represents the modified/locally altered (post embedding facial security constraints) register allocation information corresponding to unrolling SDFG#2 of kernel $[Y^1]$. Similarly, the modified register allocation information corresponding to SDFG of kernel $[Y^2]$ and $[Y^3]$ are shown in Tables 4.8–4.11, respectively. The storage variables marked in green

Table 4.5   *Encoded storage variable pairs (hardware security constraints) corresponding to facial signature fragment-2*

For bit "0"		For bit "1"	
$< V_0, V_2 >$	$< V_0, V_{14} >$	$< V_1, V_3 >$	–
$< V_0, V_4 >$	$< V_0, V_{16} >$	$< V_1, V_5 >$	–
$< V_0, V_6 >$	$< V_0, V_{18} >$	$< V_1, V_7 >$	–
$< V_0, V_8 >$	$< V_0, V_{20} >$	$< V_1, V_9 >$	–
$< V_0, V_{10} >$	–	–	–
$< V_0, V_{12} >$	–	–	–

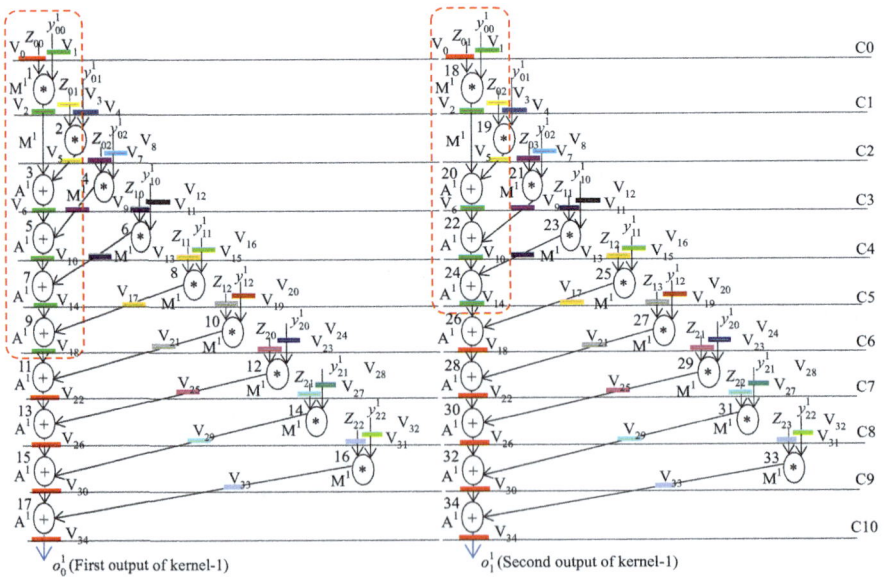

Figure 4.10   *Facial biometric embedded scheduled data flow graph of CNN convolutional IP core (corresponding to kernel-"1" ($[Y^1]$) of size "3 × 3") and twice unrolled using HLS-based approach*

Table 4.6 Register allocation information of scheduled design corresponding to convolution process with respect to feature kernel [Y^1] (for first pixel output computation)

Registers	R^1	R^2	R^3	R^4	R^5	R^6	R^7	R^8	R^9	R^{10}	R^{11}	R^{12}	R^{13}	R^{14}	R^{15}	R^{16}	R^{17}	R^{18}
C0	V_0	V_1																
C1	V_2	V_2	V_3	V_4														
C2	V_2	V_2	V_5		V_7	V_8												
C3	V_6	V_6			V_9			V_{12}										
C4	V_{10}	V_{10}					V_{11}		V_{15}	V_{16}								
C5	V_{14}	V_{14}					V_{13}		V_{17}		V_{19}	V_{20}						
C6	V_{18}	V_{18}									V_{21}		V_{23}	V_{24}				
C7	V_{22}												V_{25}		V_{27}	V_{28}		
C8	V_{26}														V_{29}		V_{31}	V_{32}
C9	V_{30}																V_{33}	
C10	V_{34}																	

Table 4.7 Register allocation information of scheduled design corresponding to convolution process with respect to feature kernel [Y^1] (for second pixel output computation).

Registers	R^1	R^2	R^3	R^4	R^5	R^6	R^7	R^8	R^9	R^{10}	R^{11}	R^{12}	R^{13}	R^{14}	R^{15}	R^{16}	R^{17}	R^{18}
C0	V_0	V_1																
C1	V_2	V_2	V_3	V_4														
C2	V_2	V_2	V_5		V_7	V_8												
C3	V_6	V_6			V_9			V_{12}										
C4	V_{10}	V_{10}					V_{11}		V_{15}	V_{16}								
C5	V_{14}	V_{14}					V_{13}		V_{17}		V_{19}	V_{20}						
C6	V_{18}										V_{21}		V_{23}	V_{24}				
C7	V_{22}												V_{25}		V_{27}	V_{28}		
C8	V_{26}														V_{29}		V_{31}	V_{32}
C9	V_{30}																V_{33}	
C10	V_{34}																	

Table 4.8 Register allocation information of scheduled design corresponding to convolution process with respect to feature kernel [Y^2] (for first pixel output computation)

Registers	R^1	R^2	R^3	R^4	R^5	R^6	R^7	R^8	R^9	R^{10}	R^{11}	R^{12}	R^{13}	R^{14}	R^{15}	R^{16}	R^{17}	R^{18}
C0	V_0	V_1																
C1	V_2	V_2	V_3	V_4														
C2	V_2	V_2	V_5		V_7	V_8												
C3	V_6	V_6			V_9		V_{11}	V_{12}										
C4	V_{10}						V_{13}		V_{15}	V_{16}								
C5	V_{14}								V_{17}		V_{19}	V_{20}						
C6	V_{18}										V_{21}		V_{23}	V_{24}				
C7	V_{22}												V_{25}		V_{27}	V_{28}		
C8	V_{26}														V_{29}		V_{31}	V_{32}
C9	V_{30}																V_{33}	
C10	V_{34}																	

Table 4.9 Register allocation information of scheduled design corresponding to convolution process with respect to feature kernel [Y^2] (for second pixel output computation)

Registers	R^1	R^2	R^3	R^4	R^5	R^6	R^7	R^8	R^9	R^{10}	R^{11}	R^{12}	R^{13}	R^{14}	R^{15}	R^{16}	R^{17}	R^{18}
C0	V_0	V_1																
C1	V_2	V_2	V_3	V_4														
C2	V_2	V_2	V_5		V_7	V_8												
C3	V_6	V_6			V_9		V_{11}	V_{12}										
C4	V_{10}	V_{10}					V_{13}		V_{15}	V_{16}								
C5	V_{14}	V_{14}							V_{17}		V_{19}	V_{20}						
C6	V_{18}										V_{21}		V_{23}	V_{24}				
C7	V_{22}												V_{25}		V_{27}	V_{28}		
C8	V_{26}														V_{29}		V_{31}	V_{32}
C9	V_{30}																V_{33}	
C10	V_{34}																	

Table 4.10   Register allocation information of scheduled design corresponding to convolution process with respect to feature kernel $[Y^3]$ (for first pixel output computation)

Registers	R^1	R^2	R^3	R^4	R^5	R^6	R^7	R^8	R^9	R^{10}	R^{11}	R^{12}	R^{13}	R^{14}	R^{15}	R^{16}	R^{17}	R^{18}
C0	V_0	V_1	–	–	–	–	–	–	–	–	–	–	–	–	–	–	–	–
C1	V_2	V_2	V_3	V_4	–	–	–	–	–	–	–	–	–	–	–	–	–	–
C2	V_2	V_2	V_5	–	–	–	–	–	–	–	–	–	–	–	–	–	–	–
C3	V_6	–	–	–	V_7	V_8	–	–	–	–	–	–	–	–	–	–	–	–
C4	V_{10}	–	–	–	V_9	–	V_{11}	V_{12}	–	–	–	–	–	–	–	–	–	–
C5	V_{14}	–	–	–	–	–	V_{13}	–	V_{15}	V_{16}	–	–	–	–	–	–	–	–
C6	V_{18}	–	–	–	–	–	–	–	V_{17}	–	V_{19}	V_{20}	–	–	–	–	–	–
C7	V_{22}	–	–	–	–	–	–	–	–	–	V_{21}	–	V_{23}	V_{24}	–	–	–	–
C8	V_{26}	–	–	–	–	–	–	–	–	–	–	–	V_{25}	–	V_{27}	V_{28}	–	–
C9	V_{30}	–	–	–	–	–	–	–	–	–	–	–	–	–	V_{29}	–	V_{31}	V_{32}
C10	V_{34}	–	–	–	–	–	–	–	–	–	–	–	–	–	–	–	V_{33}	–

Table 4.11   Register allocation information of scheduled design corresponding to convolution process with respect to feature kernel $[Y^3]$ (for second pixel output computation)

Registers	R^1	R^2	R^3	R^4	R^5	R^6	R^7	R^8	R^9	R^{10}	R^{11}	R^{12}	R^{13}	R^{14}	R^{15}	R^{16}	R^{17}	R^{18}
C0	V_0	V_1	–	–	–	–	–	–	–	–	–	–	–	–	–	–	–	–
C1	V_2	V_2	V_3	V_4	–	–	–	–	–	–	–	–	–	–	–	–	–	–
C2	V_2	V_2	V_5	–	–	–	–	–	–	–	–	–	–	–	–	–	–	–
C3	V_6	V_6	–	–	V_7	V_8	–	–	–	–	–	–	–	–	–	–	–	–
C4	V_{10}	V_{10}	–	–	V_9	–	V_{11}	V_{12}	–	–	–	–	–	–	–	–	–	–
C5	V_{14}	V_{14}	–	–	–	–	V_{13}	–	V_{15}	V_{16}	–	–	–	–	–	–	–	–
C6	V_{18}	–	–	–	–	–	–	–	V_{17}	–	V_{19}	V_{20}	–	–	–	–	–	–
C7	V_{22}	–	–	–	–	–	–	–	–	–	V_{21}	–	V_{23}	V_{24}	–	–	–	–
C8	V_{26}	–	–	–	–	–	–	–	–	–	–	–	V_{25}	–	V_{27}	V_{28}	–	–
C9	V_{30}	–	–	–	–	–	–	–	–	–	–	–	–	–	V_{29}	–	V_{31}	V_{32}
C10	V_{34}	–	–	–	–	–	–	–	–	–	–	–	–	–	–	–	V_{33}	–

color are representing the altered position of storage variables into registers and the variables marked in gray are representing the previous position of storage variables into registers. The alteration has been performed by embedding the security constraints into the design by using following rule: "two storage variables from any constraint pair cannot be assigned to the same register." Thus, the IP vendor facial biometric signature-driven secret hardware security constraints are embedded into the convolutional layer IP core design during register allocation phase of HLS process. These embedded covert hardware security constraints enable robust detective control against pirated versions. The robustness is ensured through the facial biometric-based security mechanism comprising of several security parameters, which all are unknown to an adversary, as discussed in Section 4.7.4. Therefore, facial biometric-based secured convolutional layer IP core design methodology incapacitates an adversary to replicate the security mark into pirated versions to evade their detection during piracy detection process. Therefore, it ensures the integration of only genuine IP core versions into SoCs of computing systems.

Further, it is evident from modified register allocation information corresponding to SDFG of convolution process with respect to kernels $[Y^1]$, $[Y^2]$, and $[Y^3]$ that no extra registers were required for embedding the facial security constraints into the design covertly.

4.7.6   Data path synthesis

Now, we discuss the process of performing datapath synthesis. The process has been discussed in

Generating RTL datapath using HLS: In the next design block of HLS process, datapaths corresponding to convolution process with respect to kernels $[Y^1]$, $[Y^2]$, and $[Y^3]$ are synthesized. This generates RTL design of convolutional layer IP core design capable of computing parallel pixels corresponding to each of the three kernels. The datapath block diagram corresponding to convolution process with respect to kernel $[Y^1]$ is shown in Figure 4.11(a) (for first pixel output computation) and Figure 4.11(b) (for second pixel output computation) pre-embedding facial biometric signature and Figure 4.12(a) (for first pixel output computation) and Figure 4.12(b) (for second pixel output computation) post embedding facial biometric signature, respectively, where the impact of embedding facial biometric signature on RTL design has been marked using red color boundary. The steps of generating secured RTL datapath corresponding to SDFG design of convolution process using HLS are as follows:

Step (1) *Formulation of multiplexing scheme for functional units (FUs) and registers*: Extract the information of the number of functional units (adders and multipliers) from scheduled DFG design. For example, in the scheduled DFG design corresponding to kernel $[Y^1]$, shown in Figure 4.6, all the operations are executed using 1 adder and 1 multiplier. Further, these FUs are being shared among different number of respective operations (as shown in Figure 4.6). Therefore, to enable the execution of more than one operation of same type (addition or multiplication) through a corresponding functional unit, Muxes and Demuxes are

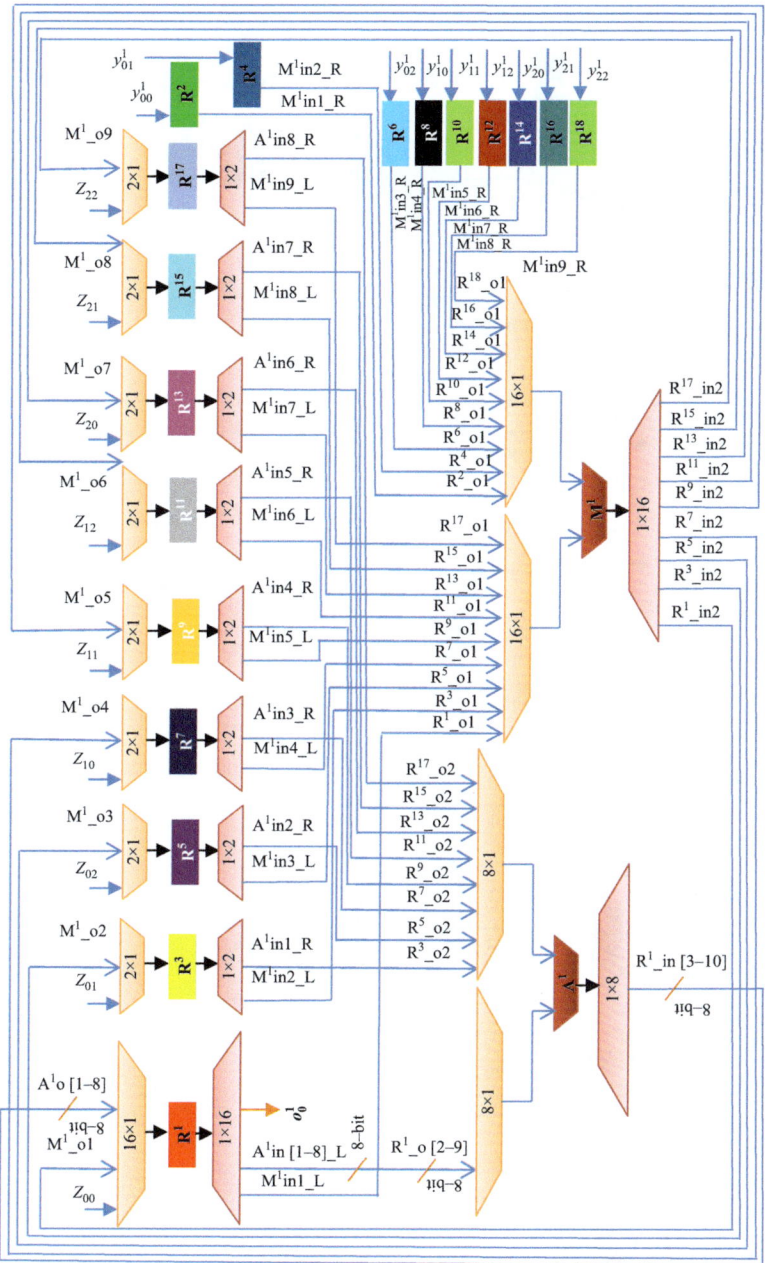

Figure 4.11   *(a) Datapath block diagram (pre embedding facial biometric signature) corresponding to convolution process w.r.t. kernel [Y¹] (for first pixel output computation). (b) Datapath block diagram (pre embedding facial signature) corresponding to convolution process w.r.t. kernel [Y¹] (for second pixel output computation).*

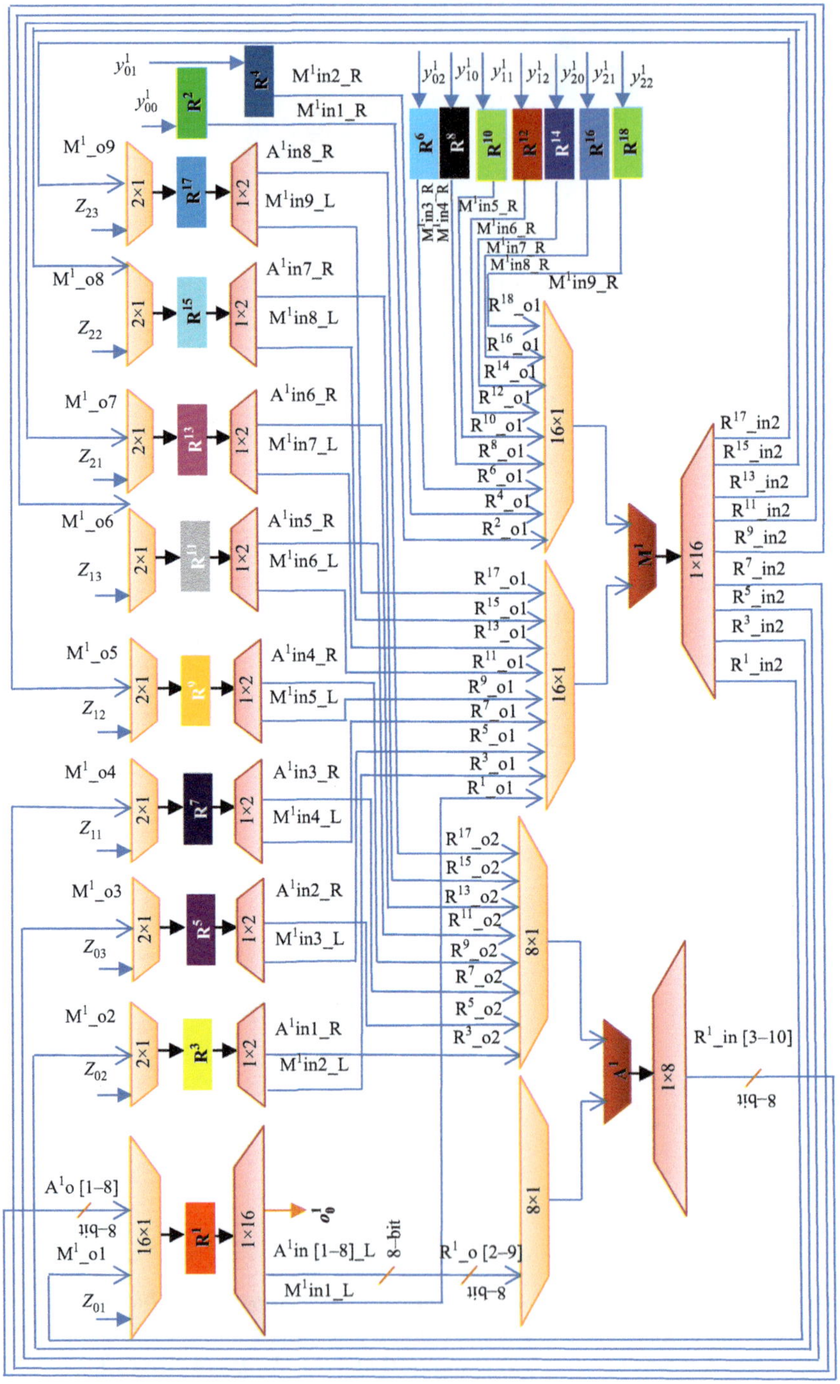

Figure 4.11 (*Continued*)

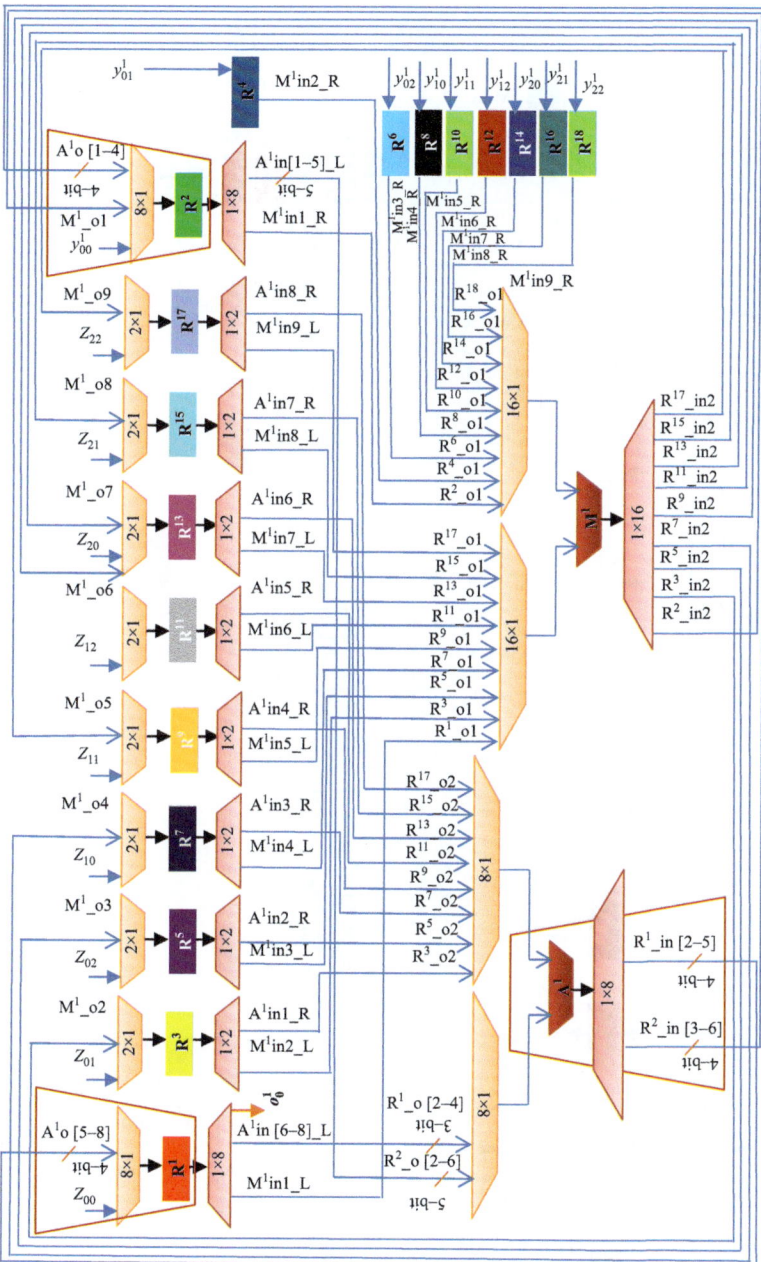

Figure 4.12　(a) Secured datapath block diagram (post embedding facial signature) corresponding to convolution process w.r.t. kernel [Y¹] (for first pixel output computation). (b) Secured datapath block diagram (post embedding facial signature) corresponding to convolution process w.r.t. kernel [Y¹] (for second pixel output computation).

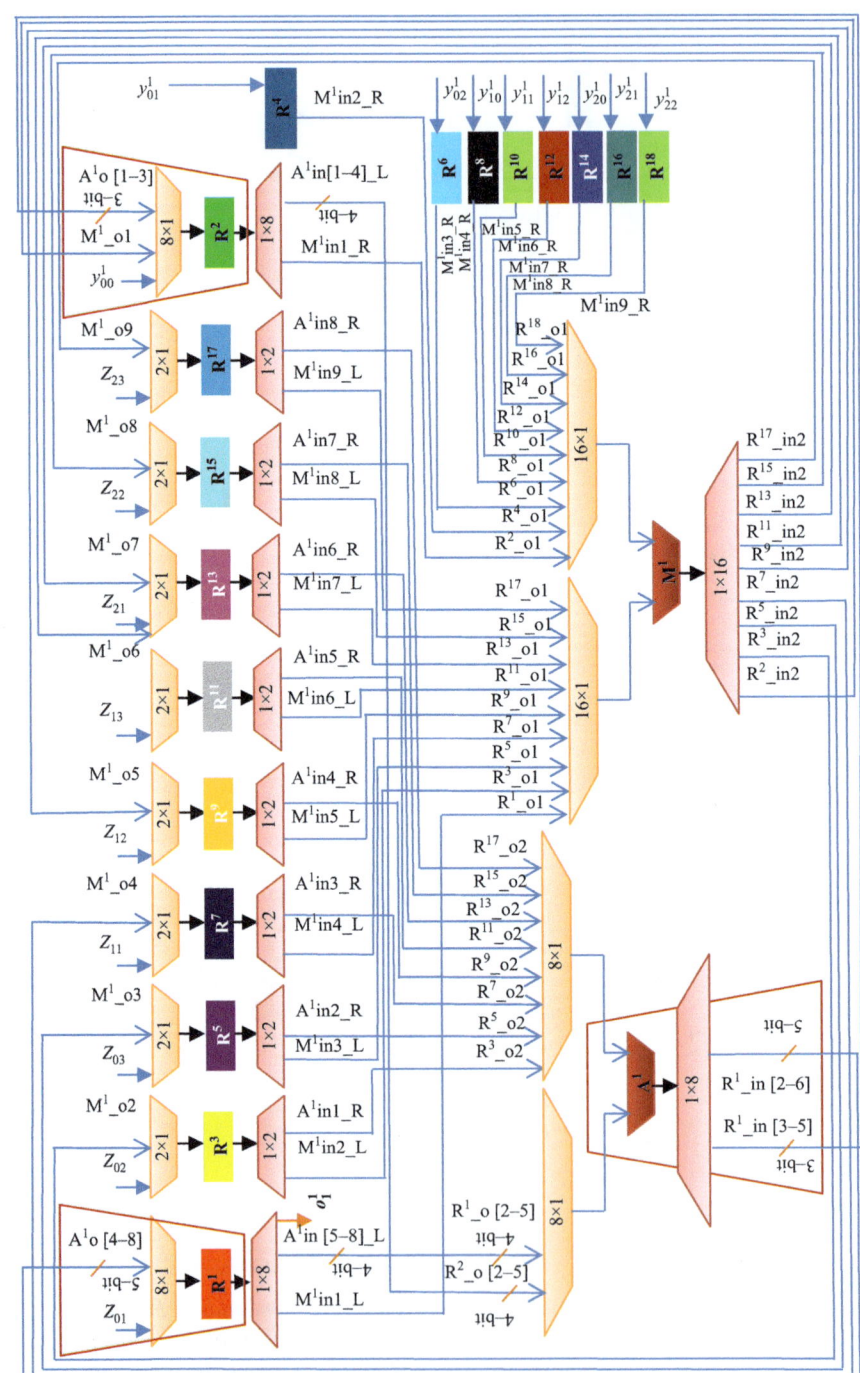

Figure 4.12 (*Continued*)

required in the datapath design, as shown in Figure 4.11(a) and (b) (pre-embedding facial signature) and Figure 4.12(a) and (b) (post embedding facial signature).

For the sake of demonstration, the multiplexing and de-multiplexing scheme corresponding to unrolling SDFG #1 pre-embedding facial signature (with respect to kernel $[Y^1]$) for FUs (adder and multiplier) is shown in Table 4.12. Further, the multiplexing scheme corresponding to registers (pre-embedding facial signature) is shown in Table 4.13(a)–(e). As evident from Table 4.12, FUs (adder and multiplier) are used for performing operations corresponding to several input pairs, therefore multiplexing scheme corresponding to these FUs require multiplexing and demultiplexing. Further, as evident from Table 4.13(a)–(e), registers (R^1, R^3, R^5, R^7, R^9, R^{11}, R^{13}, R^{15}, and R^{17}) are used for accommodating multiple storage variables during the computation process throughout the control steps C0–C10, therefore the multiplexing scheme corresponding to these registers require multiplexing and demultiplexing. In the multiplexing scheme of FUs (shown in Table 4.12), in0–in10 indicate multiplexers input (in) in order and o1–o9 indicates de-multiplexer outputs (o) in order. Further, in the multiplexing scheme of registers, as shown in Table 4.13 (a)–(e), left and right multiplexers associated with functional units (adder and multiplier) are designated as "L" and "R."

Step 2: *Data path and controller synthesis*: Post determining the multiplexing and de-multiplexing scheme corresponding to FUs and registers, datapath and controller designs are synthesized. The synthesis process therefore results into RTL data path design of CNN convolutional layer IP core.

Further, the multiplexing and de-multiplexing scheme corresponding to unrolling SDFG #1 post embedding facial biometric signature (with respect to kernel $[Y^1]$) for FUs (adder and multiplier) is shown in Table 4.14, where the

Table 4.12   *Multiplexing scheme for multiplier M^1 and adder A^1 corresponding to unrolling SDFG#1 with respect to kernel $[Y^1]$ (pre-embedding facial biometric signature)*

Control steps	M^1			A^1		
	Input1 (in)	Input2 (in)	Output (o)	Input1 (in)	Input2 (in)	Output (o)
C0	R^1_o1	R^2_o1	–	–	–	–
C1	R^3_o1	R^4_o1	R^1_in2	–	–	–
C2	R^5_o1	R^6_o1	R^3_in2	R^1_o2	R^3_o2	–
C3	R^7_o1	R^8_o1	R^5_in2	R^1_o3	R^5_o2	R^1_in3
C4	R^9_o1	R^{10}_o1	R^7_in2	R^1_o4	R^7_o2	R^1_in4
C5	R^{11}_o1	R^{12}_o1	R^9_in2	R^1_o5	R^9_o2	R^1_in5
C6	R^{13}_o1	R^{14}_o1	R^{11}_in2	R^1_o6	R^{11}_o2	R^1_in6
C7	R^{15}_o1	R^{16}_o1	R^{13}_in2	R^1_o7	R^{13}_o2	R^1_in7
C8	R^{17}_o1	R^{18}_o1	R^{15}_in2	R^1_o8	R^{15}_o2	R^1_in8
C9	–	–	R^{17}_in2	R^1_o9	R^{17}_o2	R^1_in9
C10	–	–	–	–	–	R^1_in10

Table 4.13 (a)   Multiplexing scheme for registers R^1, R^2, R^3, and R^4 corresponding to unrolling SDFG#1 with respect to kernel $[Y^1]$ (pre-embedding facial biometric signature)

Control steps	R^1		R^2		R^3		R^4	
	Input (in)	Output (o)	Input (in)	Output (o)	Input (in)	Output (o)	Input (in)	Output (o)
C0	Z_{00}	M¹in1_L	y^1_{00}	M¹in1_R	Z_{01}	–	y^1_{01}	–
C1	M¹_o1	A¹in1_L	–	–	–	M¹in2_L	–	M¹in2_R
C2	–	–	–	–	M¹_o2	A¹in 1_R	–	–
C3	A¹_o1	A¹in2_L	–	–	–	–	–	–
C4	A1_o2	A¹in3_L	–	–	–	–	–	–
C5	A1_o3	A¹in4_L	–	–	–	–	–	–
C6	A1_o4	A¹in5_L	–	–	–	–	–	–
C7	A1_o5	A¹in6_L	–	–	–	–	–	–
C8	A1_o6	A¹in7_L	–	–	–	–	–	–
C9	A1_o7	A¹in8_L	–	–	–	–	–	–
C10	A1_o8	O^1_0	–	–	–	–	–	–

Table 4.13 (b)   Multiplexing scheme for registers R^5, R^6, R^7, and R^8 corresponding to unrolling SDFG#1 with respect to kernel $[Y^1]$ (pre-embedding facial biometric signature)

Control steps	R^5		R^6		R^7		R^8	
	Input (in)	Output (o)	Input (in)	Output (o)	Input (in)	Output (o)	Input (in)	Output (o)
C0	Z_{02}	–	y^1_{02}	–	Z_{10}	–	y^1_{10}	–
C1	–	–	–	–	–	–	–	–
C2	–	M¹in3_L	–	M¹in3_R	–	–	–	–
C3	M¹_o3	A¹in2_R	–	–	–	M¹in4_L	–	M¹in4_R
C4	–	–	–	–	M¹_o4	A¹in 3_R	–	–
C5	–	–	–	–	–	–	–	–
C6	–	–	–	–	–	–	–	–
C7	–	–	–	–	–	–	–	–
C8	–	–	–	–	–	–	–	–
C9	–	–	–	–	–	–	–	–
C10	–	–	–	–	–	–	–	–

impact of implanting facial biometric signature has been emphasized using red color (representing change in their inter-connections corresponding to multiplexing scheme pre-embedding facial signature as shown in Table 4.12). Further, the multiplexing scheme corresponding to registers (post embedding facial signature) is shown in Table 4.15(a)–(e). This multiplexing scheme has been derived corresponding to unrolling SDFG#1 (generating first pixel output) post embedding IP

Table 4.13 (c)   Multiplexing scheme for registers R^9, R^{10}, R^{11}, and R^{12} corresponding to unrolling SDFG#1 with respect to kernel $[Y^1]$ (pre-embedding facial biometric signature)

Control steps	R^9		R^{10}		R^{11}		R^{12}	
	Input (in)	Output (o)	Input (in)	Output (o)	Input (in)	Output (o)	Input (in)	Output (o)
C0	Z_{11}	–	y^1_{11}	–	Z_{12}	–	y^1_{12}	–
C1	–	–	–	–	–	–	–	–
C2	–	–	–	–	–	–	–	–
C3	–	–	–	–	–	–	–	–
C4	–	M¹in5_L	–	M¹in5_R	–	–	–	–
C5	M¹_o5	A¹in4_R	–	–	–	M¹in6_L	–	M¹in6_R
C6	–	–	–	–	M¹_o6	A¹in 5_R	–	–
C7	–	–	–	–	–	–	–	–
C8	–	–	–	–	–	–	–	–
C9	–	–	–	–	–	–	–	–
C10	–	–	–	–	–	–	–	–

Table 4.13 (d)   Multiplexing scheme for registers R^{13}, R^{14}, R^{15}, and R^{16} corresponding to unrolling SDFG#1 with respect to kernel $[Y^1]$ (pre-embedding facial biometric signature)

Control steps	R^{13}		R^{14}		R^{15}		R^{16}	
	Input (in)	Output (o)	Input (in)	Output (o)	Input (in)	Output (o)	Input (in)	Output (o)
C0	Z_{20}	–	y^1_{20}	–	Z_{21}	–	y^1_{21}	–
C1	–	–	–	–	–	–	–	–
C2	–	–	–	–	–	–	–	–
C3	–	–	–	–	–	–	–	–
C4	–	–	–	–	–	–	–	–
C5	–	–	–	–	–	–	–	–
C6	–	M¹in7_L	–	M¹in7_R	–	–	–	–
C7	M¹_o7	A¹in6_R	–	–	–	M¹in8_L	–	M¹in8_R
C8	–	–	–	–	M¹_o8	A¹in 7_R	–	–
C9	–	–	–	–	–	–	–	–
C10	–	–	–	–	–	–	–	–

vendor facial signature as shown in Figure 4.10. This multiplexing scheme is exploited to synthesize the secured RTL datapath corresponding to convolutional layer IP core design. Similarly, multiplexing and de-multiplexing scheme for SDFG design (pre- and post-embedding facial biometric signature) corresponding to convolution process with respect to kernels $[Y^1]$, $[Y^2]$, and $[Y^3]$, is generated.

Table 4.13 (e)   Multiplexing scheme for registers R^{17} and R^{18} corresponding to unrolling SDFG#1 with respect to kernel $[Y^1]$ (pre-embedding facial biometric signature)

Control steps	R^{17}		R^{18}	
	Input (in)	Output (o)	Input (in)	Output (o)
C0	Z_{22}	–	y_{22}^1	–
C1	–	–	–	–
C2	–	–	–	–
C3	–	–	–	–
C4	–	–	–	–
C5	–	–	–	–
C6	–	–	–	–
C7	–	–	–	–
C8	–	M^1in9_L	–	M^1in9_R
C9	M^1_out9	A^1in8_R	–	–
C10	–	–	–	–

Table 4.14   Multiplexing scheme for multiplier M^1 and adder A^1 corresponding to unrolling SDFG#1 with respect to kernel $[Y^1]$ (post embedding facial biometric signature)

Control steps	M^1			A^1		
	Input1 (in)	Input2 (in)	Output (o)	Input1 (in)	Input2 (in)	Output (o)
C0	R^1_o1	R^2_o1	–	–	–	–
C1	R^3_o1	R^4_o1	R^2_in2	–	–	–
C2	R^5_o1	R^6_o1	R^3_in2	R^2_o2	R^3_o2	–
C3	R^7_o1	R^8_o1	R^5_in2	R^2_o3	R^5_o2	R^2_in3
C4	R^9_o1	R^{10}_o1	R^7_in2	R^2_o4	R^7_o2	R^2_in4
C5	R^{11}_o1	R^{12}_o1	R^9_in2	R^2_o5	R^9_o2	R^2_in5
C6	R^{13}_o1	R^{14}_o1	R^{11}_in2	R^2_o6	R^{11}_o2	R^2_in6
C7	R^{15}_o1	R^{16}_o1	R^{13}_in2	R^1_o2	R^{13}_o2	R^1_in2
C8	R^{17}_o1	R^{18}_o1	R^{15}_in2	R^1_o3	R^{15}_o2	R^1_in3
C9	–	–	R^{17}_in2	R^1_o4	R^{17}_o2	R^1_in4
C10	–	–	–	–	–	R^1_in5

Similarly, the secured datapaths corresponding to convolution process with respect to kernel $[Y^2]$ and $[Y^3]$ are generated. Secured datapaths corresponding to kernels $[Y^2]$ and $[Y^3]$ are shown in Figure 4.13(a) and (b) (post embedding corresponding facial biometric signature) and Figure 4.14(a) and (b) (post embedding corresponding facial biometric signature), respectively.

Table 4.15 (a)   Multiplexing scheme for registers R^1, R^2, R^3, and R^4 corresponding to unrolling SDFG#1 with respect to kernel [Y^1] (post embedding facial biometric signature)

Control steps	R^1		R^2		R^3		R^4	
	Input (in)	Output (o)	Input (in)	Output (o)	Input (in)	Output (o)	Input (in)	Output (o)
C0	Z_{00}	M^1in1_L	y_{00}^1	M^1in1_R	Z_{01}	–	y_{01}^1	–
C1	–	–	M^1_o1	A^1in1_L	–	M^1in2_L	–	M^1in2_R
C2	–	–	–	–	M^1_o2	$A^1in\,1\_R$	–	–
C3	–	–	A^1_o1	A^1in2_L	–	–	–	–
C4	–	–	$A1\_o2$	A^1in3_L	–	–	–	–
C5	–	–	$A1\_o3$	A^1in4_L	–	–	–	–
C6	–	–	$A1\_o4$	A^1in5_L	–	–	–	–
C7	$A1\_o5$	A^1in6_L	–	–	–	–	–	–
C8	$A1\_o6$	A^1in7_L	–	–	–	–	–	–
C9	$A1\_o7$	A^1in8_L	–	–	–	–	–	–
C10	$A1\_o8$	0_0^1	–	–	–	–	–	–

Table 4.15 (b)   Multiplexing scheme for registers R^5, R^6, R^7, and R^8 corresponding to unrolling SDFG#1 with respect to kernel [Y^1] (post embedding facial biometric signature)

Control steps	R^5		R^6		R^7		R^8	
	Input (in)	Output (o)	Input (in)	Output (o)	Input (in)	Output (o)	Input (in)	Output (o)
C0	Z_{02}	–	y_{02}^1	–	Z_{10}	–	y_{10}^1	–
C1	–	–	–	–	–	–	–	–
C2	–	M^1in3_L	–	M^1in3_R	–	–	–	–
C3	M^1_o3	A^1in2_R	–	–	–	M^1in4_L	–	M^1in4_R
C4	–	–	–	–	M^1_o4	$A^1in\,3\_R$	–	–
C5	–	–	–	–	–	–	–	–
C6	–	–	–	–	–	–	–	–
C7	–	–	–	–	–	–	–	–
C8	–	–	–	–	–	–	–	–
C9	–	–	–	–	–	–	–	–
C10	–	–	–	–	–	–	–	–

4.7.7   Demonstration of the methodology

The designed secured reusable CNN convolutional layer hardware IP core with embedded IP vendor facial biometric enables accurate feature/curve detection (Sengupta and Chaurasia, 2022). To do so, it employs three kernels ($N = 3$, as an example) in parallel for convolving over input image (pixel matrix). Further, each kernel in a single execution is capable of computing two output pixels concurrently corresponding to output feature map. Therefore, the secured

Table 4.15 (c)   Multiplexing scheme for registers R^9, R^{10}, R^{11}, and R^{12}
corresponding to unrolling SDFG#1 with respect to kernel $[Y^1]$
(post embedding facial biometric signature)

Control steps	R^9		R^{10}		R^{11}		R^{12}	
	Input (in)	Output (o)	Input (in)	Output (o)	Input (in)	Output (o)	Input (in)	Output (o)
C0	Z_{11}	–	y^1_{11}	–	Z_{12}	–	y^1_{12}	–
C1	–	–	–	–	–	–	–	–
C2	–	–	–	–	–	–	–	–
C3	–	–	–	–	–	–	–	–
C4	–	M^1in5_L	–	M^1in5_R	–	–	–	–
C5	M^1_o5	A^1in4_R	–	–	–	M^1in6_L	–	M^1in6_R
C6	–	–	–	–	M^1_o6	A^1in 5_R	–	–
C7	–	–	–	–	–	–	–	–
C8	–	–	–	–	–	–	–	–
C9	–	–	–	–	–	–	–	–
C10	–	–	–	–	–	–	–	–

Table 4.15 (d)   Multiplexing scheme for registers R^{13}, R^{14}, R^{15}, and R^{16}
corresponding to unrolling SDFG#1 with respect to kernel $[Y^1]$
(post embedding facial biometric signature)

Control steps	R^{13}		R^{14}		R^{15}		R^{16}	
	Input (in)	Output (o)	Input (in)	Output (o)	Input (in)	Output (o)	Input (in)	Output (o)
C0	Z_{20}	–	y^1_{20}	–	Z_{21}	–	y^1_{21}	–
C1	–	–	–	–	–	–	–	–
C2	–	–	–	–	–	–	–	–
C3	–	–	–	–	–	–	–	–
C4	–	–	–	–	–	–	–	–
C5	–	–	–	–	–	–	–	–
C6	–	M^1in7_L	–	M^1in7_R	–	–	–	–
C7	M^1_o7	A^1in6_R	–	–	–	M^1in8_L	–	M^1in8_R
C8	–	–	–	–	M^1_o8	A^1in 7_R	–	–
C9	–	–	–	–	–	–	–	–
C10	–	–	–	–	–	–	–	–

convolutional layer IP core design is capable of achieving secured accelerated hardware-based computation during convolution process in convolutional layer of CNN.

For the sake of demonstrating the process flow of convolutional layer IP core design architecture, the entire CNN working mechanism has been discussed through following phases: convolutional layer phase, pooling layer phase, and fully connected layer phase.

Table 4.15 (e)   Multiplexing scheme for registers R^{17} and R^{18} corresponding to unrolling SDFG#1 with respect to kernel $[Y^1]$ (post embedding facial biometric signature)

Control steps	R^{17}		R^{18}	
	Input (in)	Output (o)	Input (in)	Output (o)
C0	Z_{22}	–	y_{22}^1	–
C1	–	–	–	–
C2	–	–	–	–
C3	–	–	–	–
C4	–	–	–	–
C5	–	–	–	–
C6	–	–	–	–
C7	–	–	–	–
C8	–	M^{1}in9_L	–	M^{1}in9_R
C9	M^1_out9	A^{1}in8_R	–	–
C10	–	–	–	–

Convolutional layer phase: In the convolutional layer phase, a sample input image (whose features are to be detected) is convolved (*) using filter kernels $[Y^1]$, $[Y^2]$, and $[Y^3]$. As shown in Figure 4.15(a), a sample image has been accepted as an input image, which is further exploited for performing the feature (curve/edges) detection. The sample image with receptive field visualization is highlighted using yellow color boundary. Receptive field is the subregion of the image to which kernel(s) is/are applied for feature extraction. For example, a sample input image with receptive region is represented as pixel matrix (partial potion), which is of size "86 × 124," as shown in Figure 4.15(b). In the pixel matrix of sample image, receptive field/subregion is marked using cyan and brown color corresponding to parallel computations of two output pixels. Further, weights corresponding to three 2D convolutional kernels ($[Y^1]$, $[Y^2]$, and $[Y^3]$) of size "3 × 3" are defined. The weights are defined in such a way that they resembles the curve/edge feature.

So far, we have selected the sample input image and chosen the weights of all three kernel. Now, the convolution process is performed using the secured convolutional layer IP core as shown in Figure 4.15(b), which comprises of facial biometric secured RTL datapath. Consequently, the kernels detect the features from sample image that generally resembles as per the orientation of the pixels and their weight value. In output pixel matrix (feature map), the pixels that resemble the curve are non-zero or higher pixel values. Further, if the input image either partially resembles or does not resembles the object/curve corresponding to filter kernels, then output pixel value is lesser (zero, in the case of binary image). In a feature map, output pixels are the resultant of dot product between corresponding image matrix and kernel matrix. Further,

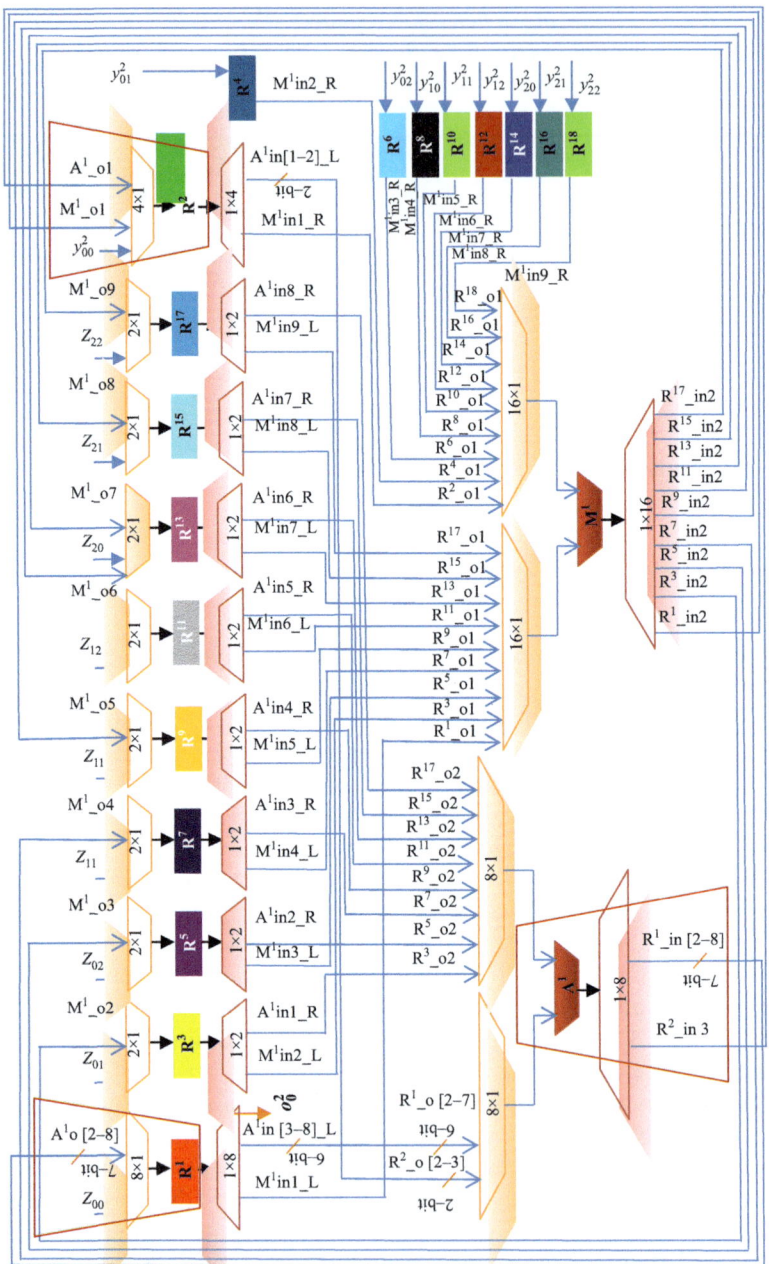

Figure 4.13   *(a) Secured datapath block diagram (post embedding facial signature)*
corresponding to convolution process w.r.t. kernel [Y²] (for first pixel
output computation). (b) Secured datapath block diagram (post
embedding facial signature) corresponding to convolution process w.r.t.
kernel [Y²] (for second pixel output computation).

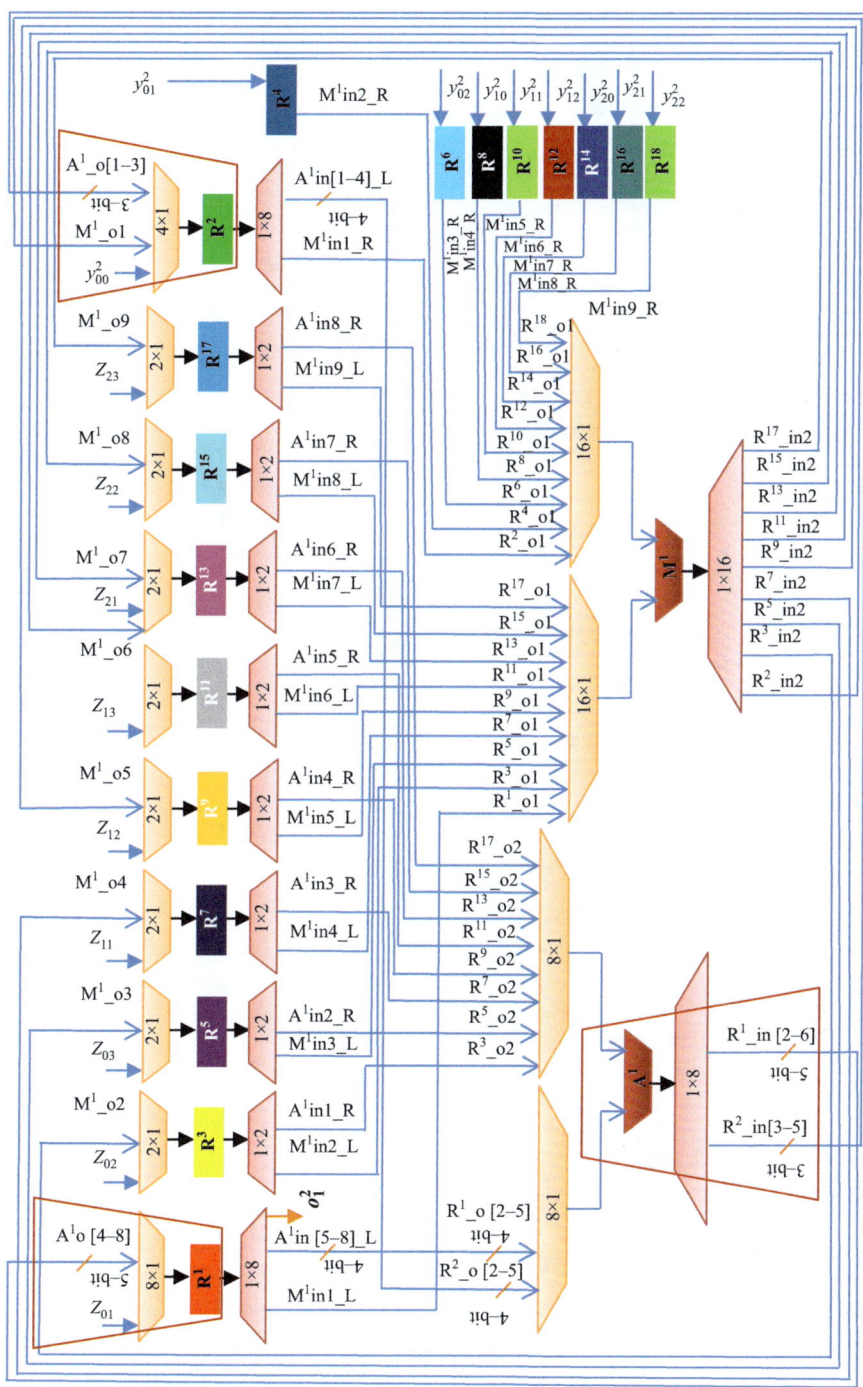

Figure 4.13   (*Continued*)

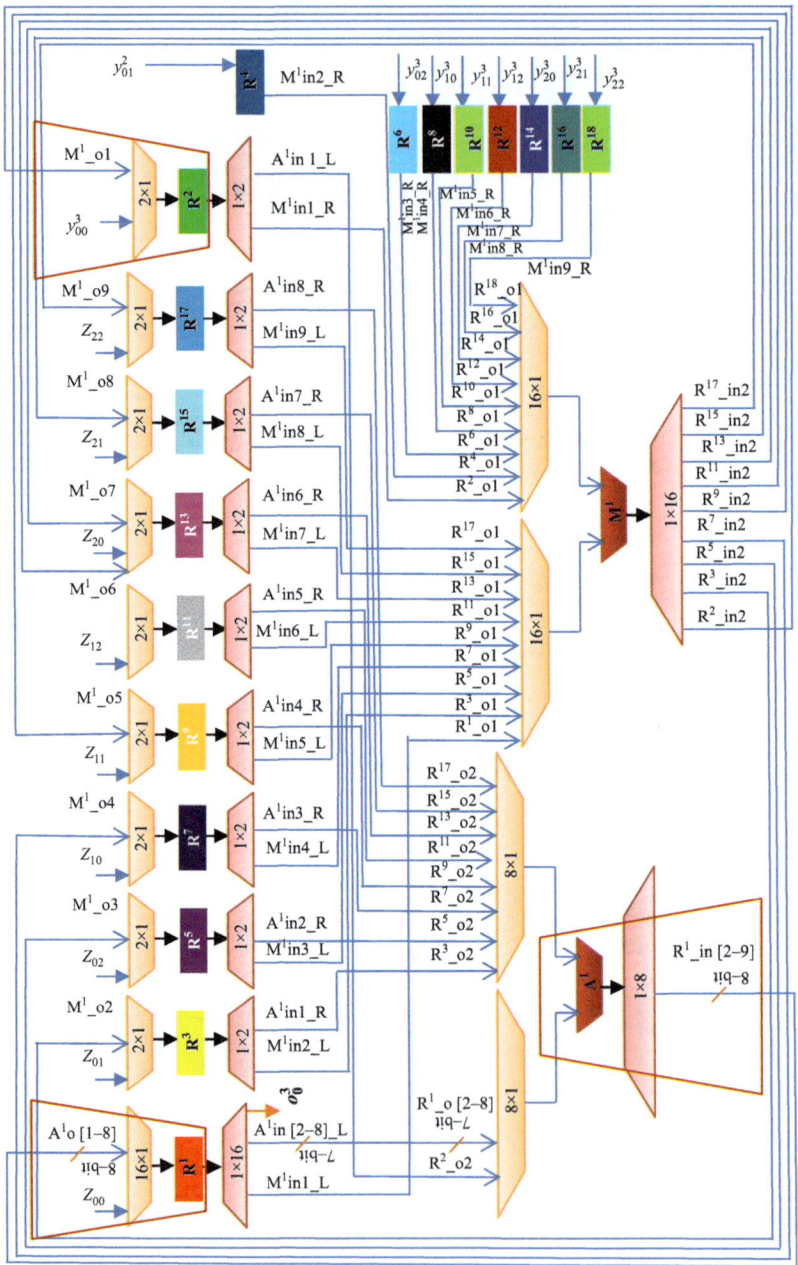

Figure 4.14   *(a) Secured datapath block diagram (post embedding facial signature)*
corresponding to convolution process w.r.t. kernel [Y³] (for first pixel
output computation). (b) Secured datapath block diagram (post
embedding facial signature) corresponding to convolution process w.r.t.
kernel [Y³] (for second pixel output computation).

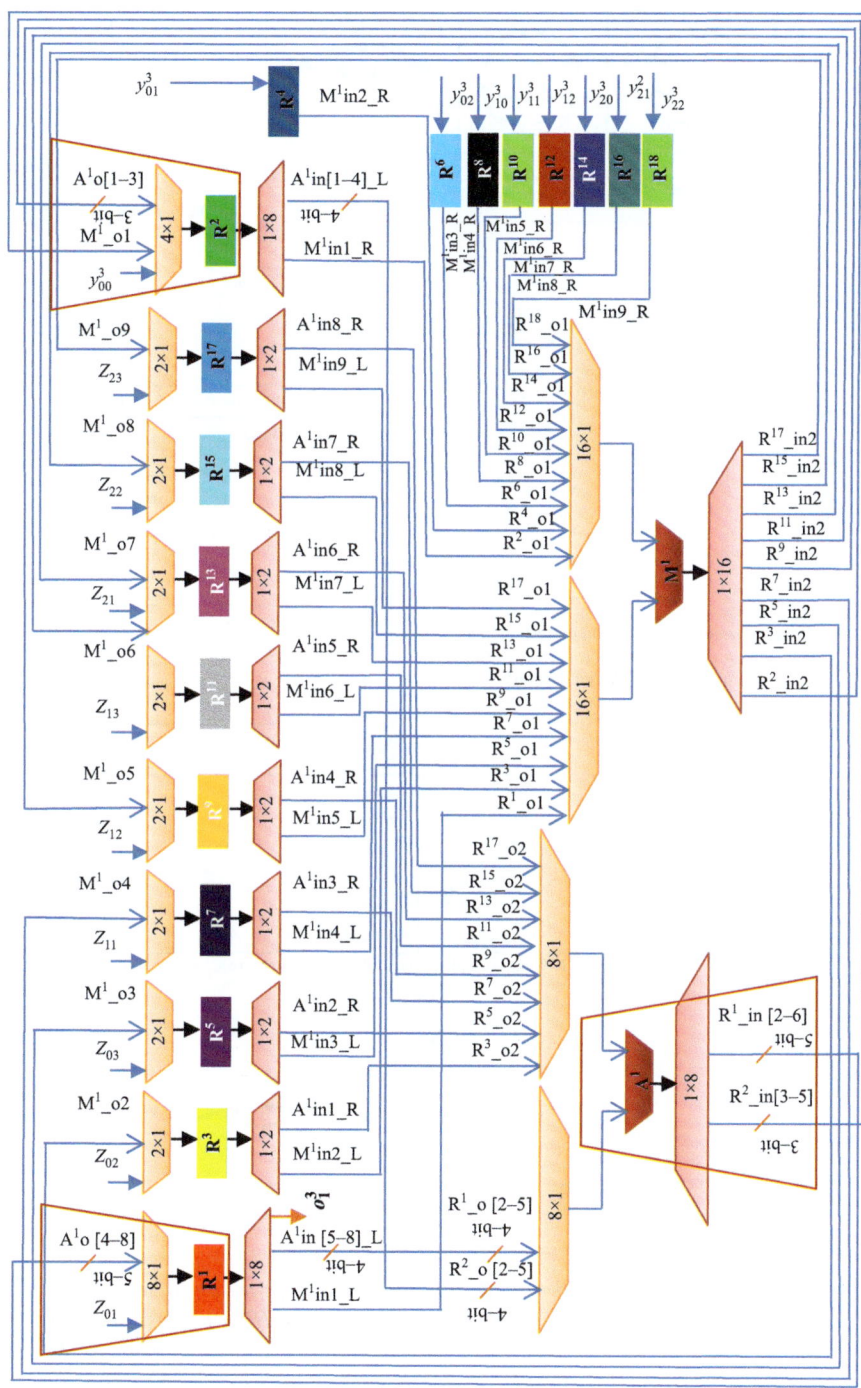

Figure 4.14 (*Continued*)

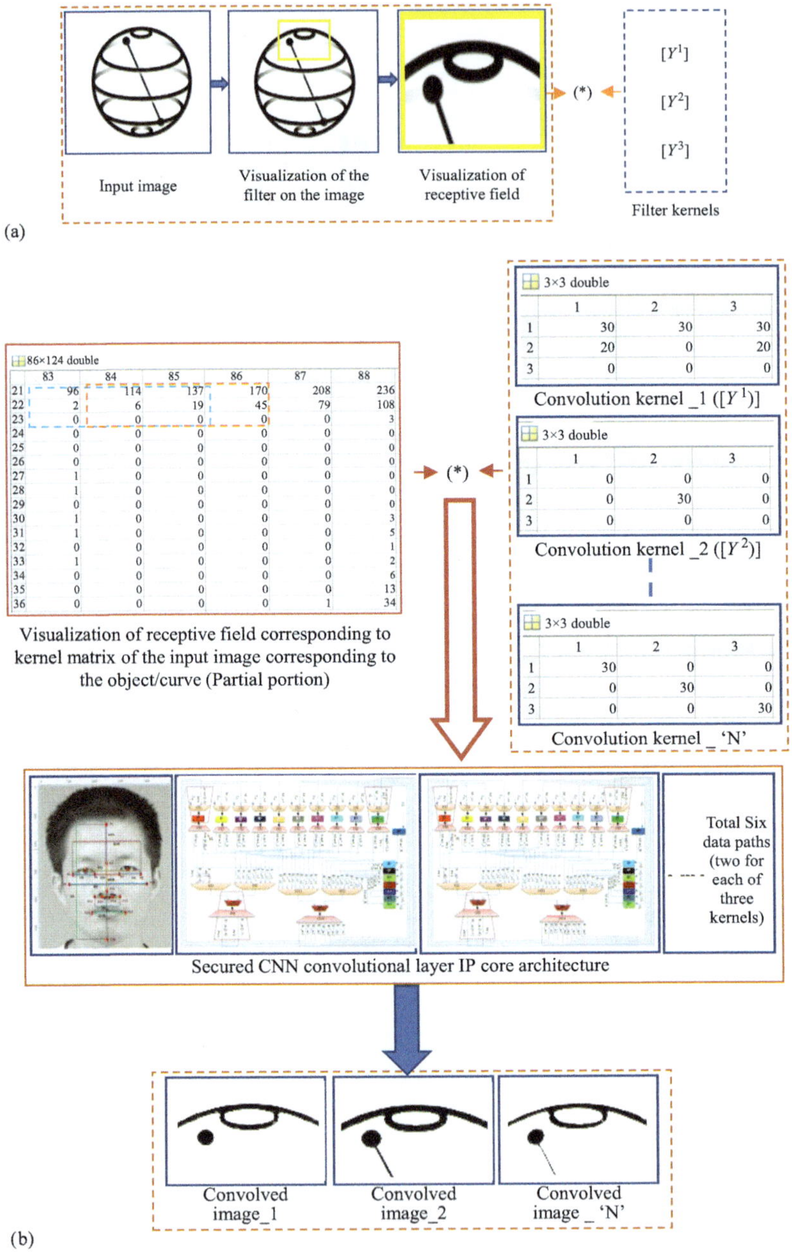

(a)

86×124 double					
	83	84	85	86	87	88
21	96	114	137	170	208	236
22	2	6	19	45	79	108
23	0	0	0	0	0	3
24	0	0	0	0	0	0
25	0	0	0	0	0	0
26	0	0	0	0	0	0
27	1	0	0	0	0	0
28	1	0	0	0	0	0
29	0	0	0	0	0	0
30	1	0	0	0	0	3
31	1	0	0	0	0	5
32	0	0	0	0	0	1
33	1	0	0	0	0	2
34	0	0	0	0	0	6
35	0	0	0	0	0	13
36	0	0	0	0	1	34

Visualization of receptive field corresponding to kernel matrix of the input image corresponding to the object/curve (Partial portion)

(*)

3×3 double		
	1	2	3
1	30	30	30
2	20	0	20
3	0	0	0

Convolution kernel _1 ([Y^1])

3×3 double		
	1	2	3
1	0	0	0
2	0	30	0
3	0	0	0

Convolution kernel _2 ([Y^2])

3×3 double		
	1	2	3
1	30	0	0
2	0	30	0
3	0	0	30

Convolution kernel _ 'N'

Secured CNN convolutional layer IP core architecture

Total Six data paths (two for each of three kernels)

Convolved image_1

Convolved image_2

Convolved image _ 'N'

(b)

Figure 4.15 (a) *Visualization of receptive field corresponding to sample input image. (b) Secure CNN convolution layer IP core design.* Note: *detailed secure RTL data path design are shown in Figures 4.12(a) and (b), 4.13(a) and (b), and 4.14(a) and (b) corresponding to kernels [Y^1], [Y^2], and [Y^3], respectively. (continues)*

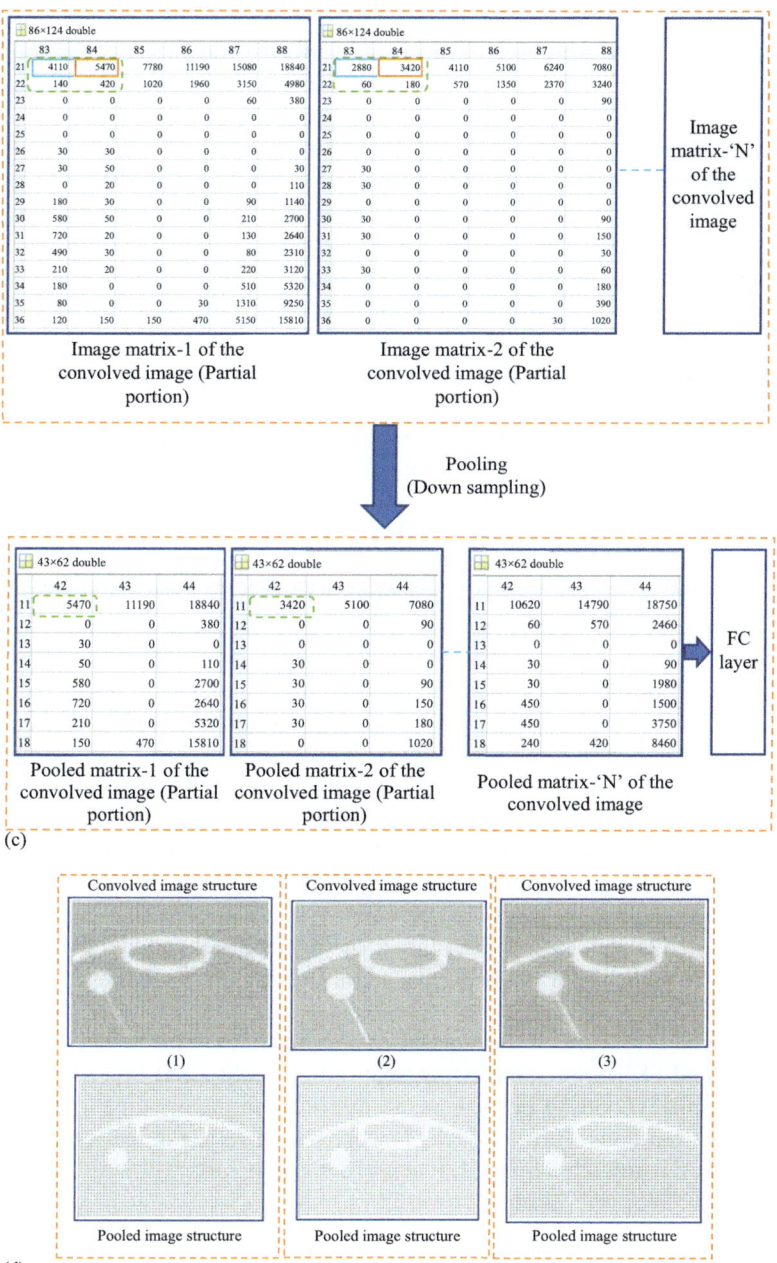

(c)

(d)

Figure 4.15   *(continued) (c) Demonstration of generated convolved image matrices corresponding to filter kernels and their respective pooled image matrices (d) Output structure of convolved images with respect to different feature kernels and their corresponding pooled images.*

convolutional layer IP core design employs the "same convolution" process, therefore, the size of convolved feature matrix post-hardware computation is also the same as input image matrix pre-padding. The generated convolved output image matrix (feature map) corresponding to filter kernels using convolutional hardware IP core is shown in Figure 4.15(c), where different convolved matrices (1, 2, and 3) are generated corresponding to kernels ($[Y^1]$, $[Y^2]$, and $[Y^3]$). As evident from figure, the dimension of convolved matrices is the same as image matrix "86 × 124." Thus, corresponding to a sample input image, convolved images are generated independently corresponding to kernels. The pixel values of these convolved images differ due to different kernel weight orientations.

Pooling layer phase: In this phase, the functionality of pooling layer has been discussed. During the pooling layer, each feature map (image matrix-generated post-convolution process) is processed independently to generate corresponding pooled image. Pooling layer is responsible for reducing the computational complexity of CNN by plummeting the spatial dimensions of the generated feature maps. In the pooling layer, the filter of size 2 × 2 with stride/shift (2) integrated with "max-pooling" technique have been employed. This therefore results into the pooled image by down sampling the feature map. The pooled images (represented in the form of pooled matrices-1, 2, and 3) corresponding to convolved images are shown in Figure 4.15(c). As evident post-performing the pooling, the size of the resultant image matrix is reduced down to "43 × 62." This is because in general pooling process with a kernel of size 2 × 2 and stride/shift = 2, down samples the image size by 2 factors. The output pixel value corresponding to "max-pooling" is represented using green color dotted boundary, which selects the maximum value from convolved image matrix corresponding to kernel of size "2 × 2." Thus, the pooled (down sampled) images corresponding to feature map images are generated.

Fully-connected layer phase: Subsequently, in this phase, the output of pooling layer (pooled images) is flattened into a vector. This is then fed into soft-max or logistic functional unit for enabling the classification or detection of an object/feature.

The output structure of convolved image matrices with respect to kernels ($[Y^1]$, $[Y^2]$, and $[Y^3]$) generated using secured convolutional layer hardware IP core and its corresponding pooled image matrices are shown in Figure 4.15(d). Thus, the secured convolutional layer IP core performs the detection of features (curves/edges) using CNN-based feature detection, after processing through FC layer.

4.8   Analysis and discussion

This section analyses the CNN convolutional hardware IP core design in terms of following aspects.

4.8.1   Analyzing the convolutional IP core design in terms of computation of pixels

The convolutional layer hardware IP core design employs three feature kernels to convolve over input sample image for performing the hardware-based detection of features. Further, each kernel is capable of computing two output pixels concurrently using loop unrolling mechanism. The comparison of discussed convolutional layer hardware IP core with respect to conventional hardware design, in terms of required number of executions with respect to convolution operation, is shown in Figure 4.16. As evident, convolutional layer hardware IP core design (Sengupta and Chaurasia, 2022) requires significantly lesser number of executions for convolution operations than conventional hardware design. This is because the secured convolutional layer hardware IP core design employs parallelism owing to unrolling of the kernel datapath.

4.8.2   Analyzing the change in resources (Muxes and Demuxes) of RTL datapath design post-implanting facial biometric signature of IP vendor

The facial biometric of IP vendor in the form of encoded signature is implanted into the design during register allocation phase of HLS (discussed earlier in Section 4.7.5). Further, an IP vendor may vary the signature strength by varying the number of selected facial features for signature generation. Thus, IP vendor-selected facial signature, post-performing secret mapping, and their encoding into security constraints is implanted into the SDFG corresponding to convolutional layer IP core design. The constraints are embedded into design by performing local alteration of storage variables among registers. This may lead change in their interconnectivity. However, if the available registers are not capable of accommodating all the generated security constraints, then, some extra registers are also

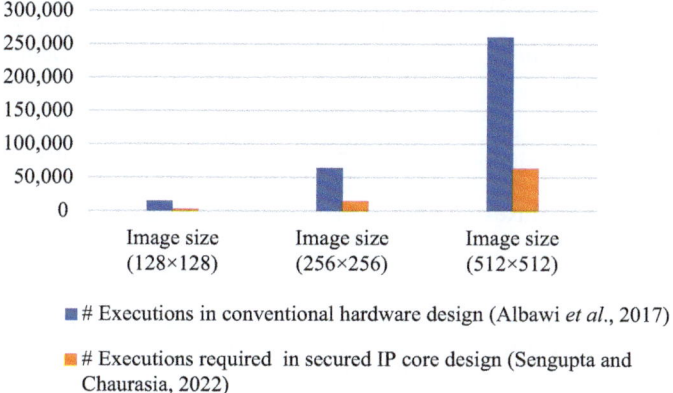

■ # Executions in conventional hardware design (Albawi *et al.*, 2017)

■ # Executions required in secured IP core design (Sengupta and Chaurasia, 2022)

Figure 4.16   Comparison of required executions for performing convolution operation (w.r.t. three kernels)

included into the design. Thus, embedding of facial security constraints may lead to structural change of the RTL datapath design without affecting the functionality. The impact of embedding IP vendor-encoded facial signature on functional units and corresponding multiplexers and de-multiplexers is shown in Table 4.16.

4.8.3   Analyzing the security strength

The security strength of convolutional hardware IP design integrated with facial biometric is analyzed in terms of: (a) authorship proof, using probability of coincidence (Xp) metrics and (b) resiliency against tampering attack using tamper tolerance (TT) metrics. The probability of coincidence metric corresponds to the probability of detecting the embedded facial security constraints of an IP vendor into an unsecured design. This therefore signifies that the Xp value of the secured IP core design must be as low as possible. In other words, lower value of Xp metric is desirable. The Xp of the secured design is computed using the following function (Koushanfar *et al.*, 2005; Sengupta and Rathor, 2019):

$$Xp = \left(1 - \frac{1}{R^n}\right)^r \tag{4.10}$$

Table 4.16   *Comparison of multiplexers and de-multiplexers of CNN convolutional IP design (pre- and post embedding encoded facial biometric signature)*

Kernel #	Pre-embedding facial biometric signature		Post embedding facial biometric signature	
	Multiplexers	**Demultiplexers**	**Multiplexers**	**Demultiplexers**
Convolutional IP datapath (Kernel-1)	#8X1Muxes=4, #16X1Muxes=6, #2X1Muxes=16	#1x8De-muxes=2, #1x16De-muxes=4, #1x2 De-muxes=16	#8X1Muxes=8, #16X1Muxes=4, #2X1Muxes=16	#1x8De-muxes=6, #1x16De-muxes=2, #1x2 De-muxes=16
Convolutional IP datapath (Kernel-2)	#8X1Muxes=4, #16X1Muxes=6, #2X1Muxes=16	#1x8De-muxes=2, #1x16De-muxes=4, #1x2 De-muxes=16	#8X1Muxes=7, #16X1Muxes=4, #2X1Muxes=16, #4X1Muxes=1	#1x8De-muxes=5, #1x16De-muxes=2, #1x2 De-muxes=16#1X4 De-muxes=1
Convolutional IP datapath (Kernel-3)	#8X1Muxes=4, #16X1Muxes=6, #2X1Muxes=16	#1x8De-muxes=2, #1x16De-muxes=4, #1x2 De-muxes=16	#8X1Muxes=6, #16X1Muxes=5, #2X1Muxes=17	#1x8De-muxes=4, #1x16De-muxes=3, #1x2 De-muxes=17

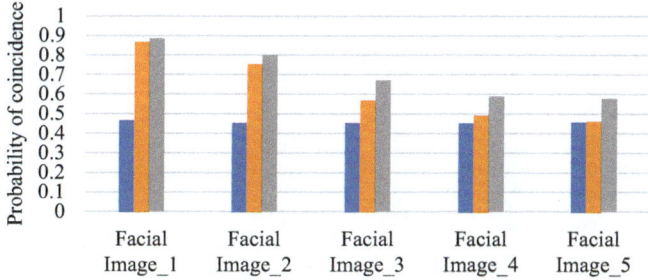

■ Xp of facial biometric secured CNN convolutional IP design approach (Sengupta and Chaurasia, 2022)

■ Xp of the digital signature based hardware security (Sengupta and Chaurasia, 2022)

■ Xp of hardware steganography approach (Sengupta and Chaurasia, 2022)

Figure 4.17   Comparison of security strength in terms of probability of coincidence

where "R^n" denotes the number of registers corresponding to SDFG design of the convolutional layer before embedding facial signature (digital evidence) and "r" denotes the number of embedded secret hardware security constraints, generated corresponding to SDFG design and encoded facial signature. The resultant "Xp" value of design methodology generating secured CNN convolutional IP core design is presented in Figure 4.17. As evident, Xp value of the facial biometric secured CNN convolutional IP design methodology (Sengupta and Chaurasia, 2022) is lesser (desirable) than the contemporary methodologies based on digital signature (Sengupta *et al.*, 2019) and hardware steganography (Sengupta and Rathor, 2019).

Further, the resiliency strength (tolerance) of the secured design against tampering attack is measured using the following function (Koushanfar *et al.*, 2005; Sengupta and Bhadauria, 2016):

$$TT = \mu^r \qquad (4.11)$$

where "μ" denotes the number of signature variables, which are two (0 and 1) in the case of facial biometric binary signature. An adversary may attempt to perform tampering of the genuine design to extract the embedded facial security constraints. The extraction of the authentic facial security constraints or a successful attempt to regenerate the authentic facial signature may lead to evasion of pirated IP versions during piracy detection process. This therefore signifies that tamper tolerance ability of the design must be as high as possible to prevent an adversary from successfully guessing (or extracting/regenerating) the exact facial signature combination. In other words, higher value of "TT" metric is desirable. The comparison of tamper tolerance ability of the facial biometric methodology with digital signature-based hardware security approach (Sengupta *et al.*, 2019) is shown in

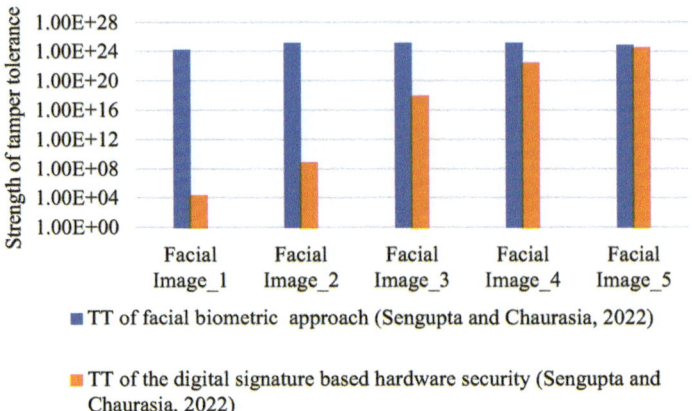

Figure 4.18   Comparison of security strength in terms of tamper tolerance ability

Figure 4.18. As evident, facial biometric methodology renders robust/higher tamper tolerance ability. This is because the number of generated covert hardware security constraints of facial biometric security methodology is higher than the digital signature-based approach.

4.8.4   Analyzing the design area

In the facial biometric secured convolutional IP methodology, security is embedded during register allocation phase of HLS. This implantation of IP vendor facial signature-driven secret hardware security constraints may lead to design overhead [in case if it is not possible to accommodate the generated hardware security constraints into the design within the available resources (registers)]. The impact of embedding facial biometric-driven secret signature into IP design area (A_d) is measured using following design metrics (Sengupta and Rathor, 2019):

$$A_d = g * (A_{adder}) + j * (A_{multiplier}) + t \quad (A_{register}) \qquad (4.12)$$

where "g," "j," and "t" denotes the number of adders, multipliers, and registers of the design, respectively, and "A" denotes the area of the respective hardware unit. The computation of design area has been performed by using a 15-nm open-cell library. The impact of the number of CNN convolutional layer kernels "N" and their unrolling factor over design area is shown in Table 4.17. As evident, more the number of kernels and more the number of times they are unrolled, it results into larger design area. In the facial biometric secured convolutional IP design methodology, three kernels are employed with unrolling factor (UF=2). Further, the embedding of facial biometric signature results into zero design cost overhead. This is because HLS-based secured IP core design methodology does not yield any design cost overhead in terms of requiring extra resources to integrate/implant

Table 4.17   *Comparison of design area (μm^2) corresponding to number of feature kernels and their unrolling factors*

	UF=4	UF=3	UF=2	UF=1
K=1	434.1	325.58	217.05	108.52
K=3	1302.32	976.74	651.16	325.58
K=5	2170.54	1627.9	1085.27	542.636
K=7	3038.76	2279.07	1519.38	759.69

facial biometric-driven secret hardware security constraints into the design. In the HLS-based methodology, the facial signature (in terms of secret hardware security constraints) is embedded into the IP design during register allocation phase. Further, it is apparent from the register allocation tables (Tables 4.6–4.11) that no extra resource was required for integrating all the facial security constraints.

4.9   Conclusion

This chapter discussed the HLS-based methodology for designing the secured reusable hardware IP core. To do so, the computationally intensive process of convolutional layer of CNN has been targeted. The generated secured CNN convolutional layer hardware IP core is capable of accelerating the pixel computation process through parallel execution of feature kernels. Further, facial biometric of IP vendor has been integrated into the design for enabling the robust detective control against IP piracy. Facial biometric signature in form of encoded naturally unique facial security constraints is embedded into the design during higher abstraction level of HLS. This therefore enables isolation of pirated IP versions during piracy detection process, thereby ensuring the integration of only authentic IP versions into SoCs of computing systems. Additionally, the HLS-based secured convolutional IP core design methodology results into zero design cost overhead for integrating facial biometric-based robust encoded facial signature (digital evidence) into the design.

At the end of this chapter, a reader gains the knowledge about the following:

- The significance of designing HLS-based secured convolutional layer IP core.
- Importance of securing IP cores against the threat of IP piracy.
- Basic functionality of CNN framework.
- Process flow of integrating facial biometric-based security into the convolutional IP during register allocation phase of HLS.
- Process flow of designing secured RTL datapath corresponding to convolution process of CNN using facial biometric.
- End to end design flow of designing CNN convolutional IP with demonstrative examples.

4.10   Questions and exercise

1. Discuss the utility of CNN framework in different domains.
2. Discuss the motivation for designing secured IPs and their importance for safe and reliable computing.
3. Discuss the motivation for designing the secured convolutional layer IP.
4. Explain the mathematical details of convolution process of CNN corresponding to a sample image and feature kernel(s).
5. Discuss the steps for designing scheduled DFG design corresponding to convolution process.
6. Discuss the HLS-based design flow for generating secured RTL datapath design corresponding to convolution process of convolutional layer.
7. Discuss the process of generating facial biometric signature from facial image of an IP vendor.
8. Discuss the inputs for the process of generating facial biometric-based encoded hardware security constraints.
9. Discuss the steps of integrating the facial biometric-based digital evidence into the register allocation phase of HLS.
10. Explain the functionality of pooling layer of CNN framework.
11. Discuss the multiplexing and demultiplexing scheme corresponding to FU resources such as adder and multiplier and storage hardware.
12. Discuss the steps for performing data path synthesis.
13. Discuss the concept of sliding window of convolution process.
14. Discuss the concept of unrolling the DFG design.
15. Discuss the concept of down sampling the image in pooling layer.
16. Discuss the factors responsible for yielding design cost overhead.
17. Analyze the security strength of the convolutional IP core design methodology.
18. Explain the security parameters responsible for enabling robust security of convolutional IP against IP piracy.

References

Albawi, S., T. A. Mohammed, and S. Al-Zawi (2017), "Understanding of a convolutional neural network," in: *Proceedings of the ICET*, pp. 1–6.

Arredondo-Velázquez, M., J. Diaz-Carmona, A. Barranco-Gutiérrez, and C. Torres-Huitzil (2020), "Review of prominent strategies for mapping CNNs onto embedded systems," *IEEE Latin America Transactions*, vol. 18, no. 5, pp. 971–982.

Bai, L., Y. Zhao, and X. Huang (2018), "A CNN accelerator on FPGA using depthwise separable convolution," *IEEE Transactions on Circuits and Systems II: Express Briefs*, vol. 65, no. 10, pp. 1415–1419.

CAD for Assurance, *IEEE Hardware Security and Trust Technical Committee*, https://cadforassurance.org/tools/ip-ic-protection/faciometric-hardware-security-tool/, accessed on January 2022.

Castillo, E., U. Meyer-Baese, A. García, L. Parrilla, and A. Lloris (2007), "IPP@HDL: efficient intellectual property protection scheme for IP cores," *IEEE Transactions on VLSI Systems*, vol. 15, no. 5, pp. 578–591.

Chang, M., Z. Pan, and J. Chen (2017), "Hardware accelerator for boosting convolution computation in image classification applications," in: *Proceedings of the GCCE*, pp. 1–2.

Davalle, D., B. Carnevale, S. Saponara, L. Fanucci, and P. Terreni (2014), "Hardware accelerator for fast image/video thinning," in: *Proc. IST,* pp. 64–67.

Gu, J., Z. Wang, J. Kuen, *et al.* (2018), "Recent advances in convolutional neural networks," *Pattern Recognition,* vol. 77, pp. 354–377, https://doi.org/10.1016/j.patcog.2017.10.013.

Guo, K., L. Sui, J. Qiu, *et al.* (2016), "Angel-eye: a complete design flow for mapping CNN onto customized hardware," in: *Proceedings of the ISVLSI*, pp. 24–29.

Haeffele, B. D. and R. Vidal (2017), "Global optimality in neural network training," in: 2017 *IEEE Conference on Computer Vision and Pattern Recognition (CVPR)*, 2017, pp. 4390–4398.

Kim, J. H., B. Grady, R. Lian, J. Brothers, and J. H. Anderson (2017), "FPGA-based CNN inference accelerator synthesized from multi-threaded C software," in: *Proceedings of the SOCC*, pp. 268–273.

Kim, T. S., J. Bae, and M. H. Sunwoo (2019), "Fast convolution algorithm for convolutional neural networks," in: *Proceedings of the AICAS*, pp. 258–261.

Kyriakos, A., V. Kitsakis, A. Louropoulos, E. -A. Papatheofanous, I. Patronas, and D. Reisis (2019), "High performance accelerator for CNN applications," in: *Proceedings of the PATMOS*, pp. 135–140.

Lemley, J., S. Bazrafkan, and P. Corcoran (2017), "Deep learning for consumer devices and services: pushing the limits for machine learning, *artificial intelligence, and computer vision,"* EEE Consumer Electronics Magazine*, vol. 6, no. 2, pp. 48–56.

Liu, Z., Dou, Y., Jiang, J., *et al.* (2017), "Throughput-optimized FPGA accelerator for deep convolutional neural networks," *ACM Transactions on Reconfigurable Technology and Systems*, vol. 10, pp. 1–23.

Ma, Y., Y. Cao, S. Vrudhula, and J. Seo (2018), "Optimizing the convolution operation to accelerate deep neural networks on FPGA," *IEEE Transactions on VLSI Systems*, vol. 26, no. 7, pp. 1354–1367.

Mahdiany, H. R., A. Hormati, and S. M. Fakhraie (2001), "A hardware accelerator for DSP system design," in: *Proceedings of the ICM*, Morocco, pp. 141–144.

Mishra, V. and A. Sengupta (2014), "MO-PSE: adaptive multi objective particle swarm optimization based design space exploration in architectural synthesis for application specific processor design," *Elsevier Journal on Advances in Engineering Software*, vol. 67, pp. 111–124, ISSN 0965-9978.

NanGate 15 nm Open Cell Library (June 2019). [Online]. Available: http:// www.nangate.com/?pageid=2328.

Plaza, S. M. and I. L. Markov (2015), "Solving the third-shift problem in IC piracy with test-aware logic locking," *IEEE Transactions on Computer-Aided Design of Integrated Circuits and Systems*, vol. 34, no. 6, pp. 961–971.

Roy, J. A., F. Koushanfar, and I. L. Markov (2008), "EPIC: ending piracy of integrated circuits," in: *Proceedings of the DATE*, Munich, pp. 1069–1074.

Schneiderman, R. (2010), "DSPs evolving in consumer electronics applications," *IEEE Signal Processing Magazine*, vol. 27, no. 3, pp. 6–10.

Sengupta, A. and R. Chaurasia (2022), "Secured convolutional layer IP core in convolutional neural network using facial biometric," *IEEE Transactions on Consumer Electronics*, vol. 68, no. 3, pp. 291–306.

Sengupta, A. and S. P. Mohanty (2016), "High-level synthesis of digital circuits in the nanoscale mobile electronics era," in: *IET Book: Nano-CMOS and Post-CMOS Electronics: Circuits and Design*, pp: 219–261.

Sengupta, A. and M. Rathor (2019), "IP core steganography for protecting DSP kernels used in CE systems," *IEEE Transactions on Consumer Electronics*, vol. 65, no. 4, pp. 506–515.

Sengupta, A. and M. Rathor (2020), "Obfuscated hardware accelerators for image processing filters—application specific and functionally reconfigurable processors," *IEEE Transactions on Consumer Electronics*, vol. 66, no. 4, pp. 386–395.

Sengupta, A. and M. Rathor (2021), "Facial biometric for securing hardware accelerators," *IEEE Transactions on VLSI Systems*, vol. 29, no. 1, pp. 112–123.

Sengupta, A. and D. Roy (2017), "Automated low-cost scheduling driven watermarking methodology for modern CAD high-level synthesis tools," *Advances in Engineering Software*, vol. 110, pp 26–33.

Sengupta, A., E. R. Kumar, and N. P. Chandra (2019), "Embedding digital signature using encrypted-hashing for protection of DSP cores in CE," *IEEE Transactions on Consumer Electronics*, vol. 65, no. 3, pp. 398–407.

Shen, Y., T. Ji, M. Ferdman, and P. Milder (2019), "Argus: An end-to-end framework for accelerating CNNs on FPGAs," *IEEE Micro*, vol. 39, no. 5, pp. 17–25.

Srivastava, H. and K. Sarawadekar (2020), "A depthwise separable convolution architecture for CNN accelerator," in: *Proceedings of the ASPCON*, pp. 1–5.

Tsiktsiris, D., D. Ziouzios, and M. Dasygenis (2018), "A portable image processing accelerator using FPGA," in: *Proceedings of the MOCAST*, pp. 1–4.

Wanhammar, L. (1999), "DSP system design," in: *Academic Press Series in Engineering, DSP Integrated Circuits*, New York, NY: Academic Press, pp. 277–356, ISBN 9780127345307, https://doi.org/10.1016/B978-012734530-7/50007-6.

Zeiler, M. and R. Fergus (2013). "Visualizing and understanding convolutional neural networks," in: *Proceedings of the ECCV*, 2014, Part I, LNCS 8689.

Zhao, Z., P. Zheng, S. Xu, and X. Wu (2019), "Object detection with deep learning: a review," *IEEE Transactions on Neural Networks and Learning Systems*, vol. 30, no. 11, pp. 3212–3232.

Chapter 5

Handling symmetrical IP core protection and IP protection (IPP) of Trojan-secured designs in HLS using physical biometrics

Anirban Sengupta[1] and Rahul Chaurasia[1]

The chapter describes the applications of biometric-based hardware security methodologies to provide symmetric protection of ownership rights for intellectual property (IP) buyer and seller. Additionally, it also presents the application of facial biometric-based hardware security for protecting Trojan-secured DSP designs against IP piracy (Chaurasia and Sengupta, 2022a, 2022b). To provide symmetric security for IP buyer and seller, first, the fingerprint biometric of IP buyer in the form of secret hardware security constraints is embedded into the digital signal processing (DSP) design during high-level synthesis (HLS). This embedded fingerprint signature IP buyer safeguards his/her rights against an adversary or an untrustworthy IP seller. Subsequently, post obtaining the IP buyer fingerprint embedded design, the facial biometric of IP seller is inserted into the design. Thus, the embedded security constraints (digital evidence) corresponding to fingerprint biometric and facial biometric signature of IP buyer and seller, respectively, enables the symmetric protection of their ownership rights. Additionally, in case of piracy, the embedded facial biometric signature also enables the robust and seamless detective control against pirated versions. Further, for protecting the Trojan-secured designs against IP piracy, first the Trojan-secured scheduled design is generated. Subsequently, the facial security constraints are inserted into the Trojan-secured design using HLS. These embedded facial security constraints enable the piracy detection control for Trojan-secured DSP designs as well. Thus, isolating pirated versions before integrating into system on chip (SoC) systems.

The organization of the chapter is as follows: Section 5.1 discusses about the introduction of the chapter; Section 5.2 presents the discussion on contemporary approaches for symmetrical IP core protection; Section 5.3 explains the process of HLS-based symmetrical IP core protection using IP seller facial biometric and IP buyer fingerprint biometric; Section 5.4 shows the protection of IP seller right's against the false claim of IP ownership; Section 5.5 shows the protection of IP buyer right's using fingerprint biometric signature; Section 5.6 explains the piracy detection process; Section 5.7 presents the process of employing facial biometric for protecting

[1]Department of Computer Science and Engineering, Indian Institute of Technology Indore, India

Trojan-secured IP cores against piracy; Section 5.8 discusses and analyses the security and design cost of symmetrical IP core protection technique and for, protecting Trojan-secured designs against piracy; Section 5.9 concludes the chapter.

5.1   Introduction

Digital signal processing (DSP) IP cores are the underlying hardware of several consumer electronics and computing systems. DSP cores such as discrete cosine transform (DCT), fast Fourier transform (FFT), Haar wavelet transform (HWT), digital filters like finite impulse response filter (FIR) and infinite impulse response filter (IIR), joint photographic expert group (JPEG), and moving picture expert group (MPEG) are some of the widely used DSP cores in digital electronic systems. DCT is used in audio, image, and video compression, FFT is used in digital video broadcasting, HWT is used in lossless image compression of camera systems, FIR and IIR are used in speech processing, noise suppression of signals, removal of attenuation of selected frequencies, audio processing devices, loud speakers, and telecommunication, JPEG is used in image and video compression and MPEG is used in audio/video processing applications. All these applications are computationally intensive in nature. Therefore, their realization as hardware IP core design is feasible for enabling high-performance digital electronics systems. In the present scenario where the development of smart cities and deployment of electronic devices in the field of health care, robotics, and Internet of Things (IoT) is thriving, the demand of DSP IP cores is increasing. Therefore, to meet the demand and supply ratio, these IP cores are designed at third-party design houses. Furthermore, especially in case of mass production, managing the entire design and fabrication process of IC design in-house increases the design and manufacturing cost.

The two major goals of an IP designer are to obtain an optimized synthesized design from the higher abstraction level and also to ensure the security of the design (Pilato *et al.*, 2018; Rai *et al.*, 2019; Wang *et al.*, 2015; Schneiderman, 2010; Plaza and Markov, 2015; Castillo et al., 2007; Sengupta et al., 2019; Colombier and Bossuet, 2015; Newbould et al., 2002; Roy *et al.*, 2008; Ni and Gao, 2005; Koushanfar, 2012; Cui and Chang, 2007; Sengupta and Mohanty, 2019; Arafin *et al.*, 2017; Mohanty and Meher, 2016; Yang *et al.*, 2021; Bao *et al.*, 2016; Ziener and Teich, 2008, Ziener and Teich, 2006). To design an IP core from higher abstraction level results in lower design cost and lesser implementation complexity. Further, integrating the security mechanism during higher abstraction also ensures the security at subsequent lower abstraction level such as RT-level (obtained using HLS), gate level/netlist level (obtained using RTL synthesis), and layout level (obtained using gate level and physical level synthesis) of an IP design process. The main purpose of symmetrical protection is to generate unique one-to-one mapped IP core between IP user and IP seller. From the standpoint of a seller, it will not only protect his/her ownership but also, he/she can identify the traitors among the list of IP buyers. Further, if a deceitful IP seller attempts to illegally redistribute the same IP core to other IP buyers, then from an honest IP buyer's standpoint, detection of such resold copies is necessary such that an honest IP buyer may protect his/her exclusive

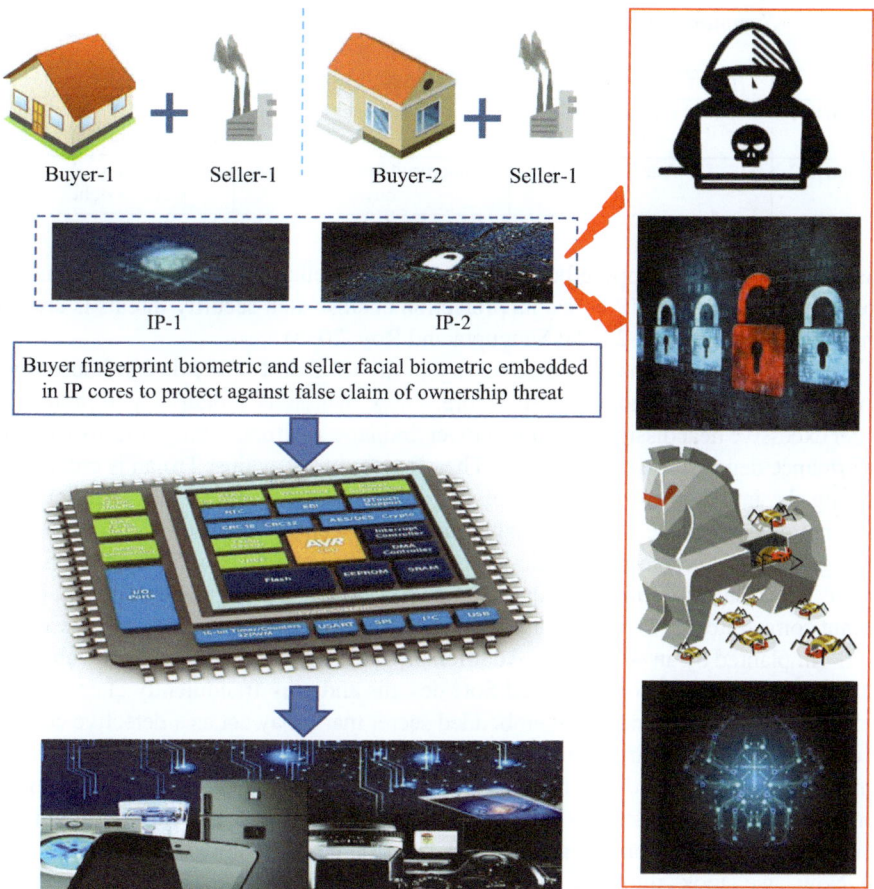

Figure 5.1 Motivation for symmetrical IP core protection

buyer rights for his/her custom design. Therefore, symmetrical IP core protection provides buyer rights as well as nullifies IP seller's ownership abuse. The motivational scenario for enabling symmetrical IP core protection is shown in Figure 5.1.

In such a case, a robust secret mark of IP seller/vendor acts as a final line of defence for proving the ownership of IP seller and nullifying the false claim of ownership from an adversary. To do so, the secret mark corresponding to IP seller is inserted into the design in the form of unique hardware security constraints. This therefore safeguards the ownership rights of IP seller against dishonest entity. On the other hand, if a deceitful IP seller attempts to illegally redistribute the same IP core to other IP buyers, then from an honest IP buyer's standpoint, detection of such resold copies is necessary such that an honest IP buyer may protect his/her exclusive buyer rights for his/her custom design, as shown in Table 5.1. For example, an IP buyer would not want redistribution of the same copy of his customized IP in the market. In such cases, buyer's rights are protected through the embedded secret mark of an IP buyer on the top of IP vendor's secret mark.

Table 5.1　Symmetric protection of IP rights for IP buyer and seller

	IP seller/ 3P IP vendor	IP buyer/ SoC integrator	Security
Scenario-1	Facial biometric of IP seller	Attacker	Protecting IP seller's right
Scenario-2	Attacker	Fingerprint biometric of IP buyer	Protecting IP buyer's right

This creates a unique mapping between IP buyer and seller. Therefore, for sustainable deployment of hardware IP cores, its protection against such security threats is crucial (Chaurasia and Sengupta, 2022a; Sengupta and Roy, 2018).

On a different dimension, an adversary in the third-party (3P) design house may also insert malicious Trojan logic 3PIP core. These malicious Trojan may imprudently trigger excessive heat dissipation, rapid power drainage, malfunctioning, and also lead to performance degradation of the device. Therefore, security against Trojan is crucial for ensuring the integrity and reliability of IP core before being integrated into SoCs of consumer electronics devices. To enable the robust security against Trojan, SoC integrator generates a Trojan-secured design by associating redundancy from another 3PIP vendor. This results in a dual modular redundant system which is passed through a fault-tolerant comparator. It, therefore, enables the detection of malicious Trojan, if present in a 3PIP (implanted by an adversary). Additionally, an adversary in the foundry (fab) may attempt to pirate such Trojan-secured SoC designs and may fraudulently claim design ownership. In such a scenario, an embedded secret mark may act as a detective control for discerning between pirated and genuine Trojan-secured SoC and also to nullify false ownership claim by an adversary. Therefore, facial biometric signature in the form of secret facial security constraints is embedded into the Trojan-secured design using HLS for enabling the protection against piracy (Chaurasia and Sengupta, 2022b).

Now, we look into the significance of biometric-based hardware security methodologies for the following: (a) protecting the rights of IP buyer and seller; (b) protecting Trojan-secured designs against IP piracy. To protect the rights of an IP buyer (in case of illegal distribution of his/her custom IP copies), his/her fingerprint biometric-driven secret signature is implanted into the design during the register allocation phase of HLS. Subsequently, to protect IP seller's right (in case of an adversary falsely claiming the ownership rights), his/her facial biometric-driven secret signature is implanted into the design obtained post embedding fingerprint signature of IP buyer. This ensures the symmetrical protection of IP rights of both the parties. This is because the embedding of fingerprint biometric corresponding to IP buyer and facial biometric of IP seller associates unique physiological features with the design to secure it.

On the other hand, non-biometric approaches for symmetrical IP core protection are based on hardware watermarking and fingerprinting (Roy and Sengupta, 2017). However, in case of non-biometric-based security methodology, an adversary may easily replicate or decode the auxiliary secret signature. This is because these security approaches do not associate unique identity of IP seller and IP buyer with the design. Thus, for embedding stronger digital evidence of both parties, biometric-based hardware security approaches render more robustness than non-biometric approaches.

Further, in case of ensuring security against piracy threat, the embedded facial biometric signature of IP seller provides the detective control against IP piracy. Facial biometric-based security mechanism includes several robust security parameters for generating the secret facial security constraints. This therefore ensures the detection and isolation of pirated versions of an IP, before their integration into the SoC system.

This chapter mainly focuses on providing a generic cognizance about the protection of rights of IP buyer and IP seller. Additionally, it discusses the piracy detection mechanism of Trojan-secured designs using facial biometric. To ensure the robust protection of IP rights of both the parties, unique fingerprint of IP buyer and facial biometric of IP seller is embedded into the design using HLS. Thus, by associating the unique identity (in terms of biometric security constraints) of both the parties, it enables the protection of their respective rights. Further, for enabling piracy detection of Trojan-secured designs, the facial biometric in the form of secret facial security constraints is inserted into Trojan-secured designs using HLS. These unique facial constraints post embedding enables the robust and seamless detection of pirated versions of Trojan-secured designs before their integration into hardware of CE systems (Chaurasia and Sengupta, 2022b).

5.2 Contemporary approaches for symmetric IP core protection

The security approaches in the literature can be classified into two major categories: symmetrical and non-symmetrical based, as shown in Figure 5.2. The non-symmetrical approaches can be further classified into two categories, depending upon the security integration in various levels of design process namely, higher abstraction level and lower abstraction level. The security methodologies that integrate security at higher abstraction level are: hardware watermarking, hardware steganography, and hardware biometrics. The watermarking approaches (Le Gal and Bossuet, 2013; Hong *et al.*, 1999; Koushanfar *et al.*, 2005; Sengupta and Bhadauria, 2016) provide the security against IP piracy using IP seller mark only. Gal and Bossuet (2013) presented an IP watermarking approach that uses mathematical relationships between numeric values as inputs and outputs at a specified time. The inserted watermark protects the sellers' right while satisfying the user constraints in terms of design latency and area. Sengupta and Bhadauria (2016) presented non-symmetrical hardware IP protection by inserting watermark in higher abstraction phase of HLS, which is based on the encoding of multi-variable signature. Multi-variable-based signature encoding offers better robustness due to complex encoding process that results in more watermarking constraints for embedding into the design. Further, it generates a low-cost solution using particle swarm optimization (PSO)-driven exploration process. PSO explores a trade-off between latency and area overhead achieved during watermarking and yields an optimal low-cost solution. These security constraints post embedding enables the piracy detection. However, it has the following shortcomings: can be easily replicated and reused by an adversary for evading piracy detection, does not offer significant robustness, and also does not provide symmetrical IP protection for buyer

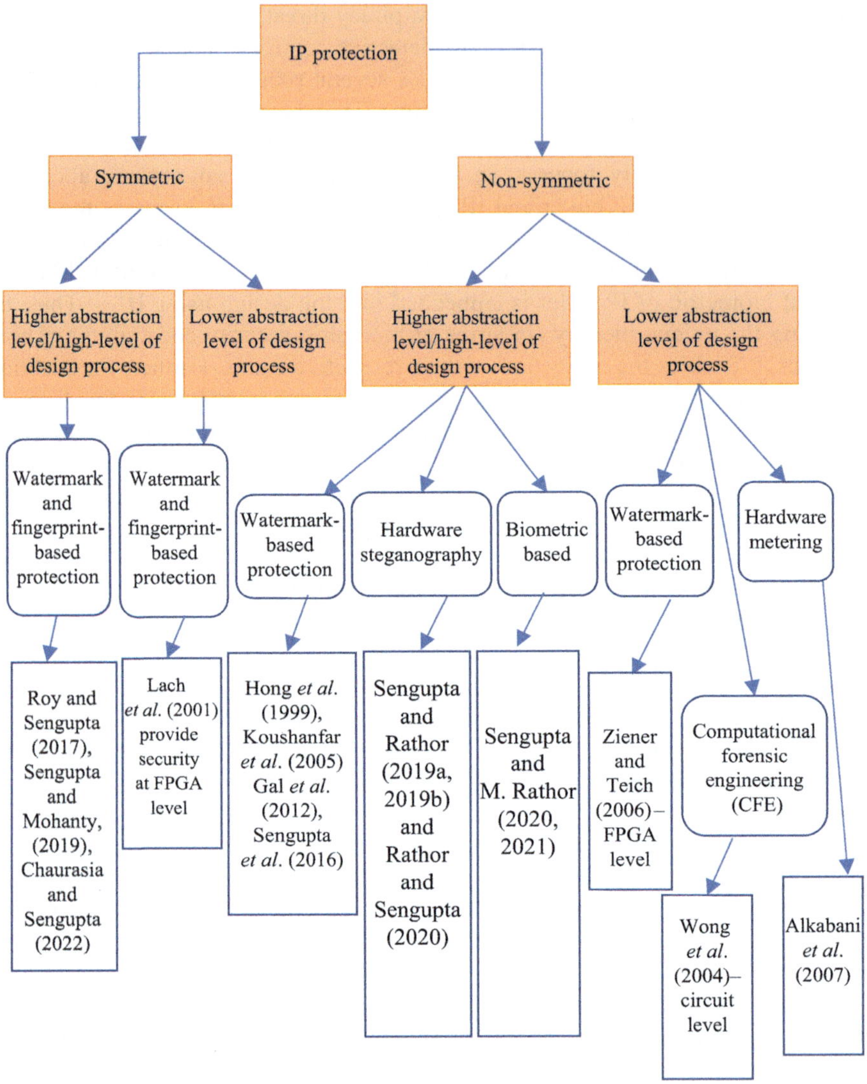

Figure 5.2   Contemporary approaches for symmetrical IP core protection

and seller. Due to above-mentioned shortcomings, these approaches therefore do not render robust security against the threats of piracy and false claim of IP ownership respectively. On the other hand, steganography-based approach (Sengupta and Rathor, 2019b) generates secret stego-constraints based on entropy threshold and encoding rules. Hardware steganography-based approach also does not provide symmetrical security. Further, like watermarking, it may also be replicated by an adversary for evading piracy detection. Digital signature-based approach enables piracy detection by embedding the encrypted digital signature obtained using SHA-512 and RSA encryption. All these approaches such as hardware watermarking, steganography, and digital signature only enable piracy detection and do not provide symmetrical security for IP buyer and seller both. Further, biometric-based approaches such as facial biometric (Sengupta and Rathor, 2021) and fingerprint biometric (Sengupta and Rathor, 2020) though provide robust security against IP piracy than hardware watermarking, steganography, and digital signature, however, have not targeted providing the symmetrical security for IP buyer and seller unlike approach (Chaurasia and Sengupta, 2022a).

Security methodologies that integrate security at lower abstraction level are based on watermarking (Ziener and Teich, 2006—FPGA level), computational forensic engineering (Wong *et al.*, 2004—circuit level), and hardware metering (Alkabani *et al.*, 2007). These approaches for IP protection also focus on seller watermark only. For example, single entity IP protection (Nie *et al.*, 2013; Ziener and Teich, 2006, 2008) through seller watermark has been presented at lower abstraction level of design process, which not only incurs overhead but also does not protect the IP buyer rights. Further, watermarking technique in Ziener and Teich (2006) embeds the signature at the netlist and bit-file level. Using this technique, seller's signature can be identified at the power supply pins of the field programmable gate array (FPGA). After measuring the supply voltage of the core, the watermark is detected using the signature detection algorithm. Nie *et al.* (2013) presented a hierarchical watermarking technique for FPGA IP protection. The seller's watermark is integrated into the netlist using a lookup table-shift register lookup (SRL) transformation, thereafter watermark is embedded into the bitstream of the same design. On the other hand, Ziener and Teich (2008) presented the detection of the signature by performing a voltage trace, which is introduced into the functional parts of the watermarked core. The voltage is sampled, analyzed, and decoded by applying the algorithm for signature detection before comparing with the original signature. All the aforementioned watermarking techniques are performed at lower abstraction level (not at behavioral/higher level). All aforesaid watermarking techniques offer non-symmetrical IP protection from the perspective of seller, thereby does not ensure protection of right from the perspective of buyer. On the other hand, Lach *et al.* (2001) presented symmetrical IP protection methodology at the lower abstraction level. The methodology offered protection from the perspective of both seller as well as buyer. This therefore protects the rights for both the entities. However, it does not protect IP cores, designed using the HLS framework.

Computational forensic engineering (CFE) is another mechanism for IP protection that offers a non-signature-based approach (Sengupta and Kachave, 2018; Wong *et al.*, 2004). It tries to judge the likelihood of the actual IP designer (tool) that has generated the IP core, in case of multiple claimants claiming IP ownership. IP metering (Alkabani and Koushanfar, 2007) is another type IP core protection mechanism (non-signature based) that identifies unauthorized duplicate versions of an IP core using an embedded ID. Koushanfar *et al.* (2012) presented hardware metering approach in which a unique ID is inserted through programming into an IP design, which helps to identify the pirated copies. The inserted ID impacts the controller design. This unique reconfigured controller at the end is programmed to provide a unique ID for every manufactured chip.

5.2.1   Symmetrical IP core protection in HLS using watermarking and fingerprinting

Roy and Sengupta (2017, 2018) presented a symmetrical IP core protection approach with multi-variable fingerprint encoding and hardware watermarking, as shown in Figure 5.3. In these approaches, along an IP seller inserting his own watermark, the fingerprint of IP buyer is also inserted into the design using HLS (for enabling symmetrical security). The combined signature in the form of seller watermark and buyer fingerprint is inserted into the design in successive phases of HLS. For example, buyer's fingerprint in the form of fingerprint security constraints is implanted during scheduling phase of HLS. Post obtaining the fingerprint-embedded scheduled design, finally the seller watermark, in the form of watermark security constraints, is embedded. Thus, embedded signature corresponding to IP seller and buyer enables the symmetric protection of IP rights. The symmetrical protection scheme by using signature of both parties i.e., fingerprint of an IP buyer and watermark of an IP seller at a higher abstraction level protect a lower-level design as well as incurs negligible overhead. Further, it also protects the IP rights of both the buyer and the seller. However, multi-variable fingerprint and watermarking approach are not as robust as that of Chaurasia and Sengupta (2022a). This is because the biometric-based security methodology embeds the

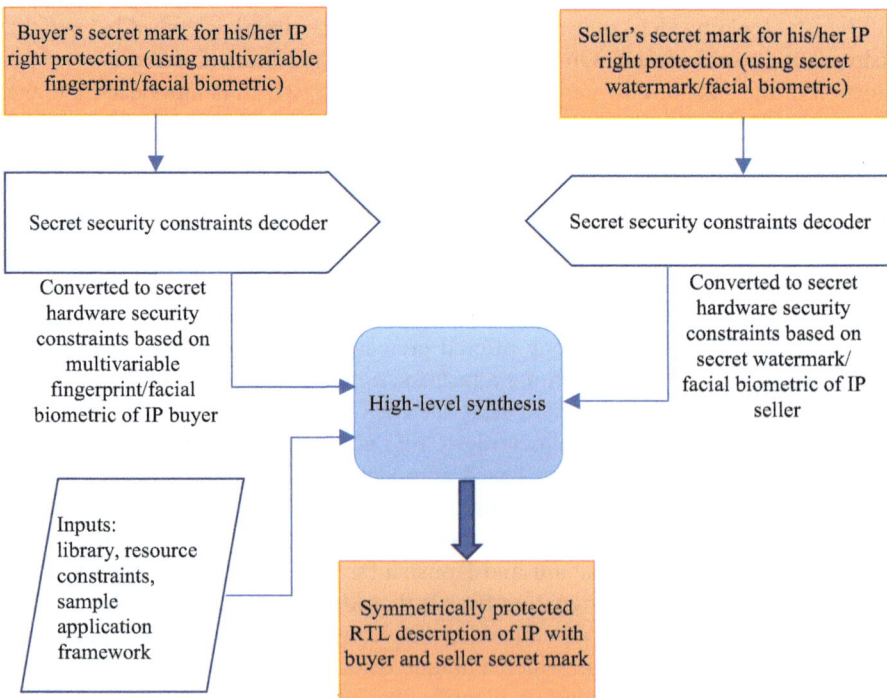

Figure 5.3   Overview of symmetrical IP core protection approach

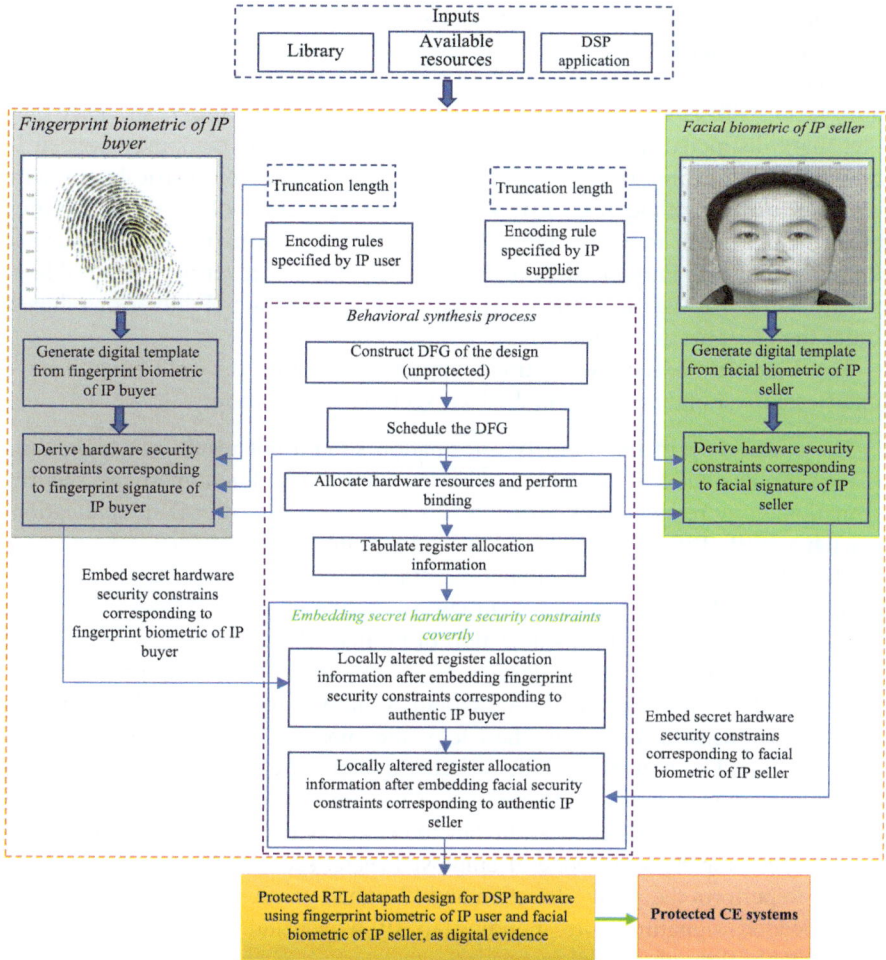

Figure 5.4   Symmetrical IP core protection using fingerprint biometric of IP user and facial biometric of IP seller

naturally unique facial biometric of IP buyer and IP seller, as shown in Figure 5.4. Additionally, the biometric-based approach results into larger security constraints than the other contemporary approaches.

In the biometric-based security methodology, first, the encoded constraints for hardware security corresponding to facial biometric features of an original IP buyer are generated. Subsequently, these constraints corresponding to facial biometric are implanted into the baseline design during register allocation phase of behavioral synthesis. Next, the constraints for hardware security corresponding to facial biometric features of original IP seller are generated. Subsequently, seller's security constraints are embedded into the resultant design obtained by post embedding buyer's facial security constraints. The embedding process of biometric security constraints corresponding to IP buyer and seller enables the symmetrical security constraints of IP seller and IP buyer uniquely without affecting the functionality of the design. Therefore, while performing the detection of illegal IP cores and to nullify false claim of ownership, both entities (IP supplier and user) can verify their secret mark distinctly. This creates an inimitable one-to-one mapping between IP buyer and IP seller in terms of the designed IP core. Thus, biometric-based symmetrical security methodology offers the robust protection of IP rights by integrating the non-replicable and unique facial biometric information of IP buyer and seller, respectively. Further, it is highly unlikely for an adversary to be able to replicate and reuse the biometric information of an authentic IP buyer or seller to acquire the IP rights deceitfully.

The biometric-based symmetrical approach satisfies the following desired properties corresponding to the embedded security information:

1. The embedding of the security information should not result into higher/significant design cost overhead. The biometric-based symmetrical security methodology renders zero design cost overhead for different DSP applications such as FIR, JPEG-codec, ARF IIR, and 8-point DCT.
2. Should be resilient against tampering attack.
3. Should be adaptable to any CAD tool.
4. Should enable the seamless detection of pirated IP cores.
5. Should enable the easy and definite establishment of ownership proof.
6. In case of multiple signatures, one should not impact the other. In the biometric-based symmetrical security methodology, the impact of biometric signature of IP buyer should be preserved while embedding biometric signature of IP seller.
7. For a genuine IP owner, the signature detection process and proving IP rights should be smooth and seamless. In the biometric-based approach, a genuine owner with the complete knowledge of employed secret security parameters should be able to perform the piracy detection and IP rights' verification seamlessly.
8. Embedded signature corresponding to IP buyer and seller should not be detectable through optical inspection, etc.

5.3   HLS-based symmetrical IP core protection using IP buyer fingerprint biometric and IP seller facial biometric

So far, we have discussed the contemporary approaches for enabling the symmetrical protection of IP rights corresponding to IP buyer and seller. Now, we discuss the demonstrative example for enabling symmetrical IP core protection using fingerprint biometric of IP buyer and facial biometric of IP seller.

5.3.1   Summary

For enabling the symmetrical protection of the IP rights for buyer and seller, first, the secret and unique security mark of IP buyer is implanted into the target DSP design. To do so, the fingerprint biometric of IP buyer is obtained and transformed into secret hardware security constraints. Subsequently, these security constraints are implanted into the target (baseline) design during register allocation phase of HLS. This results into a locally altered register allocation information corresponding to target design (post embedding fingerprint security constraints). This therefore enables the detection/tracing of illegal copies distributed by a deceitful IP seller to other IP buyers. Next, the facial biometric of IP seller is captured and transformed into encoded constraints for hardware security. Subsequently, the generated facial biometric-based constraints for hardware security corresponding to IP seller are embedded into the DSP design, obtained post embedding the fingerprint of original IP buyer (locally altered register allocation table corresponding to fingerprint signature of the original IP buyer). This therefore enables the protection for original IP seller against false claim of ownership from an adversary. The embedded biometric signature corresponding to genuine buyer and seller results into unique one-to-one mapped IP core design. Thus, ensuring the protection of their respective IP rights using their unique biometric-driven secret security mark.

5.3.2   *Deriving fingerprint security constraints of IP buyer*

To derive the secret fingerprint security constraints, following steps are followed:

- extracting the fingerprint minutiae points corresponding to fingerprint of IP buyer;
- generating fingerprint biometric digital template;
- deriving fingerprint security constraints.

Now, we discuss each of the steps for fingerprint constraints generation in details. In the fingerprint biometric, each minutiae point comprises of the unique feature of an individual. Therefore, it is crucial to locate and extract the exact/complete minutiae points. To ensure accurate minutiae point extraction, the following sub-processes are performed on the captured fingerprint image: first, we enhance the target fingerprint image as shown in Figure 5.5(a). To do so, fast Fourier transform (FFT)-based enhancement is applied on the set of pixels of fingerprint image (Sengupta and Rathor, 2020). This results into reconnection of broken ridges and separation the parallel ridges. The next subprocess is responsible for performing binarization of enhanced fingerprint image. This results in a fingerprint image with only two-pixel intensity "0" and "255" based on the threshold value. If any pixel intensity is lesser than the threshold, then its intensity value is converted into "0" (black color) and if pixel intensity is greater than the threshold, then its intensity value is converted into "255" (white color). This subprocess of binarization enables the clear visualization of minutiae points. The generated binarized image is shown in Figure 5.5(b). The third subprocess is responsible for performing thinning of the obtained binarized fingerprint image. This transforms the thickness of ridge lines to one pixel width by deleting the neighboring pixels at the edge of ridge lines. The obtained thinned fingerprint image is shown in

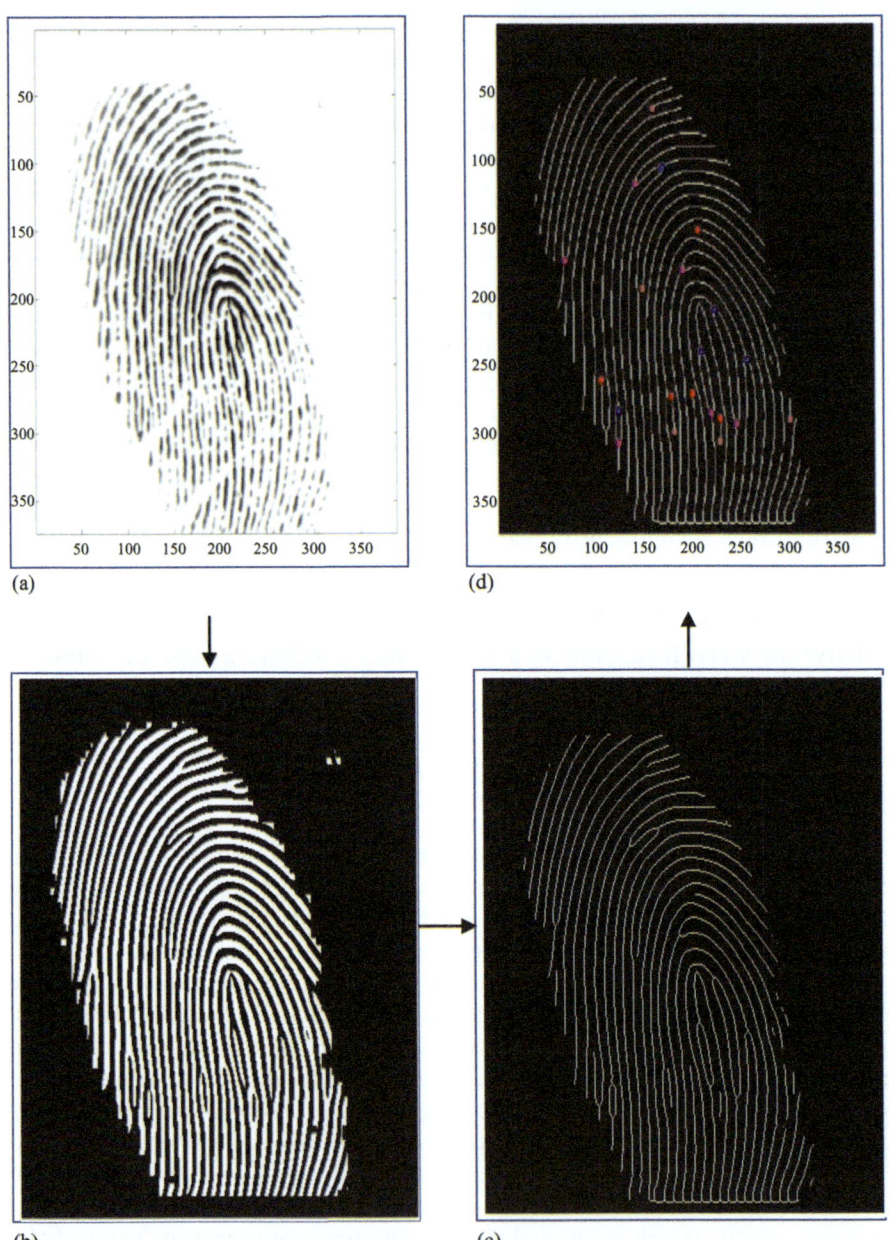

Figure 5.5   Steps for generating the fingerprint minutiae points. (a) Captured fingerprint image. (b) Enhanced fingerprint image. (c) Binarized fingerprint image. (d) Fingerprint image with minutiae.

Figure 5.5(c). The obtained thinned fingerprint image is used for generating the minutiae points. The points or locations where the ridges end abruptly or bifurcates into branches are known as the minutiae feature points. The fingerprint image with located minutiae points is shown in Figure 5.5(d).

Now, we discuss the process of fingerprint template generation. The fingerprint template comprises of the unique information corresponding to minutiae feature points of an individual. To generate fingerprint template, each minutiae point is exploited to extract their unique features. The features like coordinates (p, q), ridge angle (z), and crossing number (g) are determined corresponding to each of the minutiae points. Next, to generate fingerprint template, the features of minutiae points are transformed into their binarized form. The details of features of each minutiae point corresponding to fingerprint image are shown in Table 5.2. The final template is generated by concatenating the binarized minutiae feature points. However, there can be several possible ways for concatenating the minutiae feature points. An IP buyer may also selectively choose some of the minutiae features during final template generation. Additionally, an IP buyer may also selectively choose some of the minutiae points depending on the security-cost tradeoff. If the order of

Table 5.2   IP buyer fingerprint digital template generation

No.	p	q	Minutiae type number(z)	Angle in degree (g)	Binary (p÷q÷z÷g)
1	161	63	1	153	1010 0001-11 1111-1-1001 1001
2	171	106	3	337	1010 1011-1101010-11-1 0101 0001
3	143	118	3	130	1000 1111-111 0110-11-1000 0010
4	207	152	1	187	1100 1111-1001 1000-1-1011 1011
5	70	174	3	99	100 0110-1010 1110-11-110 0011
6	191	181	3	131	10111111-1011 0101-11-1000 0011
7	150	195	1	95	10010110-1100 0011-1-101 1111
8	224	210	3	234	1110 0000-1101 0010-11-11101010
9	210	241	3	252	1101 0010-1111 0001-11-1111 1100
10	257	247	3	247	1 0000 0001-1111 0111-11-1111 0111
11	107	262	1	262	110 1011-1 0000 0110-1-1 0000 0110
12	201	272	1	255	1100 1001-1 0001 0000-1-1111 1111
13	179	274	1	259	10110011-1 0001 0010-1-100000011
14	125	284	3	269	111 1101-1 0001 1100-11-1 0000 1101
15	220	286	3	73	1101 1100-1 0001 1110-11-1001001
16	229	290	1	264	1110 0101-100100010-1-1 0000 1000
17	301	291	1	91	1 0010 1101-1 0010 0011-1-101 1011
18	246	294	3	86	1111 0110-100100110-11-101 0110
19	182	300	1	83	1011 0110-1 0010 1100-1-101 0011
20	229	307	1	91	1110 0101-1 0011 0011-1-101 1011
21	125	308	3	88	111 1101-100110100-11-101 1000

concatenation of minutiae features by IP buyer is "p∓q∓z∓g," then the resultant fingerprint template is shown below:

"1010 0001-11 1111-1-1001 10011010 1011-1101010-11-1 0101 00011000 1111-
111 0110-11-1000 00101100 1111-1001 1000 -1-1011 1011100 0110-1010 1110-
11-110 001110111111-1011 0101 -11-1000 0011 10010110-1100 0011-1-101
11111110 0000-1101 0010-11-111010101101 0010-1111 0001-11-1111 11001
0000 0001-1111 0111-11-1111 0111110 1011-1 0000 0110-1-1 0000 01101100
1001-1 0001 0000-1-1111 111110110011-1 0001 0010-1-100000011111 1101-1
0001 1100-11-1 0000 11011101 1100-1 0001 1110-11-10010011110 0101-
100100010-1-1 0000 10001 0010 1101-1 0010 0011-1-101 10111111 0110-
100100110-11-101 01101011 0110-1 0010 1100-1-101 00111110 0101-1 0011
0011-1-101 1011111 1101-100110100-11-101 1000"

The length of generated fingerprint digital template is 538-bit size, which comprises of "277" 0s and "311" 1s.

Now, we discuss the process of generating the fingerprint biometric-based secret hardware security constraints. The fingerprint biometric-based secret hardware security constraints are generated based on IP buyer-specified signature truncation length, target-scheduled DSP design and encoding rule. For the sake of brevity, we have considered the IP buyer-specified signature truncation length of 72 bits. Therefore, the truncated fingerprint biometric signature is:

1010 0001-11 1111-1-1001 1001-1010 1011-1101010-11-1 0101 0001-1000 1111-
111 0110-11-1000 00 (72bits).

However, an IP buyer may vary the signature length depending on the size of the target application and security-design cost tradeoff. For demonstrating the symmetrical security, infinite impulse response (IIR) filter is considered as the target DSP application. The DFG (unscheduled) of IIR filter is shown in Figure 5.6(a), where {a1,a2,a3, b1,b2,b3} represents the set of input coefficient of IIR filter, {A[n], A[n-1], A[n-2], A [n-3]} are the set of previous values of input of IIR filter, {B[n-1], B[n-2], B[n-3]} represents the set of previous values of output of IIR filter, and Z[n] represents the output of IIR filter. Subsequently, the scheduled IIR filter design with its storage variable information is shown in Figure 5.6(b), where (T^0 to T^7) represents the control steps required for scheduling the design, (R^1 to R^{14}) represents the required registers corresponding to accommodating storage variables (J_0 to J_{26}) of the design. Storage variables hold the input design data. The storage variables of the design are used to form the secret hardware security constraints. Each hardware security constraint is generated by forming the pair of two storage variables. However, an IP buyer may generate different ordering of storage variables such as either sorted ascending order or sorted descending order or arrangement based on functional units or alternate arrangement of storage variables, etc. For example, the storage variables in sorted ascending order are (J_0, J_1, J_2, J_3, J_4, J_5, … …., J_{26}). Finally, by forming the pair of storage variables from the IP buyer selected storage variable order, the secret hardware security constraints are generated.

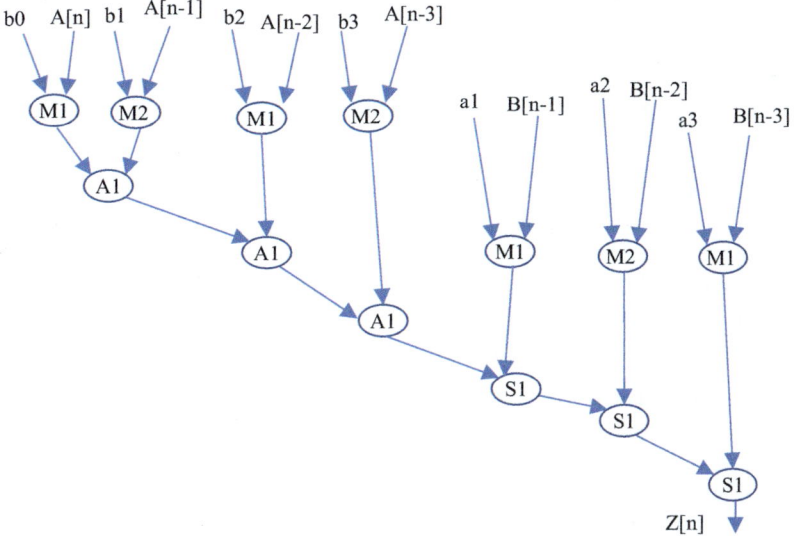

(a)

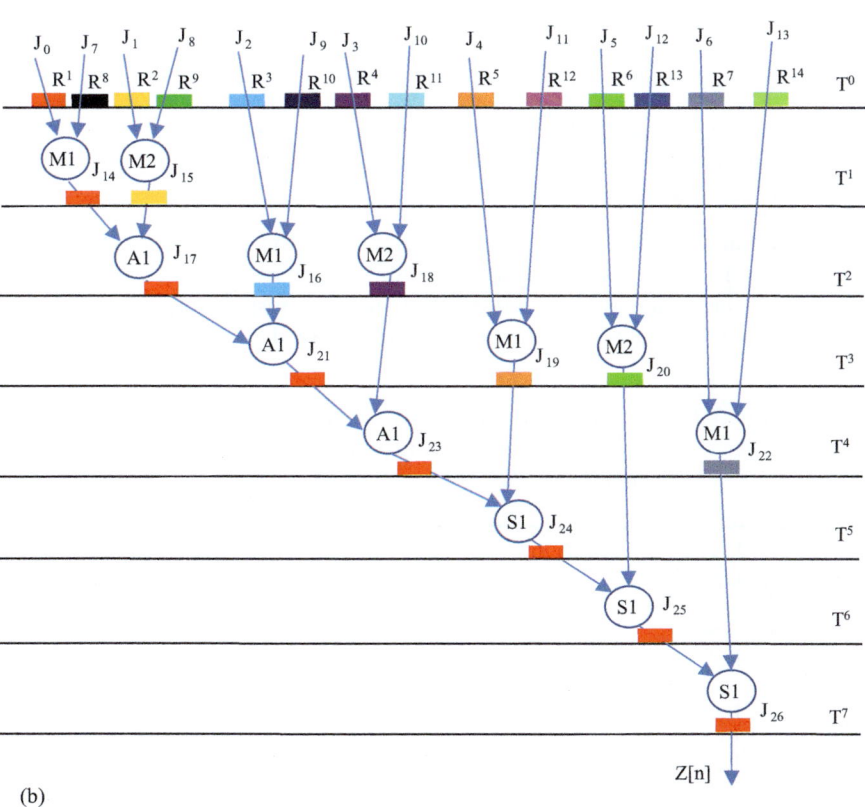

(b)

Figure 5.6   (a) Data flow graph of IIR filter design. (b) Scheduled DFG of IIR filter corresponding to resources one adder (A1), two multipliers (M1, M2), and one subtractor (S1).

Further, let us consider the IP buyer-specified encoding rules for encoding the fingerprint signature into the hardware security constraints are (Chaurasia and Sengupta, 2022):

- **Signature bit "0"**: representing storage variable pair (J_x and J_y) of even number required to be allocated to distinct registers, where J_x and J_y can be of any integer value.
- **Signature bit "1"**: representing storage variable pair between J_x and J_y, required to be allocated to distinct registers, where first variable (J_x) is 0 and second variable (J_y) can be of any integer value (except the generated pairs).

Therefore, based on the above encoding rule and the storage variable set obtained in sorted ascending order, hardware security constrains are generated. Security constraints corresponding to bit "0" and bit "1" of the fingerprint signature (72 bits) are shown in Tables 5.3 and 5.4, respectively. These generated secret constraints for hardware security, post embedding into the design, enable the protection of the IP buyer right.

5.3.3   Deriving facial security constraints of IP seller

The facial security constraints are responsible for enabling the protection of IP seller's right against the false claim of IP ownership. To derive the secret facial security constraints, following steps are followed:

(a) generating the facial image with IP seller-chosen facial features;
(b) generating facial biometric digital template;
(c) deriving facial security constraints.

Now, we discuss each of the steps for facial constraints generation in details. In the facial biometric, unique facial information comprises of various distinct features of the facial image of an IP seller. The IP seller is responsible for deciding the final set of facial features which is further transformed into the facial signature.

Table 5.3   IP buyer fingerprint biometric signature-based secret hardware security constrains corresponding to signature bit "0"

$\langle J_0, J_2 \rangle$	$\langle J_0, J_{26} \rangle$	$\langle J_2, J_{26} \rangle$	–	–
$\langle J_0, J_4 \rangle$	$\langle J_2, J_4 \rangle$	$\langle J_4, J_6 \rangle$	–	–
$\langle J_0, J_6 \rangle$	$\langle J_2, J_6 \rangle$	$\langle J_4, J_8 \rangle$	–	–
$\langle J_0, J_8 \rangle$	$\langle J_2, J_8 \rangle$	$\langle J_4, J_{10} \rangle$	–	–
$\langle J_0, J_{10} \rangle$	$\langle J_2, J_{10} \rangle$	$\langle J_4, J_{12} \rangle$	–	–
$\langle J_0, J_{12} \rangle$	$\langle J_2, J_{12} \rangle$	$\langle J_4, J_{14} \rangle$	–	–
$\langle J_0, J_{14} \rangle$	$\langle J_2, J_{14} \rangle$	–	–	–
$\langle J_0, J_{16} \rangle$	$\langle J_2, J_{16} \rangle$	–	–	–
$\langle J_0, J_{18} \rangle$	$\langle J_2, J_{18} \rangle$	–	–	–
$\langle J_0, J_{20} \rangle$	$\langle J_2, J_{20} \rangle$	–	–	–
$\langle J_0, J_{22} \rangle$	$\langle J_2, J_{22} \rangle$	–	–	–
$\langle J_0, J_{24} \rangle$	$\langle J_2, J_{24} \rangle$	–	–	–

Table 5.4   IP buyer fingerprint biometric signature-based
secret hardware security constrains
corresponding to signature bit "1"

$\langle J_0, J_1 \rangle$	$\langle J_0, J_{25} \rangle$	–
$\langle J_0, J_3 \rangle$	–	–
$\langle J_0, J_5 \rangle$	–	–
$\langle J_0, J_7 \rangle$	–	–
$\langle J_0, J_9 \rangle$	–	–
$\langle J_0, J_{11} \rangle$	–	–
$\langle J_0, J_{13} \rangle$	–	–
$\langle J_0, J_{15} \rangle$	–	–
$\langle J_0, J_{17} \rangle$	–	–
$\langle J_0, J_{19} \rangle$	–	–
$\langle J_0, J_{21} \rangle$	–	–
$\langle J_0, J_{23} \rangle$	–	–

Based on the decided facial features, generation of nodal points is performed (on the facial image). Therefore, it is crucial to locate and extract the accurate nodal point information for unique facial biometric signature generation. The process of generating the image with IP seller chosen facial features, as shown in Figure 5.7 (Part I) and Figure 5.8 (Part II), involves the following steps:

1. Capturing the face using high-resolution digital camera and placing it into a specific grid size/spacing (selected by original IP seller for secret signature generation). The captured facial image of IP seller with specific grid size and spacing, is shown in Figure 5.7(a). The IP seller-chosen grid size is 500×550.
2. Deriving the nodal points on captured facial image corresponding to the chosen facial features set by IP seller. Let us say that an IP seller has chosen the following facial features for generating the biometric information:

    1.   HFH = Height of Forehead
    2.   IPD = Inter Pupillary Distance
    3.   BOB = Bio-Ocular Breadth
    4.   IOB = Inter-Ocular Breadth
    5.   OB = Ocular Breadth
    6.   WNR = Width of Nasal Ridge
    7.   WF = Width of face
    8.   HF = Height of Face
    9.   WNB = Width of Nasal Base
    10. NB = Nasal Breadth
    11. OCW = Oral Commissure Width

    This therefore results into a facial image with the nodal points, as shown in Figure 5.7(b).
3. Assigning the naming convention on nodal points corresponding to IP seller-selected facial features. Each nodal point is designated as "Pn," where "n" is

(a)

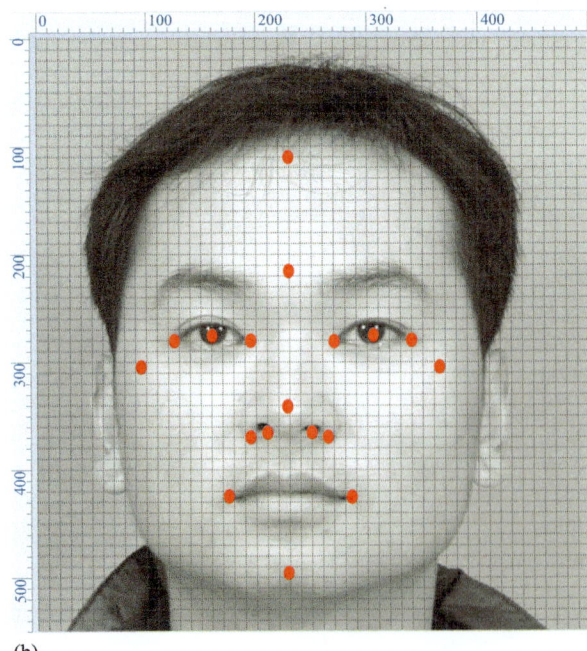

(b)

*Figure 5.7   Process for generating the image with IP seller-chosen facial features
(Part I). (a) Captured facial image of IP seller with specific grid size.
(b) Facial image with the nodal points.*

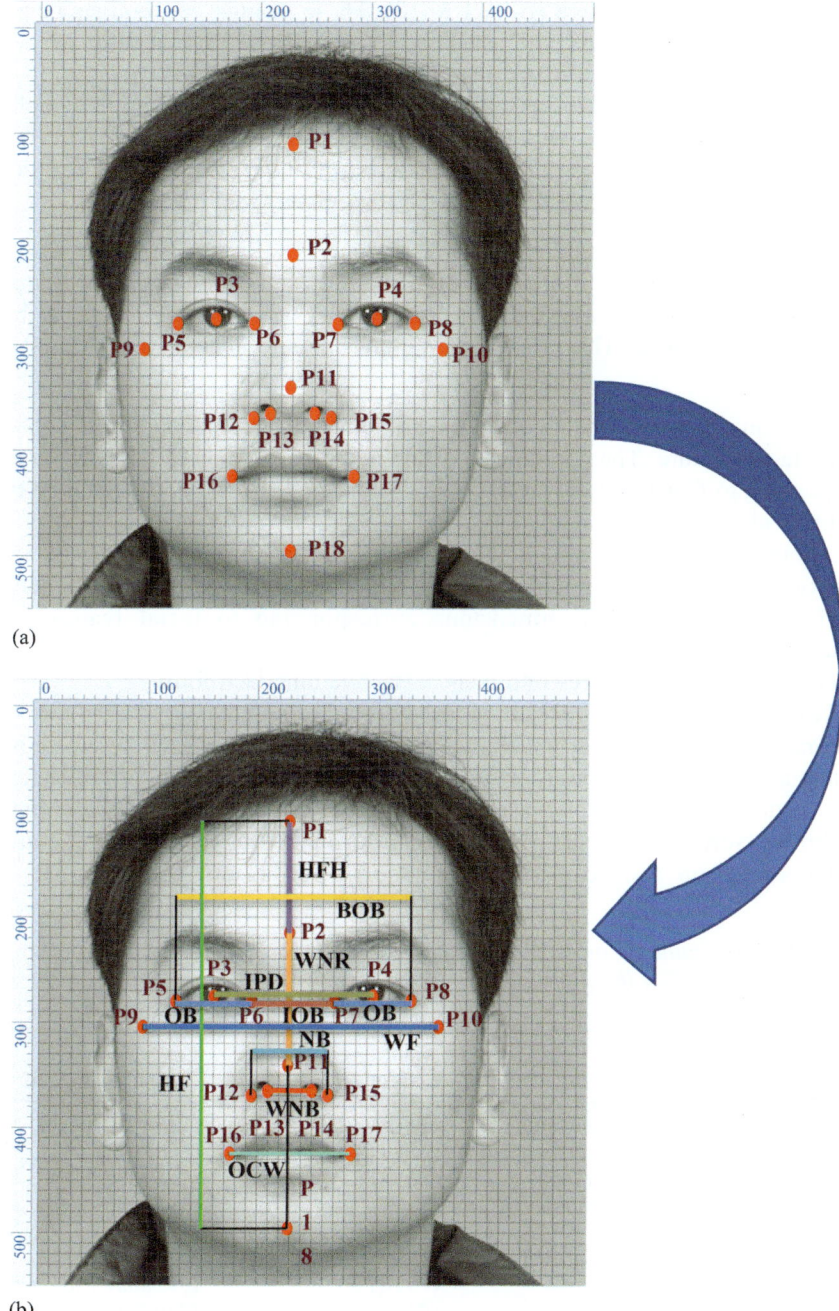

Figure 5.8 Process for generating the image with IP seller-chosen facial features (Part II). (a) Facial image with naming convention of nodal points. (b) Facial image with IP seller chosen facial features.

the number of nodal point (from 1 to 18). These naming conventions help in determining the coordinate dimensions of nodal points. The facial image with naming convention of nodal points is shown in Figure 5.8(a).

4. Generating the image with IP seller selected facial features for digital template generation. The image with IP seller-chosen facial features is shown in Figure 5.8(b), where each feature is marked using a different color. A feature is the measure of the distance between the respective nodal points pair.

So far, we have discussed the process of generating the facial image with IP seller selected facial features through designating the nodal points corresponding to captured facial image. Now, we discuss the process of generating the facial biometric digital template. The process of digital template generation involves the following steps:

1. Determine the co-ordinates of the nodal points pair corresponding to each of the facial feature. The coordinates of the nodal points pair "(Px)–(Py)" corresponding to facial feature, are shown in Table 5.5, where x and y refers to two different nodal points.
2. Determining feature dimensions by computing the Manhattan distance between the nodal points pair, for each of the features selected by IP seller. The computed feature dimensions corresponding to facial features are shown in Table 5.6.
3. Transforming the features into their respective binarized form.
4. Next, facial signature/biometric-based digital template is generated through concatenating the binarized facial features based on the IP seller-chosen feature's concatenation order. However, there can be several possible ways for concatenating the facial features. An IP seller may also selectively choose some of the facial features during final template generation depending on the security-cost tradeoff. Let us say, if the order of

Table 5.5   *Co-ordinates of each facial feature, calculated using nodal points marked on the facial image*

S. no.	Facial features	Naming convention of points $|(Px)-(Py)|$	Co-ordinates (m1, n1)–(m2, n2)
1.	HFH: Height of Forehead	(P1)–(P2)	(230, 110)–(230, 215)
2.	IPD: Inter Pupillary Distance	(P3)–(P4)	(160, 275)–(305, 275)
3.	BOB: Bio-Ocular Breadth	(P5)–(P8)	(125, 280)–(340, 280)
4.	IOB: Inter-Ocular Breadth	(P6)–(P7)	(195, 280)–(270, 280)
5.	OB: Ocular Breadth	(P5)–(P6)	(125, 280)–(195, 280)
6.	WNR: Width of Nasal Ridge	(P2)–(P11)	(230, 215)–(230, 340)
7.	WF: Width of face	(P9)–(P10)	(95, 305)–(365,305)
8.	HF: Height of Face	(P1)–(P18)	(230, 110)–(230,495)
9.	WNB: Width of Nasal Base	(P13)–(P14)	(210, 365)–(250, 365)
10.	NB: Nasal Breadth	(P12)–(P15)	(195, 370)–(265, 370)
11.	OCW: Oral Commissure Width	(P16)–(P17)	(175, 425)–(285, 425)

Table 5.6   *Determining feature dimension of IP seller-selected facial features and their binarized representation*

S. no.	Facial features	Feature dimension (Manhattan Distance) = $\|m2-m1\|+\|n2-n1\|$	Binary
1.	HFH	105	1101001
2.	IPD	145	10010001
3.	BOB	215	11010111
4.	IOB	75	1001011
5.	OB	70	1000110
6.	WNR	125	1111101
7.	WF	270	100001110
8.	HF	385	110000001
9.	WNB	40	101000
10.	NB	70	1000110
11.	OCW	110	1101110

concatenation of facial features by IP seller is:

$$f^{OB} + f^{IOB} + f^{NB} + f^{OCW} + f^{WF} + f^{BOB} + f^{HF} + f^{WNB} + f^{HFH}$$

Thus, the generated facial digital template by concatenating the IP seller-selected features in binarized form is:

"1000110100101110001101101110100001110110101111100000011010001101001"

The strength of the facial biometric digital template is 67 bits, including 33 number of 0s and 34 number of 1s. This generated digital template is further exploited for generating the secret hardware security constraints. Now, let us discuss the process of generating the facial biometric-based secret hardware security constraints. For the sake of brevity, we have considered the IP seller specified signature truncation length is 60 bits. However, an IP seller may vary the signature length depending upon the size of the target application and security-design cost tradeoff. Now, let us generate the secret constraints for hardware security corresponding to facial biometric signature of IP seller. The facial biometric-based secret hardware security constraints are generated based on IP seller-specified signature truncation length, target scheduled DSP design and encoding rule. For the sake of brevity, we have considered the IP seller-specified signature truncation length of 48 bits. The secret hardware security constraints are generated corresponding to scheduled IIR filter design (as shown in Figure 5.6(b)) by considering the storage variables in sorted ascending order are (J_0, J_1, J_2, J_3, J_4, J_5, ..., J_{26}). Further, let us consider the IP seller-specified encoding rules for encoding the facial signature into constraints for hardware security are (Chaurasia and Sengupta, 2022):

- **Signature bit "0"**: representing storage variable pair (J_x and J_y) of odd number required to be allocated to distinct registers, where J_x and J_y can be of any integer value.
- **Signature bit "1"**: representing storage variable pair (J_x and J_y) of prime number required to be allocated to distinct registers, where J_x and J_y can be of any prime integer value.

Therefore, based on the above encoding rule and the storage variable set obtained in sorted ascending order, hardware security constrains are generated. Security constraints conforming to bit "0" and bit "1" are shown in Tables 5.7 and 5.8 respectively. These generated secret hardware security constraints post embedding into the design enable the protection of IP seller right against false claim of IP ownership from an adversary.

5.3.4   Embedding the fingerprint security constraints of IP buyer in DSP design

So far, we have discussed the process of deriving secret constraints for hardware security corresponding to biometric traits of IP buyer and seller. Now we discuss the embedding of the derived security constraints into the design during register

Table 5.7   IP seller facial biometric signature-based secret hardware security constrains corresponding to signature bit "0"

$<J_1, J_3>$	$<J_3, J_5>$	$<J_5, J_9>$	–	–
$<J_1, J_5>$	$<J_3, J_7>$	$<J_5, J_{11}>$	–	–
$<J_1, J_7>$	$<J_3, J_9>$	$<J_5, J_{13}>$	–	–
$<J_1, J_9>$	$<J_3, J_{11}>$	$<J_5, J_{15}>$	–	–
$<J_1, J_{11}>$	$<J_3, J_{13}>$	$<J_5, J_{17}>$	–	–
$<J_1, J_{13}>$	$<J_3, J_{15}>$	$<J_5, J_{19}>$	–	–
$<J_1, J_{15}>$	$<J_3, J_{17}>$		–	–
$<J_1, J_{17}>$	$<J_3, J_{19}>$		–	–
$<J_1, J_{19}>$	$<J_3, J_{21}>$		–	–
$<J_1, J_{21}>$	$<J_3, J_{23}>$		–	–
$<J_1, J_{23}>$	$<J_3, J_{25}>$		–	–
$<J_1, J_{25}>$	$<J_5, J_7>$		–	–

Table 5.8   IP seller facial biometric signature-based
secret hardware security constrains
corresponding to signature bit "1"

$<J_2, J_3>$	$<J_3, J_{17}>$	$<J_7, J_{19}>$
$<J_2, J_5>$	$<J_3, J_{19}>$	$<J_7, J_{23}>$
$<J_2, J_7>$	$<J_3, J_{23}>$	$<J_{11}, J_{13}>$
$<J_2, J_{11}>$	$<J_5, J_7>$	$<J_{11}, J_{17}>$
$<J_2, J_{13}>$	$<J_5, J_{11}>$	$<J_{11}, J_{19}>$
$<J_2, J_{17}>$	$<J_5, J_{13}>$	$<J_{11}, J_{23}>$
$<J_2, J_{19}>$	$<J_5, J_{17}>$	$<J_{13}, J_{17}>$
$<J_2, J_{23}>$	$<J_5, J_{19}>$	$<J_{13}, J_{19}>$
$<J_3, J_5>$	$<J_5, J_{23}>$	–
$<J_3, J_7>$	$<J_7, J_{11}>$	–
$<J_3, J_{11}>$	$<J_7, J_{13}>$	–
$<J_3, J_{13}>$	$<J_7, J_{17}>$	–

allocation phase of HLS process. In this subsection, mainly we discuss the process of embedding the secret constraints for hardware security corresponding to the finger-print biometric of IP buyer. The embedding process includes the following steps:

(a) Allocating the hardware resources available in the module library, to respective operations (multiplication, addition, and subtraction) of the scheduled DFG design based on the resource constraints specified by IP designer. The scheduled DFG was obtained earlier (in Section 5.3.2 as shown in Figure 5.6(b)). Subsequently binding is performed.

(b) Constructing the register allocation table corresponding to baseline design (without the embedded biometric signature of IP buyer and seller). The register allocation table corresponding to IIR filter is shown in Table 5.9. Register allocation table comprises of the details of required registers (R^1 to R^{14}), available storage variables (J_0 to J_{26}) in the design and their assignment to particular register and required control steps (T^0 to T^7).

(c) Now, performing the embedding of IP buyer fingerprint biometric signature-based hardware security constraints (generated earlier, as shown in Tables 5.3 and 5.4.) into the scheduled DSP design. The embedding is performed by locally altering the storage variables among the available registers into the register allocation information, corresponding to fingerprint-based hardware security constraints. The constraints embedding block accepts the following two inputs such as (i) generated security constraints corresponding to target scheduled design and fingerprint biometric signature of IP buyer and (ii) register allocation information. The register allocation table is used for embedding the hardware security constraints by locally altering (modifying) the register assignments using the following constraint rule such that "two storage variables in a pair cannot be assigned to the same register." The register allocation information post embedding the IP buyer fingerprint-based hardware security constraints corresponding to signature bit "0" is shown in

Table 5.9 Register allocation information corresponding to scheduled IIR digital filter design (before embedding fingerprint biometric signature of IP buyer)

CS	R^1	R^2	R^3	R^4	R^5	R^6	R^7	R^8	R^9	R^{10}	R^{11}	R^{12}	R^{13}	R^{14}
T^0	J_0	J_1	J_2	J_3	J_4	J_5	J_6	J_7	J_8	J_9	J_{10}	J_{11}	J_{12}	J_{13}
T^1	J_{14}	J_{15}	J_2	J_3	J_4	J_5	J_6	–	–	J_9	J_{10}	J_{11}	J_{12}	J_{13}
T^2	J_{17}	–	J_{16}	J_{18}	J_4	J_5	J_6	–	–	–	–	J_{11}	J_{12}	J_{13}
T^3	J_{21}	–	–	J_{18}	J_{19}	J_{20}	J_6	–	–	–	–	–	–	J_{13}
T^4	J_{23}	–	–	–	J_{19}	J_{20}	J_{22}	–	–	–	–	–	–	–
T^5	J_{24}	–	–	–	–	J_{20}	J_{22}	–	–	–	–	–	–	–
T^6	J_{25}	–	–	–	–	–	J_{22}	–	–	–	–	–	–	–
T^7	J_{26}	–	–	–	–	–	–	–	–	–	–	–	–	–

Table 5.10 Register allocation information corresponding to scheduled IIR digital filter design (post embedding IP buyer fingerprint biometric constraints for bit "0")

CS	R^1	R^2	R^3	R^4	R^5	R^6	R^7	R^8	R^9	R^{10}	R^{11}	R^{12}	R^{13}	R^{14}
T^0	J_0	J_1	J_2	J_3	J_4	J_5	J_6	J_7	J_8	J_9	J_{10}	J_{11}	J_{12}	J_{13}
T^1	J_{15}	J_{14}	J_2	J_3	J_4	J_5	J_6	–	–	J_9	J_{10}	J_{11}	J_{12}	J_{13}
T^2	J_{17}	–	–	J_{16}	J_4	J_5	J_6	J_{18}	–	–	–	J_{11}	J_{12}	J_{13}
T^3	J_{21}	–	–	–	J_{19}	J_{20}	J_6	J_{18}	–	–	–	–	–	J_{13}
T^4	J_{23}	–	–	–	J_{19}	J_{20}	J_{22}	–	–	–	–	–	–	–
T^5	–	J_{24}	–	–	–	J_{20}	J_{22}	–	–	–	–	–	–	–
T^6	J_{25}	–	–	–	–	–	J_{22}	–	–	–	–	–	–	–
T^7	–	J_{26}	–	–	–	–	–	–	–	–	–	–	–	–

Table 5.10, where the variables marked in red color indicate the altered position of variables post embedding the security constraints based on the embedding rule. Similarly, IP buyer fingerprint-based hardware security constraints corresponding to signature bit "1" are embedded into the design. The

Table 5.11   Register allocation information corresponding to scheduled IIR digital filter design (post embedding fingerprint biometric signature of IP buyer for bit "0" and "1")

CS	R^1	R^2	R^3	R^4	R^5	R^6	R^7	R^8	R^9	R^{10}	R^{11}	R^{12}	R^{13}	R^{14}
T^0	J_0	J_1	J_2	J_3	J_4	J_5	J_6	J_7	J_8	J_9	J_{10}	J_{11}	J_{12}	J_{13}
T^1	–	J_{14}	J_2	J_3	J_4	J_5	J_6	J_{15}	–	J_9	J_{10}	J_{11}	J_{12}	J_{13}
T^2	–	J_{17}	–	J_{16}	J_4	J_5	J_6	J_{18}	–	–	–	J_{11}	J_{12}	J_{13}
T^3	–	J_{21}	–	–	J_{19}	J_{20}	J_6	J_{18}	–	–	–	–	–	J_{13}
T^4	–	J_{23}	–	–	J_{19}	J_{20}	J_{22}	–	–	–	–	–	–	–
T^5	–	J_{24}	–	–	–	J_{20}	J_{22}	–	–	–	–	–	–	–
T^6	–	J_{25}	–	–	–	–	J_{22}	–	–	–	–	–	–	–
T^7	–	J_{26}	–	–	–	–	–	–	–	–	–	–	–	–

register allocation information post embedding the fingerprint security constraints conforming to both bit "0" and "1" are shown in Table 5.11. The variables marked in red color indicate the altered position of storage variables post embedding the IP buyer fingerprint biometric signature-driven security constraints. As evident from the register allocation (Tables 5.10 and 5.11), no extra register was required to accommodate the locally altered storage variables corresponding to embedding the IP buyer fingerprint security constraints. Subsequently, the scheduled DSP design post embedding the IP buyer fingerprint biometric-based secret hardware security constraints is obtained, as shown in Figure 5.9.

5.3.5   Embedding the facial security constraints of IP seller in IP buyer fingerprint biometric-embedded DSP design

In this subsection, mainly we discuss the process of embedding the secret constraints for hardware security corresponding to facial biometric of IP seller. The embedding process includes the following steps:

(a) performing embedding of the IP seller facial biometric signature-based hardware security constraints (generated earlier, as shown in Tables 5.7 and 5.8.) into the IP buyer fingerprint biometric embedded register allocation table (obtained in Table 5.11). The embedding of corresponding facial biometric-based hardware security constraints is performed by locally altering

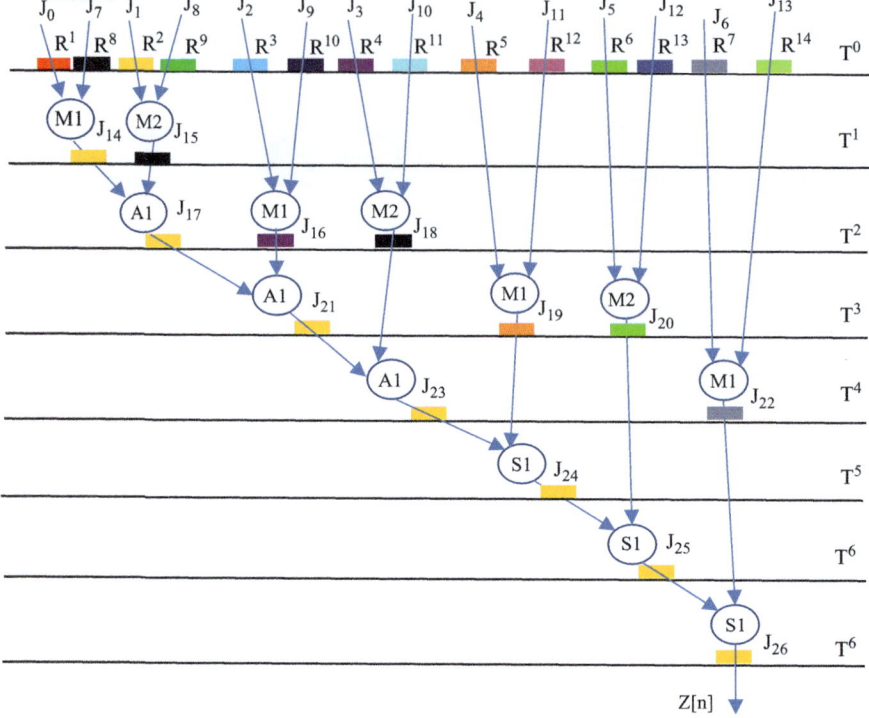

Figure 5.9   Scheduled DFG of IIR filter post embedding IP buyer fingerprint biometric signature

the storage variables among the available registers into the register allocation information of the design, obtained post embedding fingerprint signature. The constraints embedding block accepts the following two inputs: (i) generated security constraints corresponding to the facial biometric signature of IP seller with respect to target scheduled design and (ii) register allocation information post embedding IP buyer fingerprint signature. The embedding process therefore results into a secured DSP design with integrated facial biometric of IP seller. The register allocation information post embedding the IP seller facial biometric-based hardware security constraints corresponding to signature bit "0" is shown in Table 5.12, where the storage variables marked in green color indicates the altered position of variables post embedding the facial security constraints corresponding to signature bit "0." Similarly, IP seller facial biometric-based hardware security constraints corresponding to signature bit "1" are embedded into the design as shown in Table 5.13, where the variables marked in green color indicate the altered positions of storage variables post embedding the facial biometric security constraints corresponding to both bit "0" and "1" of the IP seller facial biometric signature-driven security constraints. As evident from the register allocation Tables 5.12 and 5.13, no extra

Table 5.12 Register allocation information corresponding to scheduled IIR digital filter design (post embedding facial biometric signature of IP seller into the design obtained with embedding IP buyer fingerprint biometric for bit "0")

CS	R^1	R^2	R^3	R^4	R^5	R^6	R^7	R^8	R^9	R^{10}	R^{11}	R^{12}	R^{13}	R^{14}
T^0	J_0	J_1	J_2	J_3	J_4	J_5	J_6	J7	J_8	J_9	J_{10}	J_{11}	J_{12}	J_{13}
T^1	–	J_{14}	J_2	J_3	J_4	J_5	J_6	J_{15}	–	J_9	J_{10}	J_{11}	J_{12}	J_{13}
T^2	–	–	J_{17}	J_{16}	J_4	J_5	J_6	J_{18}	–	–	–	J_{11}	J_{12}	J_{13}
T^3	–	–	J_{21}	–	J_{19}	J_{20}	J_6	J_{18}	–	–	–	–	–	J_{13}
T^4	–	–	J_{23}	–	J_{19}	J_{20}	J_{22}	–	–	–	–	–	–	–
T^5	–	J_{24}	–	–	–	J_{20}	J_{22}	–	–	–	–	–	–	–
T^6	–	–	J_{25}	–	–	–	J_{22}	–	–	–	–	–	–	–
T^7	–	J_{26}	–	–	–	–	–	–	–	–	–	–	–	–

Table 5.13 Register allocation information corresponding to scheduled IIR digital filter design (post embedding facial biometric signature of IP seller into the design obtained with embedding IP buyer fingerprint biometric 0 and 1)

CS	R^1	R^2	R^3	R^4	R^5	R^6	R^7	R^8	R^9	R^{10}	R^{11}	R^{12}	R^{13}	R^{14}
T^0	J_0	J_1	J_2	J_3	J_4	J_5	J_6	J_7	J_8	J_9	J_{10}	J_{11}	J_{12}	J_{13}
T^1	–	J_{14}	J_2	J_3	J_4	J_5	J_6	J_{15}	–	J_9	J_{10}	J_{11}	J_{12}	J_{13}
T^2	–	–	–	J_{16}	J_4	J_5	J_6	J_{18}	J_{17}	–	–	J_{11}	J_{12}	J_{13}
T^3	–	–	J_{21}	–	J_{19}	J_{20}	J_6	J_{18}	–	–	–	–	–	J_{13}
T^4	–	–	–	–	J_{19}	J_{20}	J_{22}	–	J_{23}	–	–	–	–	–
T^5	–	J_{24}	–	–	–	J_{20}	J_{22}	–	–	–	–	–	–	–
T^6	–	–	J_{25}	–	–	–	J_{22}	–	–	–	–	–	–	–
T^7	–	J_{26}	–	–	–	–	–	–	–	–	–	–	–	–

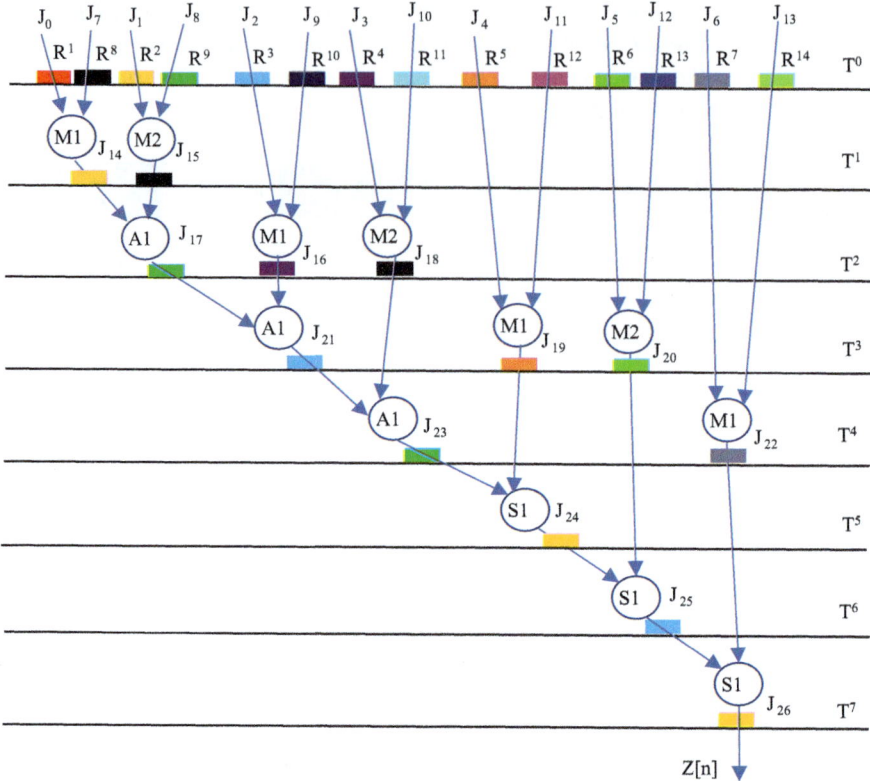

Figure 5.10   Scheduled DFG of IIR filter post embedding IP seller facial biometric signature

registers were needed to accommodate the locally altered storage variables corresponding to embedding the IP seller facial biometric-driven encoded constraints for hardware security. Thus, the register allocation information post embedding facial biometric of IP seller into the IP buyer fingerprint-embedded design is obtained. Subsequently, the scheduled DSP design post embedding the IP seller facial biometric-based secret hardware security constraints is obtained, as shown in Figure 5.10.

5.4   Protecting an IP seller (vendor) right against false ownership claim using facial biometric signature

In case if an adversary from offshore design house or at foundry level of IP design process, fraudulently claims IP ownership, then the IP seller facial biometric-based security constraints into the DSP design offers a highly robust and seamless verification of true IP ownership. The facial biometric of IP seller is only captured once before the embedding process and the corresponding facial biometric image (with grid size and spacing) is safely stored for IP ownership verification. There is no need to recapture the facial biometric again for verification process. The existing pre-stored facial biometric image is used to regenerate the facial signature and its corresponding hardware security constraints to prove IP ownership and nullify fraudulent claim of IP ownership. The same features and their dimensions of the facial biometric can be identified and computed accurately from the pre-stored facial image for proving IP ownership. Since the facial biometric is only captured once and hence factors such as slight tilt of camera, variation in resolution, difference in cropping size do not have any impact on the IP ownership verification. The verification process is independent of recapturing of the facial biometric information. There will be no differences in extracted biometric data as second time capturing of the biometric information is not required. The pre-stored facial image (with grid size and spacing) is sufficient to prove IP ownership of original IP seller. Further, in case of IP ownership, to nullify the false claim of IP ownership and awarding the IP ownership to genuine IP seller, the positions of authentic facial signature bits are matched bit by bit with the biometric template embedded into the design under test. In case of complete matching, IP ownership is awarded to the genuine IP seller.

The embedded facial signature (corresponding to IP seller) protects his/her right against false claim of IP ownership. The embedded IP seller facial signature renders following security properties:

- exact regeneration of IP seller facial biometric is impossible for an adversary as the regeneration involves the following security parameters (which are all unknown to an adversary) such as (i) specific grid size and spacing for subjecting the facial image for accurate feature extraction; (ii) number of facial nodal points based on the IP seller-selected facial features for signature generation; (iii) concatenation order of features; (iv) final signature length used for secret hardware security constraints generation; (v) ordering of storage variables; (vi) secret encoding rule corresponding to facial signature bits (0 and 1).
- Immune against key-based attack as it does not involve any secret key for generating the facial security constraints.
- Stronger than non-biometric-based approaches such as IP watermarking and IP steganography. This is because the facial biometric associates the unique identity of an IP seller with the design to seamlessly prove his/her IP ownership.

Therefore, the involvement of several complex security information during facial signature generation and implantation makes it almost impossible for an adversary to assert fraudulently claim IP ownership.

5.5 Protecting an IP buyer' right using fingerprint biometric signature

So far, we have discussed, the mechanism for protecting the rights of an IP seller against false IP ownership claim using his/her facial biometric, embedded into the design. However, the protection of IP buyer rights is also equally important to safeguard him against a rogue adversary. In case if a deceitful IP seller or an adversary in the untrustworthy design house attempts to illegally re-distribute the same IP core to other IP buyers, then from an honest IP buyer's standpoint, detection of such resold copies is necessary such that an honest IP buyer may retain his/her exclusive buyer rights for his/her custom design.

To enable the protection of IP buyer's right, his/her fingerprint biometric-driven secret hardware security constraints are covertly inserted into the design. These embedded fingerprint security constraints corresponding to an original IP buyer enable the detection of redistributed/resold copies. However, the captured fingerprint of IP buyer is only captured once before the embedding process and the corresponding fingerprint biometric image is safely stored for further performing the detection of illegally resold copies and verification of IP buyer's right. There is no need to recapture the fingerprint biometric again for verification process. The existing pre-stored fingerprint biometric image is used to regenerate the fingerprint signature and its corresponding hardware security constraints to detect illegally distributed IP copies. The detection process is independent of recapturing of the fingerprint biometric information. There will be no differences in extracted biometric data as second time capturing of the biometric information is not required. Further, in case of detection of illegally distributed copies of custom IPs from a rogue IP seller, the positions of authentic fingerprint signature bits are matched bit by bit with the digital template embedded into the design under test. The embedded fingerprint signature corresponding to original IP buyer renders several robust security properties, which makes it impossible for an adversary to regenerate exactly the same fingerprint signature for evading the detection of illegally resold IPs. The embedded IP buyer fingerprint signature renders following security properties (which are all unknown to an adversary):

(a) Exact regeneration of IP buyer fingerprint biometric is impossible for an adversary as the regeneration involves the following security parameters such as (i) type and number of minutiae points chosen for signature generation; (ii) concatenation order of the features of a minutiae point; (iii) concatenation order of minutiae points for final signature generation; (iv) truncation length of the signature; (v) secret encoding rule by IP buyer for generating secret hardware security constraints
(b) Non-vulnerable to key leakage or theft unlike watermarking and steganography-based hardware security approaches, where an adversary can easily compromise the employed encoding rules and secret stego keys respectively.
(c) No storage required for fingerprint driven secret security constraints as they are derived based on the IP buyer specified security parameters, as shown in (a) i-v.
(d) Robustness against brute force attack for regenerating the security constraints corresponding to original IP buyer because an adversary cannot prove the generated security constraints as his/her own. This is because IP buyer fingerprint security constraints are associated with his/her unique identity and specified robust secret security parameters while generating and embedding

them into the design. The aforementioned security properties make it almost impossible for an adversary to evade the detection of illegally distributed IPs. Thus, the embedded fingerprint biometric signature of IP buyer enables the protection of his/her IP right.

5.6   Detecting IP piracy before integration into SoC systems

An adversary present in untrustworthy design house may attempt to pirate the IP cores without the knowledge of IP seller. These pirated cores may contain the malicious logic which may cause safety and reliability hazards to end consumer. Therefore, it is crucial to detect and isolate such pirated cores before their integration into SoC systems. IP seller facial biometric in the form of embedded secret constraints for hardware security enables the robust detective control against pirated IP cores. The piracy detection process is shown in Figure 5.11. While performing the IP piracy detection, the presence of authentic security constraints as per the facial biometric signature of the original IP seller is verified. To perform the verification, the embedded constraints from the target design under test are extracted. If they completely match with the security constraints of the original IP seller, then the target IP design is considered as genuine otherwise it may be considered a pirated design.

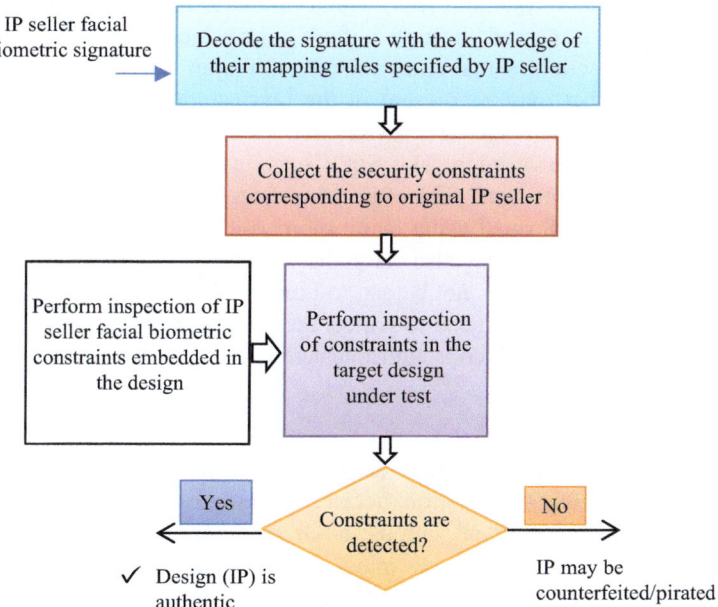

Figure 5.11   Detection of pirated versions of IP cores

5.7   Employing facial biometric for protecting Trojan-secured SoC design against piracy

So far, we have discussed the application of physical biometrics for enabling the symmetric protection of the respective rights for intellectual property (IP) buyer and seller. Now we discuss the application of facial biometric-based hardware security methodology for protecting Trojan-secured DSP designs against IP piracy (Chaurasia and Sengupta, 2022b).

5.7.1   Threat model

Trojan-secured SoC design indicates a design that has the capability to either detect functional difference due to hidden functional Trojan or capability of Trojan isolation. In this chapter, we discuss Trojan-secured SoC design that has the capability to detect functional Trojan in 3PIP as well as application of facial biometric signature, post embedding into the Trojan-secured designs that enable robust security against external threats. The following points highlight the threats handled through this approach (Chaurasia and Sengupta, 2022b):

- An adversary in the 3PIP design house may implant the malicious Trojan in the micro-IPs. Therefore, it is crucial to detect and isolate such malicious Trojan or Trojan-infected designs before their integration into the SoC of the CE systems. To ensure this a SoC integrator generates a Trojan-secured design by associating redundancy from distinct 3PIP vendor(s). This results into a dual modular redundant system at least which produces distinct output from separate units that is passed through a fault tolerant comparator. It, therefore, enables the detection of malicious functional Trojan.
- An adversary that may be present in the untrustworthy foundry (fab) house may attempt to pirate such Trojan-secured design created by SoC integrator and the integration of such fake SoCs may lead to security and reliability hazards to end consumer. Therefore, the detection of such pirated Trojan-secured SoC is performed using facial biometric signature.
- An adversary in the fab may also claim the ownership of the Trojan-secured SoC design which actually does not belong to him/her. Therefore, from the SoC integrator's standpoint, it is crucial to protect the Trojan-secured design against such false ownership threat which is authenticated using facial biometric signature.

5.7.2   Summary

Hardware Trojans in 3PIP cores may cause different payloads such as leakage of secret data, performance degradation, accelerate aging of design components, and change in computed functional output. Such malicious behavior due to secret backdoor Trojan in pirated IP cores may result into safety concern for end consumers. One of the critical Trojan that exist in 3PIPs is functional Trojan that affects functional output behavior. Therefore, Trojan-secured designs are helpful for detecting such functional Trojan, before their possible integration into a CE system. Further, during fabrication process of such Trojan-secured SoC designs, an adversary present in the offshore foundry may attempt to pirate the design or claim false IP ownership. Therefore, protection of such Trojan-secured designs against IP piracy from an SoC integrator's perspective is equally important. Embedded security constraints derived from facial biometric of SoC integrator has seamlessly provide detection capability against such pirated versions (if potentially performed in the external fab).

The overview of the mechanism for generating Trojan-secured DSP design and its protection against IP piracy is highlighted in Figure 5.12. To safeguard Trojan-secured DSP designs against IP piracy, first the DSP designs are equipped with Trojan detection capability. To enable the Trojan detection capability, first, we construct the Trojan-secured scheduled design using distinct multi-vendor policy. The Trojan-secured design is constructed by creating a sister unit of the original DSP design by duplicating its operations and subsequently passing both units through fault-tolerant comparator. Next, the facial biometric signature corresponding to facial image of the SoC integrator is generated. The facial biometric signature comprises of the unique facial features of SoC integrator. Finally, this generated facial signature, in the form of covert secret security constraints, is embedded into the design during high-level synthesis process. Thus, the embedded facial biometric signature ensures the detection of pirated Trojan-secured designs before their integration into CE systems. Now, we discuss the process of generating and embedding the facial signature into the DSP design with Trojan detection capability.

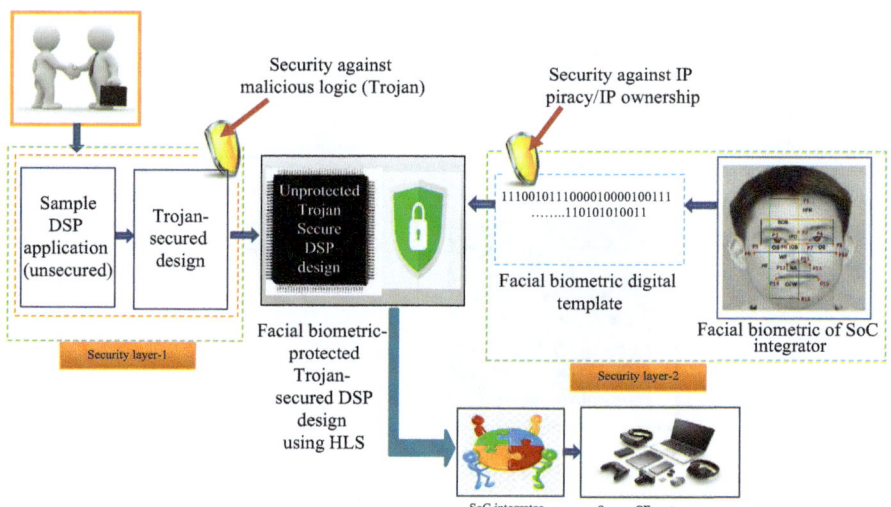

Figure 5.12   Overview of mechanism for protecting Trojan-secured design against IP piracy

5.7.3 Designing Trojan-secured design architecture

Now, we discuss the process of generating the Trojan-secured design architecture. As shown in Figure 5.13, the process of generating a Trojan-secured design architecture corresponding to a DSP algorithm accepts the following primary inputs: behavioral description of DSP application, library which comprises the details of IP vendors, and resource constraints. To generate Trojan detectable design architecture, the first phase of Trojan aware security block creates a redundant CDFG (corresponding to the DSP application). To do so, first the

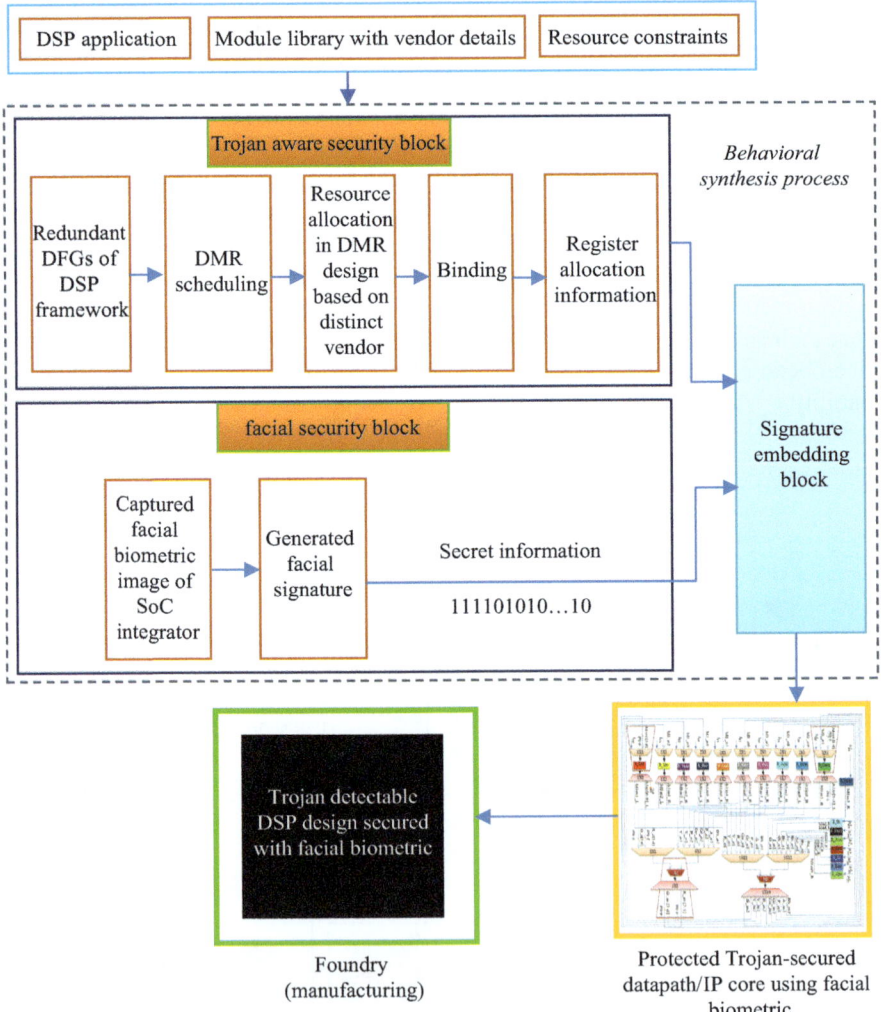

Figure 5.13   Design flow of security methodology for protecting Trojan-secured design against IP piracy using facial biometric

behavioral description of the input DSP application is transformed into its respective CDFG. This results into the formation of CDFG of the DSP design and is termed as original unit as shown in Figure 5.14, where each node is representing an operation and edges are representing their functional dependency. Subsequently, the operations of the original unit are replicated to generate a sister unit, as shown in Figure 5.15. These two units combinedly are called as dual modular redundant

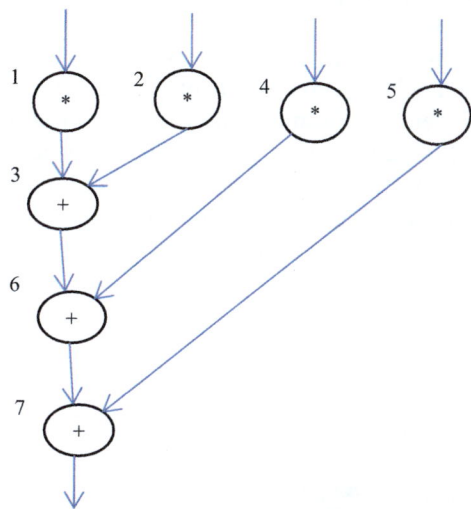

Figure 5.14   CDFG of DCT application (4-point)

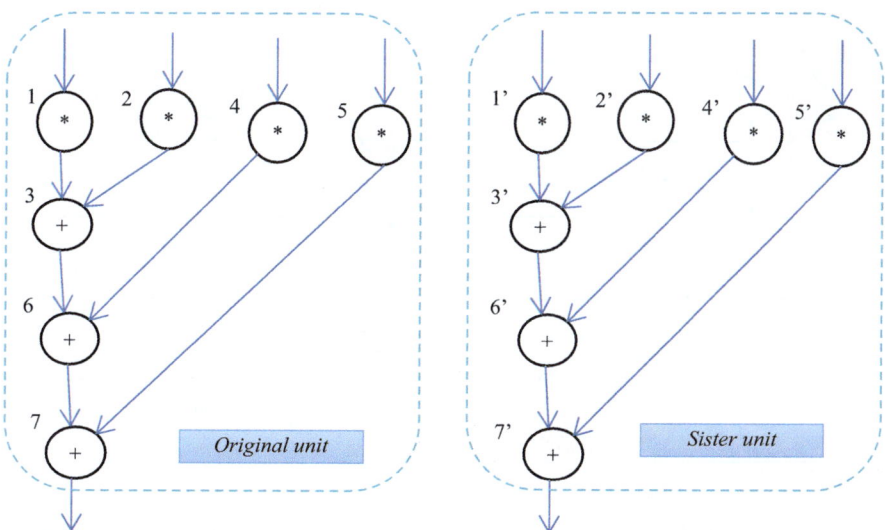

Figure 5.15   Redundant CDFGs of DCT 4-point application (original and sister unit)

(DMR) design. In the next phase, the generated DMR design is scheduled. The scheduling is performed based on the scheduling algorithm, resource constraints, and the available dependency information of operational nodes of the design. LIST scheduling algorithm has been used for scheduling the Trojan-secured design (Sengupta *et al.*, 2017). The scheduled Trojan-secured design is shown in Figure 5.16, where the storage variables of the design are represented as J_0 to J_{21}, the storage elements/registers (designated as different colors) are represented as R^1 to R^8, and the control steps required to schedule the design are represented as T^0 to T^7. Further, the two distinct vendors are designated as S1 and S2, respectively, and the corresponding operation numbers are denoted as (1, 2, ….,7) and (1', 2', ….,7'), respectively. In this methodology for generating Trojan-secured design, a distinct multi-vendor allocation policy is used, where the resources from a particular vendor are allocated to a single unit of the DMR design. This is effective for providing Trojan detection compared to assigning multi-vendor resources simultaneously within a single unit (either original or sister unit) of DMR logic. In the next phase, allocation of the hardware resources is performed, to respective

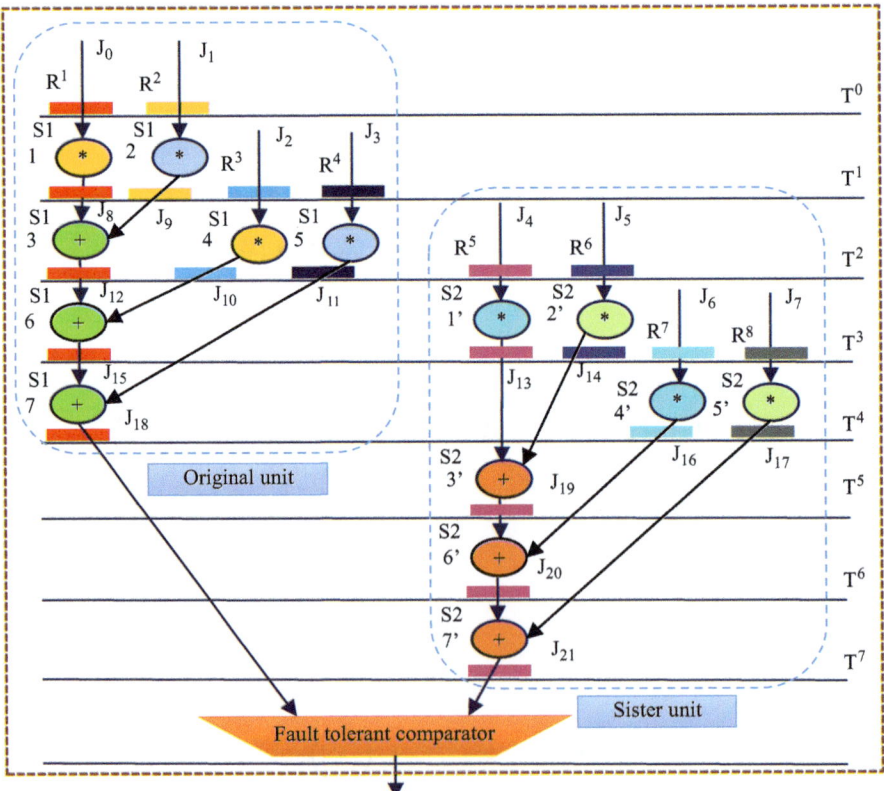

Figure 5.16   Scheduled CDFG of Trojan-secured DSP design corresponding to 4-point DCT design architecture (baseline design)

operations (multiplication, addition, etc.) of the scheduled CDFG design based on the resource constraints specified by SoC integrator and subsequently their binding is performed. Next, we construct the register allocation table corresponding to Trojan-secured design (without the embedded biometric signature of SoC integrator). The register allocation table corresponding to Trojan-secured 4-point DCT is shown in Table 5.16. Register allocation table comprises of the details of required registers (R^1 to R^8), available storage variables (J_0 to J_{21}) in the design and their assignment to particular register and required control steps (T^0 to T^7).

This register allocation information is subjected to embedding of the facial biometric signature of SoC integrator. However, before we explain the embedding process, let us discuss the facial biometric signature generation technique.

5.7.4 Generating facial signature-driven secret constraints for hardware security

To generate the facial biometric signature corresponding to captured facial image of the SoC integrator, the following steps are followed:

(a) Capture the facial image with high-quality camera and subject it to a specific grid size and spacing for accurate feature extraction, as shown in Figure 5.7(a).
(b) Derive the facial nodal points on the captured image corresponding to chosen facial features set for signature generation, as shown in Figure 5.7(b). The details of facial features are discussed in Section 5.3.
(c) Assign the naming convention on nodal points, as shown in Figure 5.8(a).
(d) Generate the image with chosen facial features for facial biometric signature generation, as shown in Figure 5.8(b).
(e) Determine the co-ordinates of the nodal points pair corresponding to each of the facial feature, as shown in Table 5.5.
(f) Determine feature dimensions by computing the Manhattan distance between the nodal points pair, as shown in Table 5.6.
(g) Generate the binarized signature corresponding to each feature dimension and subsequently concatenate them as per SoC integrator specified concatenation order, for generating the facial biometric signature.

Let us consider that the chosen concatenation order for generating the facial biometric signature is:

$$f^{OB} + f^{IOB} + f^{NB} + f^{OCW} + f^{WF} + f^{BOB} + f^{HF} + f^{WNB} + f^{HFH} + f^{IPD} + f^{WNR}$$

Then the generated facial biometric signature is as follows:

"1000110100101110001101101110100001110110101111100000011010001101001100100011111101".

The size of the facial biometric digital template is 82 bits, including 39 number of 0s and 43 number of 1s. Thus, the above facial biometric signature is generated as the output of facial security block (shown in Figure 5.13). However, the strength of the signature can be varied by SoC integrator by removing or adding any chosen facial features and selecting the final truncation length. It therefore offers several possible combinations of facial biometric signature, which provide flexibility for choosing any combination of facial signature bits for embedding into the Trojan-secured design.

Now, let us discuss the process for generating the facial security constraints corresponding to generated biometric signature from the captured facial image of SoC integrator. The facial biometric-based secret constraints for hardware security are generated based on SoC integrator specified signature truncation length, encoding rules, and target scheduled Trojan-secured DSP design. For the sake of brevity, we have considered the truncation length corresponding to facial signature of SoC integrator of 74-bit size. The secret hardware security constraints are generated corresponding to scheduled Trojan-secured 4-point DCT design (as shown in Figure 5.16) by considering the storage variables in sorted ascending order are (J_0, J_1, J_2, J_3, J_4, J_5, ..., J_{21}). Further, let us consider the SoC integrator specified encoding rules for performing the generation of encoded constraints for hardware security are (Chaurasia and Sengupta, 2022b):

- **Signature bit "0"**: representing storage variable pair (J_x and J_y) of even number required to be allocated to distinct registers, where J_x and J_y can be of any integer value.
- **Signature bit "1"**: representing storage variable pair (J_x and J_y) of odd number required to be allocated to distinct registers, where J_x and J_y can be of any integer value.

Therefore, based on the above encoding rule and the storage variable set obtained in sorted ascending order, hardware security constrains are generated corresponding to facial signature of SoC integrator (post truncation). Security constraints conforming to bit "0" and bit "1" are shown in Tables 5.14 and 5.15, respectively.

Table 5.14   Facial biometric signature-based secret hardware security constrains corresponding to Trojan-secured scheduled DSP design and signature bit "0"

$<J_0, J_2>$	$<J_2, J_8>$	$<J_4, J_{16}>$	$<J_8, J_{14}>$	–
$<J_0, J_4>$	$<J_2, J_{10}>$	$<J_4, J_{18}>$	$<J_8, J_{16}>$	–
$<J_0, J_6>$	$<J_2, J_{12}>$	$<J_4, J_{20}>$	–	–
$<J_0, J_8>$	$<J_2, J_{14}>$	$<J_6, J_8>$	–	–
$<J_0, J_{10}>$	$<J_2, J_{16}>$	$<J_6, J_{10}>$	–	–
$<J_0, J_{12}>$	$<J_2, J_{18}>$	$<J_6, J_{12}>$	–	–
$<J_0, J_{14}>$	$<J_2, J_{20}>$	$<J_6, J_{14}>$	–	–
$<J_0, J_{16}>$	$<J_4, J_6>$	$<J_6, J_{16}>$	–	–
$<J_0, J_{18}>$	$<J_4, J_8>$	$<J_6, J_{18}>$	–	–
$<J_0, J_{20}>$	$<J_4, J_{10}>$	$<J_6, J_{20}>$	–	–
$<J_2, J_4>$	$<J_4, J_{12}>$	$<J_8, J_{10}>$	–	–
$<J_2, J_6>$	$<J_4, J_{14}>$	$<J_8, J_{12}>$	–	–

Table 5.15 *Facial biometric signature-based secret hardware*
security constrains corresponding to Trojan-secured
scheduled DSP design and signature bit "1"

$<J_1, J_3>$	$<J_3, J_9>$	$<J_5, J_{17}>$	–
$<J_1, J_5>$	$<J_3, J_{11}>$	$<J_5, J_{19}>$	–
$<J_1, J_7>$	$<J_3, J_{13}>$	$<J_5, J_{21}>$	–
$<J_1, J_9>$	$<J_3, J_{15}>$	$<J_7, J_9>$	–
$<J_1, J_{11}>$	$<J_3, J_{17}>$	$<J_7, J_{11}>$	–
$<J_1, J_{13}>$	$<J_3, J_{19}>$	$<J_7, J_{13}>$	–
$<J_1, J_{15}>$	$<J_3, J_{21}>$	$<J_7, J_{15}>$	–
$<J_1, J_{17}>$	$<J_5, J_7>$	$<J_7, J_{17}>$	–
$<J_1, J_{19}>$	$<J_5, J_9>$	$<J_7, J_{19}>$	–
$<J_1, J_{21}>$	$<J_5, J_{11}>$	$<J_7, J_{21}>$	–
$<J_3, J_5>$	$<J_5, J_{13}>$	$<J_9, J_{11}>$	–
$<J_3, J_7>$	$<J_5, J_{15}>$	$<J_9, J_{13}>$	–

5.7.5   Embedding the extracted facial security constraints into Trojan-secured design

Trojan secured design and generating the facial security constraints corresponding to facial biometric of SoC integrator. Now, we discuss the embedding of facial signature-driven secret constraints into register allocation of Trojan-secured design. The inserted authentic facial biometric signature aids the detection of pirated versions of Trojan-secured SoCs before it can be integrated into CE systems. Additionally, it also safeguards an SoC integrator against the potential threat of false claim of IP ownership from an adversary that may be present in the offshore foundry house. As shown in Figure 5.13, the signature embedding block accepts the following two inputs: (a) register allocation information corresponding to scheduled Trojan-secured design and (b) extracted secret constraints for hardware security corresponding to facial biometric of SoC integrator and scheduled Trojan-secured design. The register allocation table is used for embedding the hardware security constraints by locally altering (modifying) the register assignments. The register allocation information post embedding the SoC integrator's facial biometric-based hardware security constraints corresponding to signature bit "0" and "1" is shown in Table 5.17, where the variables marked in red color indicate the altered position of variables post embedding the security constraints. As evident from the register allocation Table 5.17, no extra registers were required to accommodate the locally altered storage variables corresponding to embedding the facial security constraints of SoC integrator. Subsequently, the scheduled Trojan-secured DSP design post embedding the secret constraints for hardware security corresponding to facial biometric of SoC integrator is obtained as shown in Figure 5.17. Next, a Trojan-secured RTL design (datapath and controller) with embedded facial biometric of SoC integrator is obtained using HLS. This secured RTL design is subjected to gate level synthesis and layout level synthesis before it proceeds to the external foundry for fabrication. This facial signature embedded

Table 5.16   *Register allocation information corresponding to scheduled Trojan-secured 4-point DCT design (before embedding facial biometric signature)*

CS	R^1	R^2	R^3	R^4	R^5	R^6	R^7	R^8
T^0	J_0	J_1	–	–	–	–	–	–
T^1	J_8	J_9	J_2	J_3	–	–	–	–
T^2	J_{12}	–	J_{10}	J_{11}	J_4	J_5	–	–
T^3	J_{15}	–	–	J_{11}	J_{13}	J_{14}	J_6	J_7
T^4	J_{18}	–	–	–	J_{13}	J_{14}	J_{16}	J_{17}
T^5	–	–	–	–	J_{19}	–	J_{16}	J_{17}
T^6	–	–	–	–	J_{20}	–	–	J_{17}
T^7	–	–	–	–	J_{21}	–	–	–

Table 5.17   *Register allocation information corresponding to scheduled Trojan-secured 4-point DCT design (post embedding facial biometric signature of SoC integrator)*

CS	R^1	R^2	R^3	R^4	R^5	R^6	R^7	R^8
T^0	J_0	J_1	–	–	–	–	–	–
T^1	J_9	J_8	J_2	J_3	–	–	–	–
T^2	–	–	J_{11}	J_{10}	J_4	J_5	–	J_{12}
T^3	J_{15}	–	J_{11}	–	J_{13}	J_{14}	J_6	J_7
T^4	–	J_{18}	–	–	J_{13}	J_{14}	J_{17}	J_{16}
T^5	–	–	–	–	J_{19}	–	J_{17}	J_{16}
T^6	–	–	–	–	–	J_{20}	J_{17}	–
T^7	–	–	–	–	J_{21}	–	–	–

Trojan-secured design offers the detection of any possible pirated version as well as is also capable to nullify the false claim of ownership (in case of a potential adversary claiming fraud ownership in the foundry during fabrication stage). This therefore ensures the integration of only authentic Trojan-secured SoCs into the CE systems, thereby providing safety and integrity of end consumer.

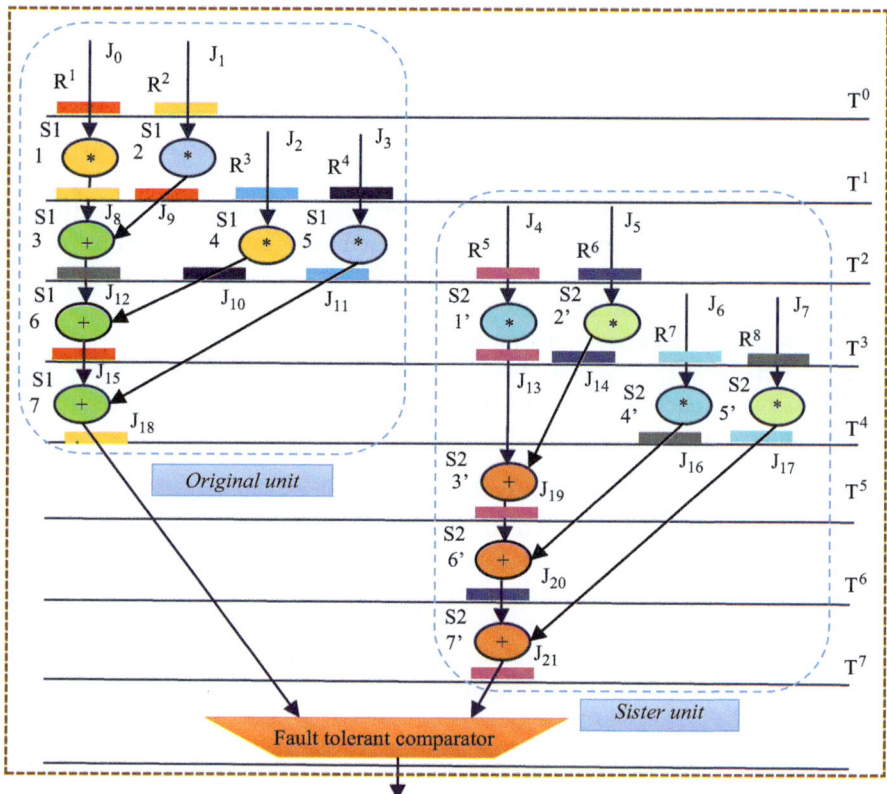

Figure 5.17   SDFG of Trojan-secured DMR design corresponding to 4-point DCT design architecture (post embedding facial biometric)

5.8   Analysis and discussion

This section analyzes the physical biometric-based hardware security methodologies for: (a) enabling symmetrical protection of respective rights for IP buyer and seller using fingerprint biometric and facial biometric, respectively (Chaurasia and Sengupta, 2022a; Sengupta and Rathor, 2020) and (b) protecting Trojan-secured DSP designs against IP piracy using facial biometric approach (Chaurasia and Sengupta, 2022b).

5.8.1   *Analyzing security and design cost of symmetric IP core protection using facial and fingerprint biometric for DSP applications*

An IP buyer may vary his/her fingerprint signature by selecting or dropping minutiae points from the fingerprint image, corresponding to the captured fingerprint image. The more the number of selected minutiae feature points, the more is the size of the generated fingerprint biometric signature. This therefore enables the generation of a greater number of fingerprint biometric-based secret hardware security constraints to be implanted into the design. Thus, larger the biometric digital template, the stronger is the digital evidence. The fingerprint security constraints are generated corresponding to fingerprint image chosen from Alilou (2019).

Similarly, an IP seller may vary his/her facial signature by selecting or dropping the facial features from the generated image, corresponding to the captured facial image. The more the number of selected facial features, the more is the size of the facial biometric signature and thus stronger is the digital evidence. The facial security constraints are generated corresponding to facial image chosen from the dataset (Multimedia Laboratory Datasets, 2020). Now, we discuss the impact of employing physical biometric on security strength and embedding design cost.

5.8.1.1   Security strength analysis

The security strength is analyzed in terms of probability of coincidence (Xp) metric. Xp designates the probability of coincidently detecting the secret hardware security constraints of an authentic design in an unsecured version. Therefore, in case of proving the ownership, lesser value of Xp is desirable (as low as possible). The lesser the value of Xp, the stronger is the definitive proof of ownership. The value of Xp is computed as follows (Sengupta and Rathor, 2020):

$$Xp = \left(1 - \frac{1}{R^n}\right)^l \tag{5.1}$$

where "R^n" denotes the strength of registers/colors used for compiling the storage variables of the design during register allocation, before implanting secret constraints and "l" denotes the number of secret hardware security constraints (storage variable pairs generated corresponding to DSP design and biometric signature) to be implanted into the design. The "Xp" value achieved in symmetrical protection of DSP designs using physical biometric (Chaurasia and Sengupta, 2022a) and multivariable fingerprinting with watermarking approach (Roy and Sengupta, 2017), corresponding to IP buyer signature and different DSP designs, is shown in Figure 5.18. As evident from the figure, physical biometric-based approach renders lesser value of Xp (desirable) than multivariable fingerprinting with watermarking approach. This is because the strength of secret security constraints generated using physical biometric of an IP buyer is larger than hardware security constraints generated using multivariable fingerprint signature of an IP buyer.

The Xp value achieved using facial biometric and watermarking signature of IP seller, corresponding to different DSP designs, is shown in Figure 5.19. As evident from the figure, facial biometric signature renders lesser value of Xp (desirable) than watermarking signature corresponding to an IP seller.

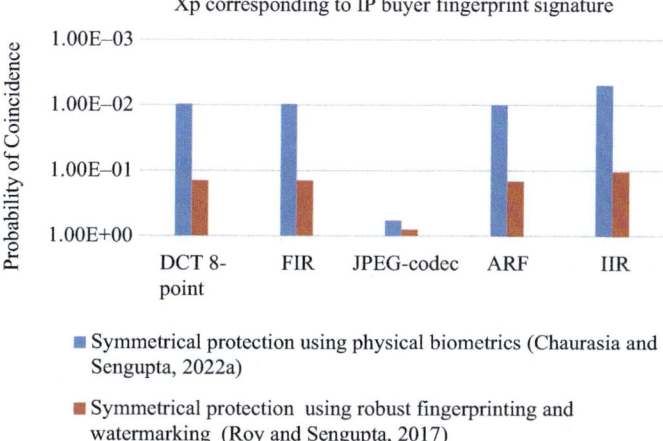

Figure 5.18 Comparison of probability of coincidence corresponding to IP buyer

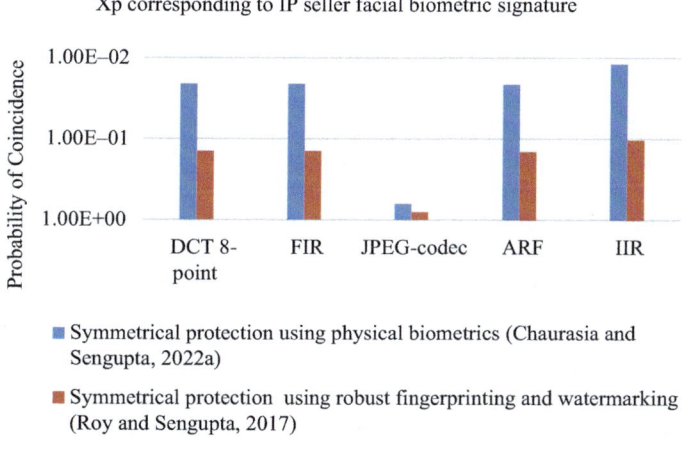

Figure 5.19 Comparison of probability of coincidence corresponding to IP seller

5.8.1.2 Design cost analysis

To enable symmetrical protection of IP rights of buyer and seller, secret signature in the form of encoded hardware security constraints is embedded into the design. This may impact the design cost overhead of target DSP design. Therefore, it is desirable to have lesser design cost, post embedding the security constraints into the design. In the physical biometric-based symmetrical protection approach, first we embed the biometric signature of IP buyer and subsequently embed the biometric signature of IP seller during register allocation phase of HLS without affecting actual functionality of the target hardware design. The resulting design cost of embedding secret security mark is evaluated using the following design metric (Sengupta and Rathor, 2020):

$$S_{cf}(C_i) = u_1 \frac{A_d}{A_{max}} + u_2 \frac{L_d}{L_{max}} \qquad (5.2)$$

where "C_i" denotes the architectural constraints (adder and multiplier) used for designing the DSP hardware, A_d and L_d are the area of the design and the latency, respectively, A_{max} and L_{max} denote the maximum area of the design and latency, u_1 and u_2 are the weighing factors of normalized design area and latency in the cost function. The assumed values by the IP vendor corresponding to $u1$ and $u2$ are 0.5. The design cost of symmetrical protection using physical biometric approach, post embedding fingerprint biometric of IP buyer and subsequently embedding facial biometric of IP seller, is shown in Table 5.18, where "*," "+," and "−" represent the adder, multiplier, and subtractor logic (hardware resources), respectively. As evident from the table, the approach (Chaurasia and Sengupta, 2022a) results into zero design cost overhead. This is because no extra hardware resources (registers) were required during the embedding of IP buyer and seller physical biometric driven secret security constraints into the hardware design.

5.8.2   Analyzing security and design cost overhead of facial biometric embedded Trojan-secured DSP design

An SoC integrator may vary the facial signature by selecting or dropping the facial features from his/her captured facial image. The more the number of selected facial features, the more is the size of the facial biometric signature and thus stronger is the digital evidence for protecting the Trojan-secured SoC design. Now, we discuss the impact of employing facial biometric on Trojan-secured designs in terms of security strength and embedding design cost.

Table 5.18   *Design cost of symmetrical protection using physical biometric approach (post embedding the fingerprint biometric of IP buyer and subsequently embedding the facial biometric of IP seller into the DSP design)*

DSP bench-marks	No. of registers (R)	Resource configuration	Design cost of baseline design	Design cost after embedding fingerprint biometric of IP buyer	Design cost after embedding facial biometric of IP seller	% Design cost overhead
DCT-8point	16	1(+), 2(*)	0.447	0.447	0.447	0.00%
FIR filter	16	1(+), 3(*)	0.5697	0.5697	0.5697	0.00%
JPEG-codec	129	3(+), 3(*)	0.2178	0.2178	0.2178	0.00%
ARF	16	2(+), 4(*)	0.4121	0.4121	0.4121	0.00%
IIR filter	14	1(+),2(*),1(−)	0.5247	0.5247	0.5247	0.00%

5.8.2.1 Security analysis

The security of the approach (Chaurasia and Sengupta, 2022b) is analyzed in terms of probability of coincidence and probability of attacker to guess the exact signature combination. First, we discuss the achieved probability of coincidence corresponding to Trojan-secured designs, embedded with facial biometric signature of the SoC integrator. The resultant value of Xp for Trojan-secured 4-point DCT design corresponding to varying facial features is shown in Figure 5.20. As evident from the figure, increasing the number of facial features results into a greater strength of facial signature yielding larger number of facial security constraints. This, therefore, generates stronger digital evidence and thus results into lesser Xp value (using (5.1)). Further, achieved Xp of Trojan-secured design corresponding to varying facial signature is also compared with hardware steganography (Rathor and Sengupta, 2020) and watermarking (Sengupta and Bhadauria, 2016) approaches, as shown in Table 5.19. As evident, the facial biometric approach results into lesser probability of coincidence. This is because the facial biometric corresponding to SoC integrator results into a greater number of facial security constraints. Additionally, the comparison of Xp corresponding to different DSP applications and contemporary approaches is shown in Table 5.20. As evident, the Xp of facial biometric protected Trojan-secured design (corresponding to any particular application) is lesser than other contemporary hardware security approaches.

Now, we discuss the probability of an attacker to guess the exact facial biometric signature combination (G_p) using brute force attack. This is computed using the following metric (Chaurasia and Sengupta, 2022b):

$$G_p = \frac{1}{v^l} \qquad\qquad (5.3)$$

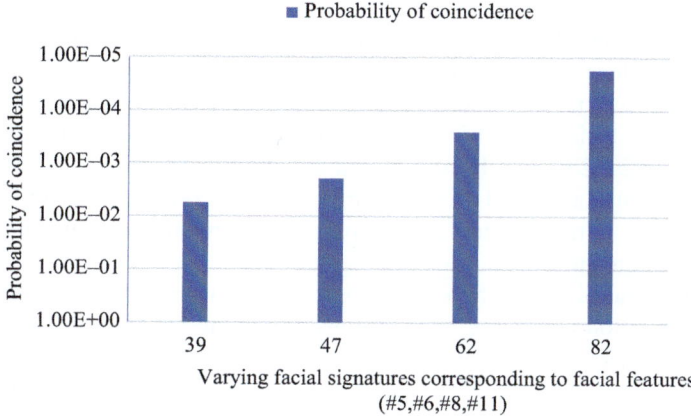

Figure 5.20   Comparison of probability of coincidence for Trojan-secured 4-point DCT corresponding to varying facial features

Table 5.19   Xp comparison of Trojan-secured 4-point DCT design for varying hardware security constraints

Chaurasia and Sengupta (2022b)		Hardware steganography (Rathor and Sengupta, 2020)		Hardware watermarking (Sengupta and Bhadauria, 2016)	
Hardware security constraints	Xp	Hardware security constraints	Xp	Hardware security constraints	Xp
40	4.70E−03	13	1.76E−01	15	1.30E−01
47	1.80E−03	20	6.90E−02	30	1.80E−02
62	2.50E−04	50	1.20E−03	60	3.30E−04
78	2.90E−05	63	2.20E−04	70	8.70E−05

Table 5.20   Xp comparison corresponding to different DSP applications

Benchmarks	(Chaurasia and Sengupta, 2022b) Xp	Hardware steganography (Rathor and Sengupta, 2020) Xp	Hardware watermarking (Sengupta and Bhadauria, 2016) Xp
4-point DCT	2.90E−05	2.20E−04	8.70E−05
FIR	6.50E−03	1.70E−02	1.00E−02
MPEG	5.80E−02	1.00E−01	7.80E−02
JPEG-sample	1.80E−02	3.90E−02	2.70E−02
ARF	6.50E−03	1.70E−02	1.00E−02

where "v" represents the type of signature bits, "v_l" represents the complete signature space indicating the exhaustive possible combinations of signature. The probability for an attacker to successfully guess the exact signature combination corresponding to facial biometric for protecting Trojan-secured designs (Chaurasia and Sengupta, 2022b) and Hardware watermarking (Sengupta and Bhadauria, 2016) is shown in Table 5.21. As evident, for an attacker, guessing of facial biometric signature is more complex and has lesser probability than guessing the watermarking signature.

Further, facial biometric approach employs several security parameters during signature generation unlike the hardware watermarking approach, which makes it impossible for an attacker to evade the piracy detection by regenerating the security constraints. Therefore, facial biometric-based protection of Trojan-secured design offers robust security and ensures the integration of only authentic Trojan-secured SoC designs in the CE systems (by enabling the isolation of pirated ones). Additionally, facial biometric also associates the unique identity of the SoC

Table 5.21   Probability for an attacker to guess the exact signature combination

Facial biometric for protecting Trojan-secured DSP design (Chaurasia and Sengupta, 2022b)			Hardware watermarking (Sengupta and Bhadauria, 2016)		
$v = 2$	l	G_p	$v = 2$	l	G_p
	40	9.10E−13		15	3.00E−05
	47	7.10E−15		30	9.30E−10
	62	2.16E−19		60	8.67E−19
	78	3.30E−24		70	8.47E−22

Table 5.22   Design cost of protecting Trojan-secured design using facial biometric of SoC integrator

DSP benchmarks	No. of storage variable	No. of registers	Design cost of generating the Trojan-secured designs	Design cost of Trojan-secured design embedded with facial biometric signature	% Overhead
DCT	22	8	0.53	0.53	0.0
FIR	62	16	0.39	0.39	0.0
JPEG-sample	90	20	0.64	0.64	0.0
ARF	72	16	0.37	0.37	0.0

integrator with the design unlike watermarking approach. This, therefore, ensures the seamless verification of ownership rights, in case if an adversary from offshore foundry attempts to falsely claim the ownership during fabrication phase.

5.8.2.2   Design cost analysis

The design cost of generating the Trojan-secured SoC and protecting it by embedding the facial biometric signature is shown in Table 5.22. As evident, the embedding of facial biometric signature into Trojan-secured design results in zero design cost overhead. This is because embedding of the facial security constraints corresponding to facial image of the SoC integrator does not incur any overhead in terms of requiring extra registers to accommodate the secret security constraints into the design. Thus, it ensures the robust security of Trojan-secured DSP designs against IP piracy at negligible or zero design cost overhead.

5.9   Conclusion

This chapter discussed the applications of physical biometric for enabling robust security of DSP hardware. In the first application, fingerprint and facial biometric corresponding to IP buyer and seller, respectively, are used for providing the symmetrical protection of their IP rights. The embedded biometric signature into the design corresponding to IP buyer and seller provides robust symmetrical security in terms of lesser probability of coincidence at zero design cost overhead. Further, in the next application, facial biometric is used for protecting Trojan-secured SoCs against IP piracy. The embedded facial signature corresponding to SoC integrator provides: (a) the robust security of Trojan-secured designs in terms of lower probability of coincidence; (b) lower probability of guessing the exact signature combination (security constraints) by an attacker. The security is achieved at zero overhead in terms of design cost. The robustness of physical biometric-based hardware security approach lies in the fact that the naturally unique biometric features (corresponding to facial image and fingerprint) are exploited for generating the robust secret security constraints (digital evidence) for embedding into the target design. Further, unlike non-biometric-based hardware security approaches, an adversary cannot easily compromise, regenerate, or replicate the unique biometric signature to assert false IP rights as well as evade IP piracy detection.

At the end of this chapter, a reader gains the knowledge of the following:

- Applications of physical biometrics for securing DSP hardware designs.
- Importance of HLS for designing secure DSP hardware designs.
- The significance of physical biometric-based hardware security for ensuring robust symmetrical IP core protection.
- Design flow of generating Trojan-secured SoC design.
- Facial biometric-based hardware security methodology for securing Trojan-secured designs, in terms of enabling the detective control against IP piracy.
- Facial and fingerprint biometric digital template generation and their encoding into respective secret hardware security constraints for embedding into the target hardware application.
- Comparative study of different contemporary hardware security approaches used for IP core protection.

5.10   Questions and exercise

1. What is symmetrical protection of IP cores?
2. How is symmetrical IP protection different from non-symmetrical protection?
3. What is Trojan payload?
4. Explain DMR logic.
5. Explain multi-vendor allocation policy.
6. How does multi-vendor allocation policy assist in Trojan detection?
7. What is Trojan security?
8. What is multivariable fingerprinting and how is it different from finger-printing biometric?
9. Explain the threat model in Trojan security.
10. Explain the threat model in IP core protection.

11. Who are the threat actors in symmetrical IP protection?
12. How many types of Trojans are there?
13. Explain at least three applications of biometric signature for hardware security
14. Explain facial biometric pairing for symmetrical IP protection.
15. Explain the encoding rule used in multivariable fingerprinting for protecting IP buyer's right.
16. Discuss the design flow of protecting Trojan-secured SoC designs against piracy.
17. Discuss the responsible factors or security parameters for ensuring robust hardware security.
18. Discuss the embedding of facial biometric signature into Trojan-secured SoC designs.
19. Explain the symmetrical protection corresponding to 4-point DCT application.
20. Discuss the process of generating Trojan-secured SoC corresponding to IIR filter design.

References

CAD for Assurance, IEEE Hardware Security and Trust Technical Committee, https://cadforassurance.org/tools/ip-icprotection/faciometric-hardware-security-tool/, accessed on March 2022.

Multimedia Laboratory Datasets, Available: http://mmlab.ie.cuhk.edu.hk/datasets.html, last accessed on June 2020.

15 nm open cell library. [Online], Available: https://si2.org/open-cell-library/, last accessed on January 2020.

Alilou, V. K. (December 2019). FingerPrint Matching: A simple approach MATLAB Central File Exchange. [Online]. Available: https://www.mathworks.com/matlabcentral/fileexchange/44369-fingerprint-matching-a simple-approach.

Alkabani, Y., F. Koushanfar, and M. Potkonjak (2007), "Remote activation of ICs for piracy prevention and digital right management," in: *Proceedings of IEEE/ACM International Conference of Computer-Aided Design*, November 2007, pp. 674–677.

Arafin, M. T., A. Stanley, and P. Sharma (2017), "Hardware-based anticounterfeiting techniques for safeguarding supply chain integrity," in: *2017 IEEE International Symposium on Circuits and Systems (ISCAS)*, 2017, pp. 1–4.

Bao, C. D. Forte, and A. Srivastava (2016), "On reverse engineering-based hardware trojan detection," *IEEE Transactions on Computer-Aided Design of Integrated Circuits and Systems*, vol. 35, no. 1, pp. 49–57.

Castillo, E., U. Meyer-Baese, A. García, L. Parrilla, and A. Lloris (2007), "IPP@HDL: efficient intellectual property protection scheme for IP cores," *IEEE Transactions on VLSI Systems*, vol. 15, no. 5, pp. 578–591.

Chaurasia, R. and A. Sengupta (2022a), "Symmetrical protection of ownership right's for IP buyer and IP vendor using facial biometric pairing," in: *2022 IEEE International Symposium on Smart Electronic Systems (iSES)* Warangal, India, pp. 272–277.

Chaurasia, R. and A. Sengupta (2022b), "Protecting Trojan secured DSP cores against IP piracy using facial biometrics," in: *2022 IEEE 19th India Council International Conference (INDICON)* Kochi, India, pp. 1–6.

Colombier, B. and L. Bossuet (2015), "Survey of hardware protection of design data for integrated circuits and intellectual properties," *IET Computers & Digital Techniques*, vol. 8, no. 6, pp. 274–287.

Cui, A. and C. Chang (2007), "Watermarking for IP protection through template substitution at logic synthesis level," in: *Proceedings of ISCAS*, New Orleans, LA, 2007, pp. 3687–3690.

Hong, I. and M. Potkonjak (1999), "Behavioral synthesis techniques for intellectual property security," in: *Proceedings 1999 Design Automation Conference* New Orleans, LA, pp. 849–854.

Koushanfar, F., I. Hong, and M. Potkonjak (2005), "Behavioral synthesis techniques for intellectual property protection," in: *ACM Transactions on Design Automation of Electronic Systems*, vol. 10, no. 3, pp. 523–545.

Koushanfar, F. (2012), "Hardware metering: a survey," in: *Introduction to Hardware Security and Trust*, New York, NY: Springer, pp. 103–122.

Koushanfar, F., S. Fazzari, C. McCants, *et al.* (2012), "Can EDA combat the rise of electronic counterfeiting?," in: DAC Design Automation Conference, San Francisco, CA, 2012, pp. 133–138.

Lach, J., W.H. Mangione-Smith, and M. Potkonjak (2001), "Fingerprinting techniques for field-programmable gate array intellectual property protection," *IEEE Transactions on Computer-Aided Design of Integrated Circuits and Systems*, vol. 20, no. 10, pp. 1253–1261.

Mohanty, B. K. and P. K. Meher (2016), "A high-performance FIR filter architecture for fixed and reconfigurable applications," *IEEE Transactions on VLSI Systems*, vol. 24, no. 2, pp. 444–452.

Newbould, R. D., J. D. Carothers, and J. J. Rodriguez (2002), "Watermarking ICs for IP protection," *Electronics Letters*, vol. 38, no. 6, pp. 272–274.

Ni, M. and Z. Gao (2005), "Detector-based watermarking technique for soft IP core protection in high synthesis design level," in: *Proceedings of CCS*, Hong Kong, 2005, pp. 1348–1352.

Nie, T., L. Zhou, and Y. Li (2013), "Hierarchical watermarking method for FPGA IP protection," *IETE Technical Review,* vol. 30, no. 5, pp. 367–374.

Pilato, C., S. Garg, K. Wu, R. Karri, and F. Regazzoni (2018), "Securing hardware accelerators: a new challenge for high-level synthesis," *IEEE Embedded Systems Letters*, vol. 10, no. 3, pp. 77–80.

Plaza, S. M. and I. L. Markov (2015), "Solving the third-shift problem in IC piracy with test-aware logic locking," *IEEE Transactions on Computer-Aided Design of Integrated Circuits and Systems*, vol. 34, no. 6, pp. 961–971.

Rai, S., A. Rupani, P. Nath, and A. Kumar (2019), "Hardware watermarking using polymorphic inverter designs based on reconfigurable nanotechnologies," in: *2019 IEEE Computer Society Annual Symposium on VLSI (ISVLSI)*, 2019, pp. 663–669.

Rathor, M. and A. Sengupta (2020), "IP core steganography using switch based key-driven hash-chaining and encoding for securing DSP kernels used in CE systems," *IEEE Transactions on Consumer Electronics*, vol. 66, no. 3, pp. 251–260.

Roy, D. and A. Sengupta (2017), "Low overhead symmetrical protection of reusable IP core using robust fingerprinting and watermarking during high level synthesis", *Elsevier Journal of Future Generation Computer Systems*, vol. 71, pp. 89–101.

Roy, D. and A. Sengupta (2018), "Reusable intellectual property core protection for both buyer and seller", in: *Proceedings of 36th IEEE International Conference on Consumer Electronics (ICCE) 2018*, Las Vegas, January 2018, pp. 1–3, doi:10.1109/ICCE.2018.8326059.

Roy, J. A., F. Koushanfar, and I. L. Markov (2008), "EPIC: ending piracy of integrated circuits," in: *2008 Design, Automation and Test in Europe*, pp. 1069–1074.

Schneiderman, R. (2010), "DSPs evolving in consumer electronics applications," *IEEE Signal Processing Magazine*, vol. 27, no. 3, pp. 6–10.

Sengupta, F. and M. Rathor (2019), "IP core steganography for protecting DSP kernels used in CE systems," in: *IEEE Transactions on Consumer Electronics*, vol. 65, no. 4, pp. 506–515.

Sengupta, A. (2020), Frontiers in Securing IP Cores – Forensic Detective Control and Obfuscation Techniques, *The Institute of Engineering and Technology (IET)*, ISBN-10: 1-83953-031-6, ISBN-13: 978-1-83953-031-9.

Sengupta, A. and S. Bhadauria (2016), "Exploring low cost optimal watermark for reusable IP cores during high level synthesis," *IEEE Access*, vol. 4, pp. 2198–2215.

Sengupta, A. and S. Mohanty (2019), "Symmetrical protection of DSP IP core and integrated circuits using fingerprinting and watermarking", in: *IET Book: IP Core Protection and Hardware-Assisted Security for Consumer Electronics*, USA: The Institution of Engineering and Technology (IET), ISBN: 978-1-78561-799-7, e-ISBN: 978-1-78561-800-0.

Sengupta, A. and D. Kachave (2018), "Forensic engineering for resolving ownership problem of reusable IP core generated during high level synthesis", *Elsevier Journal of Future Generation Computer Systems*, vol. 80, pp. 29–46.

Sengupta, A. and M. Rathor (2019a), "Crypto-based dual-phase hardware steganography for securing IP cores," *IEEE Letters of the Computer Society*, vol. 2, no. 4, pp. 32–35.

Sengupta, A. and M. Rathor (2019b), "IP core steganography for protecting DSP kernels used in CE systems," *IEEE Transactions on Consumer Electronics*, vol. 65, no. 4, pp. 506–515.

Sengupta, A. and M. Rathor (2020), "Securing hardware accelerators for CE systems using biometric fingerprinting," *IEEE Transactions on VLSI Systems*, vol. 28, pp. 1979–1992, doi: 10.1109/TVLSI.2020.2999514.

Sengupta, A. and M. Rathor (2021), "Facial biometric for securing hardware accelerators," *IEEE Transactions on VLSI Systems,* vol. 29, no. 1, pp. 112–123. doi: 10.1109/TVLSI.202 0.3029245.

Sengupta, A. and D. Roy (2018), "Reusable intellectual property core protection for both buyer and seller," in: *2018 IEEE International Conference on Consumer Electronics (ICCE)*, 2018, pp. 1–3, doi:10.1109/ICCE.2018.8326059.

Sengupta, A., S. Bhadauria, and S. P. Mohanty (2017), "TL-HLS: methodology for low cost hardware trojan security aware scheduling with optimal loop unrolling factor during high level synthesis," *IEEE Transactions on Computer-Aided Design of Integrated Circuits and Systems*, vol. 36, no. 4, pp. 655–668.

Sengupta, A., E. R. Kumar, and N. P. Chandra (2019), "Embedding digital signature using encrypted-hashing for protection of DSP cores in CE," *IEEE Transactions on Consumer Electronics*, vol. 65, no. 3, pp. 398–407.

Wang, X., Y. Zheng, A. Basak, and S. Bhunia (2015), "IIPS: infrastructure IP for secure SoC design," *IEEE Transactions on Computers*, vol. 64, no. 8, pp. 2226–2238.

Wong, J. L., D. Kirovski, and M. Potkonjak (2004), "Computational forensic techniques for intellectual property protection," *IEEE Transactions on Computer-Aided Design of Integrated Circuits and Systems*, vol. 23, no. 6, pp. 987–994.

Yang, J., Y. Zhang, Y. Hua, J. Yao, Z. Mao, and X. Chen (2021), "Hardware Trojans detection through RTL features extraction and machine learning," in *2021 Asian Hardware Oriented Security and Trust Symposium (AsianHOST)*, 2021, pp. 1–4.

Ziener, D. and J. Teich (2006), "FPGA core watermarking based on power signature analysis," in: *Proceedings of IEEE International Conference of Field Program. Technol. (FPT)*, December 2006, pp. 205–212.

Ziener, D. and J. Teich (2008), "Power signature watermarking of IP cores for FPGAs," *Journal of Signal Processing Systems*, vol. 51, no. 1, pp. 123–136.

Chapter 6

Palmprint biometrics vs. fingerprint biometrics vs. digital signature using encrypted hash: qualitative and quantitative comparison for security of DSP coprocessors

Anirban Sengupta[1] and Aditya Anshul[1]

Security of data and computation-intensive application-specific integrated circuits (ASICs) or reusable intellectual property cores is essential because of their wide applicability in the modern digital ecosystem. Integration of these ASICs while designing a complex computation system helps in the parallel and smooth execution of data and computation-intensive functions (such as image pixel computations and digital data filtering). This chapter presents a discussion and qualitative comparative analysis between three different recent hardware security methodologies, i.e., fingerprint-based hardware security methodology, palmprint-based hardware security methodology, and digital signature using encrypted hash-based digital signature methodologies, respectively. Moreover, this chapter also describes the design flow of the security methodologies in detail. Further, this chapter analyzes the security strength of the security methodologies using security metrics such as tamper tolerance and probability of coincidence.

6.1   Introduction

The evolution of a rapidly developing modern society can only be realized with the integration of various ubiquitous computing devices such as mobile phones, smartwatches, tablets, sensors (IoT), and digital cameras. These computing devices are intended to perform different data-intensive functions such as compression and decompression of files (audio, video, and image files), digital data filtering (noise removal from data), and various complex mathematical computations on digital data. Nowadays, they are used almost everywhere, from number plate recognition and automation at tolls, advanced military operations, biometric fingerprinting, automation of vehicle driving, and robotics to medical imaging systems and some real-time mission-critical applications. The above-mentioned

[1]Department of Computer Science and Engineering, Indian Institute of Technology Indore, India

computing devices contain various ASICs or reusable intellectual property (IP) cores in the form of digital signal coprocessors (DSP), which are intended to perform a particular function. Some examples of these DSP applications are the following: finite (FIR) and infinite (IIR) impulse response filters (commonly used in audio video systems for signal attenuation), discrete cosine transform (DCT), and Haar wavelet transform (HWT) (widely used for lossy compression of image, audio, and video files), fast Fourier transform (FFT) (commonly used in the field of digital video broadcasting), joint photographic expert group (JPEG) (widely used in digital camera systems for image and video compression), etc. These DSP applications are designed as a dedicated reusable IP core using a high-level synthesis (HLS) framework (Pilato *et al.*, 2018; Schneiderman, 2010).

The globalization of the chip design industry has introduced the involvement of multiple offshore entities (different third-party IP vendors) in designing a particular reusable IP core. This happens because some places can afford greater technical skills at lower cost while some cheaper labor cost in addition to the need for design cost optimization and meeting time-to-market demands. But the involvement of multiple third-party vendors in designing IP cores makes it susceptible to various hardware security threats and attacks, such as IP piracy (counterfeiting and cloning), hardware trojan insertion that affects functionality, and fraudulent claim of IP ownership. Pirated IPs are unreliable as they have not gone through rigorous testing compared to genuine ones. Further, they can cause severe damage, such as producing incorrect output (can be proved fatal also in the case of wrong medical diagnosis in case of medical imagining), leaking sensitive information (threat to end consumer), causing excessive heating, and tarnishing the esteem and reputation of both IP vendor and designer. Therefore, detecting and segregating pirated IP cores from authentic ones is crucial before integrating them into the final electronic and computing devices. The next paragraph discusses different hardware security techniques for detective control of pirated IP cores before integration into the system-on-chip of such electronic/computing devices (Colombier and Bossuet, 2015; Yasin *et al.*, 2016).

The different methodologies in the field of hardware security or IP protection (IPP) include hardware watermarking (Le Gal and Bossuet, 2012; Koushanfar *et al.*, 2005; Sengupta and Bhadauria, 2016), hardware steganography (Sengupta and Rathore, 2019), digital signature-based approach (Sengupta *et al.*, 2019), and various biometric-based security approaches (Chaurasia *et al.*, 2022; Sengupta and Rathor, 2020; Sengupta *et al.*, 2021). The approaches mentioned are based on a detective control mechanism, indicating that they embed some authentic secret information into the IP core's design file. The same authentic secret information is used for detecting pirated or counterfeited IP cores from genuine ones and determining the correct ownership of IP. The hardware watermarking mechanism involves several signature variables, their combination/sizes, and mapping rules to generate hardware security signatures and constraints. Similarly, hardware steganography exploits the register allocation table (RAT) of the DSP application itself along with the IP designer-selected threshold entropy value to generate hardware security constraints, which are further embedded into the design of the DSP IP core. Both approaches become vulnerable if the respected signature variables, combination/size, mapping rule, and threshold entropy value get compromised. Moreover, watermarking and steganography-based security approach does not integrate a natural identity with the DSP IP core. Similarly, digital signature using encrypted hash methodology uses SHA-512 and RSA cryptosystem to generate a security signature. Despite providing greater security than hardware watermarking and steganography, the digital signature-based approach also fails to integrate a natural identity with the DSP IP core and becomes vulnerable in case of a compromised RSA key. Integrating IP vendors/designer's natural identity with the IP core enhances the robustness of the security. In such a scenario, in such a scenario, it is very hard or impossible for the adversary to pirate or fraudulently claim ownership of the IP core. Therefore, the recent biometric-based hardware security approaches support the integration of the IP vendor's and designer's natural identity with the DSP IP core and provide more robust security compared to hardware watermarking, steganography, and digital signature-based approach in terms of greater tamper tolerance and lower probability of coincidence.

This chapter discusses a qualitative and quantitative comparison between palmprint biometrics (Chaurasia *et al.*, 2021; Sengupta *et al.*, 2021), fingerprint biometrics (Sengupta and Rathor, 2020), and a digital signature-based security approach (Sengupta *et al.*, 2019) in terms of hardware security. The fingerprint and palmprint biometrics methodologies involve various parameters such as the selection of grid size and spacing (explained in detail in Sections 6.3.2 and 6.4.1), generation of minutiae and nodal points on fingerprint and palmprint image, respectively, selection and concatenation order of generated features, and mapping rules (to generate hardware security constraints from the signature). Therefore, if an adversary gets access to a biometric image, it will be impossible for him/her to generate an exact signature (or hardware security constraints) without having complete knowledge of the above-mentioned parameters. Further, as biometrics are the unique natural identity of a person, it will provide complete immunity from the fraudulent claim of IP ownership. Figure 6.1 illustrates the overview of biometric-based hardware security methodologies.

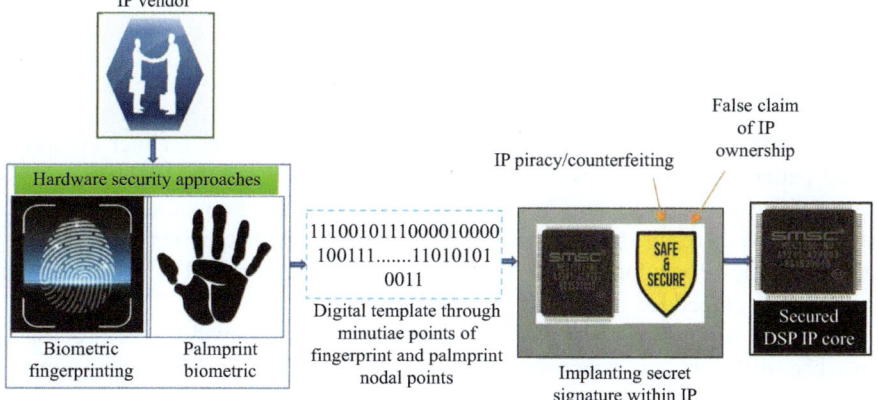

Figure 6.1   Overview of the biometric-based hardware security approach

6.2   Threat model

The reusable IP cores from third-party IP vendors may not always be trustworthy. It is crucial to isolate counterfeited IPs from authentic IP cores to ensure the integration of only genuine ones in electronic computing systems at the SoC integrator level. The counterfeited IPs may contain malicious logic as they are not rigorously tested and may cause reliability and safety hazard (Rizzo *et al.*, 2019). Therefore, the detection of such pirated IP cores becomes essential. Further, an adversary may fraudulently claim IP ownership. Therefore, it is also important to ensure the protection of IP vendor rights to nullify the false claim of IP ownership (Islam *et al.*, 2020).

6.3   Fingerprint biometric for IPP of DSP coprocessors:

6.3.1   Summary of fingerprint biometric

The fingerprint-based hardware security approach (Sengupta and Rathor, 2020) accepts an IP vendor fingerprint (captured through an optical scanner) along with the control data flow graph (CDFG) of the respective DSP application and yields a fingerprint biometric secured DSP IP core design. The secured DSP IP core is immune to hardware threats such as piracy and fraudulent claim of IP ownership. At first, a signature (digital template) is generated using captured fingerprint and its respective fingerprint biometric algorithm. Subsequently, the generated signature is converted into its respective hardware security constraints based on IP vendor-selected encoding/embedding rules. Next, the CDFG is scheduled based on IP vendor-selected resource constraints and the LIST scheduling algorithm to generate the DSP application's scheduled data flow graph (SDFG). Further, their corresponding RAT (RAT) and colored interval graph (CIG) are generated using the SDFG of the DSP application. Subsequently, the security constraints corresponding to the fingerprint biometric-based signature is embedded into the design via the CIG framework of HLS. The details of the fingerprint biometric-based hardware security approach are discussed in detail in the next subsection (Sengupta and Rathor, 2020).

6.3.2   Details of fingerprint biometric methodology

The IP vendor's fingerprint captured with the aid of an optical scanner comprises distinct patterns called valleys and ridges. Minutiae points are the bifurcation point where a ridge line divides into branch ridges, and they are known as characteristic features of a fingerprint image. Therefore, the unique minutiae points corresponding to the authentic IP vendor's respective fingerprint image are used as a security feature, which is embedded into the IP core design file (Sengupta and Rathor, 2020).

First, the captured fingerprint biometric image is subjected to pre-processing step. The different pre-processing steps involved are (a) FFT enhancement, (b) binarization, and (c) thinning. All these pre-processing steps remove unnecessary noise present in fingerprint biometric images and increase the captured image's quality. The thinned fingerprint biometric image is used to extract the minutiae points, which is further used to generate the covert hardware security constraints for embedding. Figure 6.2 depicts an IP vendor fingerprint's binarized and thinned image with its respective generated minutiae points (Sengupta and Rathor, 2020).

Further, the fingerprint biometric image containing minutiae points is subjected to IP vendor-selected grid size and spacing to determine the coordinate values of a particular minutiae point. There are four parameters corresponding to a single minutiae point: (i) x-axis coordinate, (ii) y-axis coordinate, (iii) ridge angle (clockwise angle from the horizontal axis), and (iv) minutiae type. For example, ridge ending minutiae points are denoted with $c=1$, and ridge bifurcation minutiae points are denoted with $c=3$. Table 6.1 shows the respective values of four parameters corresponding to different generated minutiae points of the IP vendor fingerprint image (shown in Figure 6.2). All the parameters values are determined for all present minutiae points (however, for demonstration, only six minutiae points

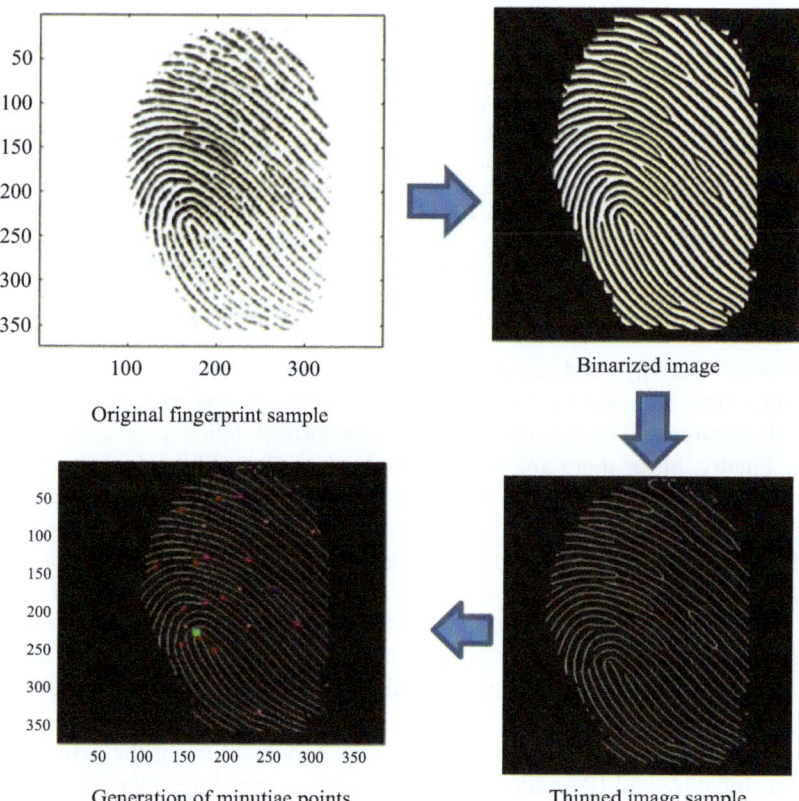

Figure 6.2   Different steps of the fingerprint biometric approach to generate minutiae points

Table 6.1   Binary representation of all parameters corresponding to each minutiae points in IP vendor-selected concatenation fashion

No.	x	y	Minutiae type number	Angle in degree	Binary	Bits
1	216	46	3	29	110110001011101111101	21
2	190	49	1	205	10111110110001111001101	23
3	146	64	1	187	100100101000000110111011	24
4	247	80	1	40	1111011110100001101000	22
5	173	86	1	21	101011011010110110101	21
6	302	93	1	48	10010111010111011110000	23

Table 6.2   Determined values of four parameters corresponding to different minutiae points of fingerprint

No.	x	y	Minutiae type number	Minutiae type name	Minutiae type color	Angle in radian	Angle in degree
1	216	46	3	Ridge bifurcation	Pink	0.503045247761871	28.8
2	190	49	1	Ridge ending	Red	3.58273130505241	205.3
3	146	64	1	Ridge ending	Red	3.26835482366637	187.3
4	247	80	1	Ridge ending (short ridge)	Red (brown)	0.700240286597619	40.1
5	173	86	1	Ridge ending (short ridge)	Red (brown)	0.366566654990681	21.0
6	302	93	1	Ridge ending (short ridge)	Red (brown)	0.837181054258808	48.0

are shown for brevity) and the decimal value is further converted into their corresponding binary equivalents. Table 6.2 depicts the binary equivalents (in IP vendor-selected concatenated order) of four parameters corresponding to different minutiae points. Finally, all of them are concatenated as per IP vendor-selected concatenation rule to generate fingerprint biometric-based digital signature template. Based on the concatenation order selected, the obtained digital template is: "*11011000101110111110110110111110110001111001101100100101000000110111 0111111011110100001101000101011011010101101101010110010111010111011110 000.*" There are a total of 134 digits in the above palmprint signature template. However, we have only used 68 digits for demonstration while embedding (explained in the next sub-section), with the *first forty 0's* and *twenty-eight 1's* as the signature bits.

Next, the hardware security constraints are generated using the generated fingerprint signature and IP vendor-selected constraint encoding/embedding rule:

An artificial edge between <even, even> storage variables (nodes) pair in CIG is added if the bit is "0." Otherwise, an edge is added between <odd, odd> storage variable pairs (Sengupta and Rathor, 2020).

Thus, obtained 68 hardware security constraints corresponding to 28 0s and 40 1s are: <*M0, M2*>, <*M0, M4*>, <*M0, M6*>, <*M0, M8*>, <*M0, M10*>, <*M0, M12*>, <*M0, M14*>, <*M0, M16*>, <*M0, M18*>, <*M0, M20*>, <*M0, M22*>, <*M0, M24*>, <*M0, M26*>, <*M2, M4*>, <*M2, M6*>, <*M2, M8*>, <*M2, M10*>, <*M2, M12*>, <*M2, M14*>, <*M2, M16*>, <*M2, M18*>, <*M2, M20*>, <*M2, M22*>, <*M2, M24*>, <*M2, M26*>, <*M4, M6*>, <*M4, M8*>, <*M4, M10*> and <*M1, M3*>, <*M1, M5*>, <*M1, M7*>, <*M1, M9*>, <*M1, M11*>, <*M1, M13*>, <*M1, M15*>, <*M1, M17*>, <*M1, M19*>, <*M1, M21*>, <*M1, M23*>, <*M1, M25*>, <*M3, M5*>, <*M3, M7*>, <*M3, M9*>, <*M3, M11*>, <*M3, M13*>, <*M3, M15*>, <*M3, M17*>, <*M3, M19*>, <*M3, M21*>, <*M3, M23*>, <*M3, M25*>, <*M5, M7*>, <*M5, M9*>, <*M5, M11*>, <*M5, M13*>, <*M5, M15*>, <*M5, M19*>, <*M5, M21*>, <*M5, M23*>, <*M5, M25*>, <*M7, M9*>, <*M7, M11*>, <*M7, M13*>, <*M7, M15*>, <*M7, M17*>, <*M7, M19*>, <*M7, M21*>.

Finally, all the generated covert security constraints are embedded into the register allocation design of the DSP application using general HLS framework. The target DSP application's register allocation information (represented as CIG) is harnessed to embed extra artificial edges (security constraints). For example, if the added edge is between the same-colored storage variables (register), then either the colors of the register are swapped, or a new colored register is allocated to accommodate the security constraints. In the end, a secured DSP IP core register transfer level (RTL) datapath design is produced as the output, which contains covert security constraints as invisible digital evidence (Sengupta and Rathor, 2020).

6.3.3 Embedding of fingerprint biometric signature on FIR filter

Figure 6.3 depicts the scheduled DFG (using IP vendor-selected resource constraints) of the FIR filter which also shows the register allocation of storage variables (M0–M26). The target DSP IP is scheduled using two multipliers and one adder (IP vendor selected). Similarly, Figure 6.4 illustrates the modified register allocation of storage variables in the SDFG of FIR filter after embedding fingerprint biometric-based hardware security constraints. Table 6.3 depicts the RAT of storage variables of FIR before (with black and indigo color) and after (with black and red color) implanting fingerprint biometric signature. After generating the secret security constraints (hardware) from the fingerprint biometric signature, these constraints are embedded into the target DSP IP design during the HLS process. A colored interval graph (CIG) framework corresponding to the respective design is used to implant the additional constraints. These extra security constraints are embedded in the form of additional edges into the CIG. Figures 6.5 and 6.6 illustrate the CIG of the FIR filter before and after embedding secret hardware constraints (fingerprint-based). Additional implanted edges (constraints) are shown with red-colored edges. The security constraints are mentioned in Section 6.3.2, corresponding to the 40 zeros and 28 ones, and they are embedded as additional edges into the CIG of the FIR filter.

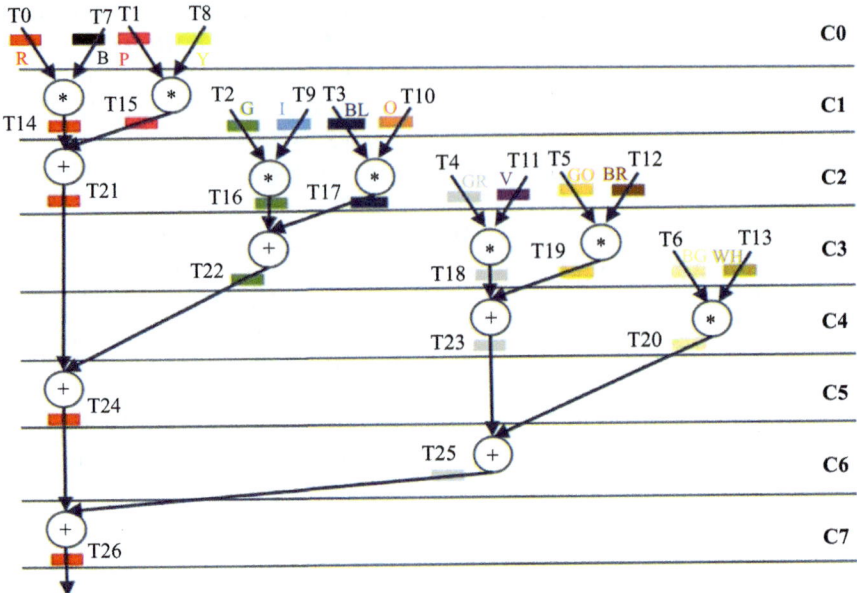

Figure 6.3   Schedule data flow graph 6-order FIR using IP vendor-selected one adder and two multiplier

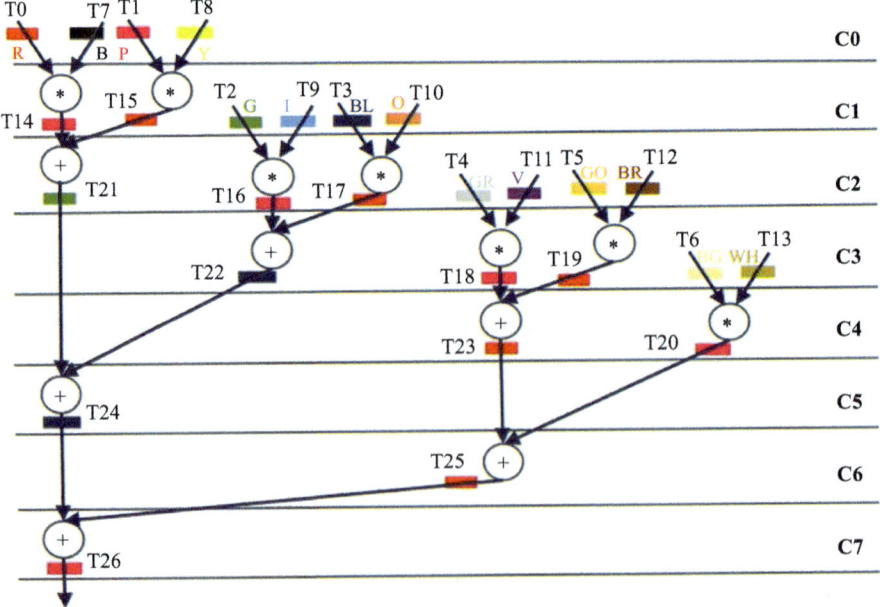

Figure 6.4   Schedule data flow graph 6-order FIR using IP vendor-selected one adder and two multiplier after embedding fingerprint biometric-based hardware security constraints

Table 6.3 RAT before and after embedding hardware security constraints corresponding to sharpening filter

		C0	C1	C2	C3	C4	C5	C6	C7
Red(R)	M0	M14/M15	M21/M17	M21/M19	M21/M23	M24/M23	M24/M25	M26	
Pink (P)	M1	M15/M14	−/M16	−/M18	−/M20	−/M20	−	M26	
Green (G)	M2	M2	M16/M21	M22/M21	M22/M21	−/M24	−	−	
Blue (BL)	M3	M3	−	−/M22	M22	−	−	−	
Gray (G)	M4	M4	M4	M18	M23	M23	M25/M24	−	
Gold (GO)	M5	M5	M5	M19	−	−	−	−	
Beige(BG)	M6	M6	M6	M6	M20	M20	−	−	
Black (B)	M7	−	−	−	−	−	−	−	
Yellow (Y)	M8	−	−	−	−	−	−	−	
Indigo (I)	M9	−	−	−	−	−	−	−	
Orange (O)	M10	−	−	−	−	−	−	−	
Violet (V)	M11	−	−	−	−	−	−	−	
Brown (BR)	M12	−	−	−	−	−	−	−	
Wheat (WH)	M13	−	−	−	−	−	−	−	

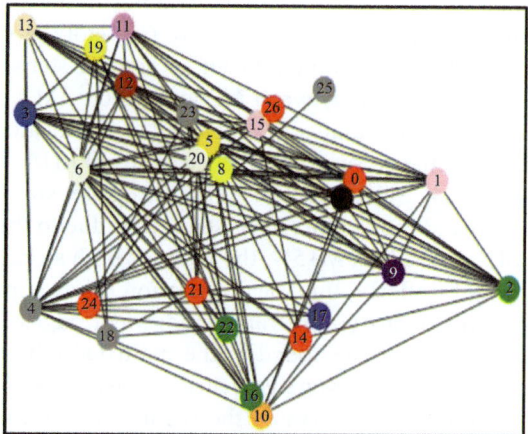

Figure 6.5 CIG of FIR before embedded fingerprint-based hardware security constraints

While embedding, it can be possible that these extra constraints (additional edges) are added between the nodes having the same color. However, we cannot embed an edge between two storage variables (nodes) of similar color. An edge between the same color nodes shows that both nodes (storage variables) are assigned to be implemented on the same register (color) within the same control step, which is not practically possible. Therefore, two methods are followed to resolve the raised conflicts: (i) first, local alteration (swapping) between the register allocation of storage variables is performed; for example, let us compare

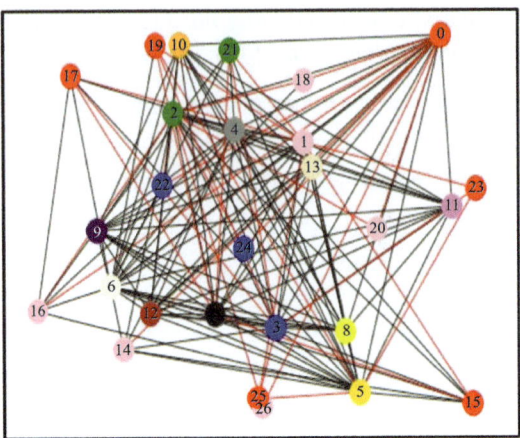

Figure 6.6   CIG of FIR after embedded fingerprint-based hardware security constraints

the register allocation information of storage variables in the SDFG of FIR filter from Figures 6.3 and 6.4, respectively. In Figure 6.3, storage variables M2 and M16 are allocated on the green color register in control steps 1 and 2, respectively. But, due to an extra edge between M2 and M16 (<M2, M16>) obtained from hardware security constraints, M16 cannot be allocated on the green-colored register. So, now the color of storage variables of M16 is changed to pink from green in Figure 6.4. Subsequently, M14, and M15 storage variables are earlier allocated to red and pink colored registers. But, due to the additional edge between <M0, M14> and < M1, M15>, the color of both storage variables M14 and M15 are swapped. Similarly, this local alteration (or swapping) corresponds to each embedded extra security constraint (additional edges). (ii) Sometimes, different registers (color) must be added in the modified CIG of the respective DSP IP in order to satisfy the hardware security constraints. This occurs when local alteration (swapping) cannot satisfy the security constraints. Similarly, all the above-mentioned secret hardware security constraints are embedded using these two rules. Finally, a modified RAT, post embedding of fingerprint biometric signature, is obtained corresponding to the target DSP IP core. Subsequently, RTL design is produced with the embedded fingerprint biometric signature after data path synthesis of HLS (Sengupta and Rathor, 2020).

6.3.4   Detection and validation of fingerprint biometrics for detective control against IP piracy and nullifying fraud claim of IP ownership

The detection and validation of fingerprint biometrics for detective control against IP piracy and nullifying fraud claim of IP ownership requires a fingerprint biometric image of an authentic IP vendor and fingerprint signature generation algorithm. This is because, alongside the fingerprint biometric image, the following security parameters are also required for regenerating the exact signature (to validate and prove ownership): (a) grid size and spacing, (b) generation of minutiae points, (c) minutiae coordinates, (d) concatenation order of features, and (e) constraints embedding rules (Sengupta and Rathor, 2020).

During the detection process and ownership proof, the fingerprint signature template is regenerated using the existing (pre-captured saved) fingerprint biometric image and fingerprint biometric signature generation algorithm. Subsequently, hardware security constraints are generated corresponding to regenerated fingerprint biometric signature. Further, storage variable register allocation information is extracted using IP core RTL datapath (IP design under examination) and the generated hardware security constraints are matched with it. In case of complete matching, the design is considered authentic; otherwise, the design may be a counterfeit (due to the absence of a genuine IP vendor's fingerprint signature).

Moreover, it is impossible for the adversary to fraudulently claim ownership of the IP, as they cannot match their signature with the embedded hardware security constraints of a genuine IP vendor's fingerprint biometric signature. However, a genuine IP vendor can easily match their fingerprint signature and the embedded storage variable register allocation information extracted using RTL datapath of the IP under examination (Sengupta and Rathor, 2020).

6.4   Palmprint biometric for IPP of DSP coprocessors

6.4.1   Summary of approach

At first, a unique palmprint image of the original IP vendor (representing its natural unique identity) is captured with the help of a high-resolution (quality) digital camera to generate its respective palmprint signature (Chaurasia *et al.*, 2022; Sengupta *et al.*, 2021). Second, hardware security constraints are generated by mapping palmprint signature to its equivalent hardware security constraints based on the IP vendor-specified mapping rule. Finally, the obtained hardware security constraints are added/implanted into the target DSP IP core design during HLS. During HLS, the target DSP IP core's algorithmic representation (such as high-level C/C++ code) is transformed into its scheduled and hardware allocation design based on IP vendor-selected resource architecture (constraints) and module library. Then, the previously obtained palmprint signature is implanted into the schedule design during the register allocation phase of the HLS framework, which yields a palmprint signature embedded RTL design of the target DSP IP core. The embedded authentic palmprint signature-based design makes it easy to differentiate between an original and a counterfeited IP during the counterfeit detection process. The details of counterfeit detection process are explained in Section 6.4.3 (Chaurasia *et al.*, 2022).

The palmprint-based approach (Chaurasia *et al.*, 2022) offers the following advantages:

(a) The used palmprint signature (obtained from the original IP vendor) is unique. It provides an authentic secret mark to the target DSP IP core used during the IP counterfeit detection.
(b) Any extremely reliable person in the IP vendor's firm can be carefully selected to obtain the palmprint biometric signature. The process is contactless as the stored image of palmprint can be used for detection. So, if, unfortunately, that person meets with an accident or leaves the company, then it does not affect the overall authentication process.
(c) Extraction of palmprint signature constraints is easier than facial biometric.
(d) The palmprint biometric-based approach provides a more significant variation in features than the facial biometric technique, resulting in a model with a higher tamper tolerance and lower probability of coincidence.
(e) Additionally, the palmprint-based approach has various benefits over state-of-the-art: the palmprint methodology is contact-less, non-vulnerable, non-replicable, unlike stego-constraints and digital signature, and is independent of secret key also.

The complete palmprint biometric-based hardware security technique is depicted in Figure 6.7.

The steps involved are (Sengupta *et al.*, 2021):

(i) At first, the palmprint image is captured with the help of a high-resolution camera, and the captured image is subjected to grid size and spacing. Grid size and spacing are required to create precise nodal points that facilitate seamless authentication, as these spacing and grid sizes will help regenerate the exact nodal points. As this approach is contactless, there is no need to use an optical scanner. The image of the palmprint used for detection/validation is the same one used for generating the original palmprint signature-based hardware security constraints.

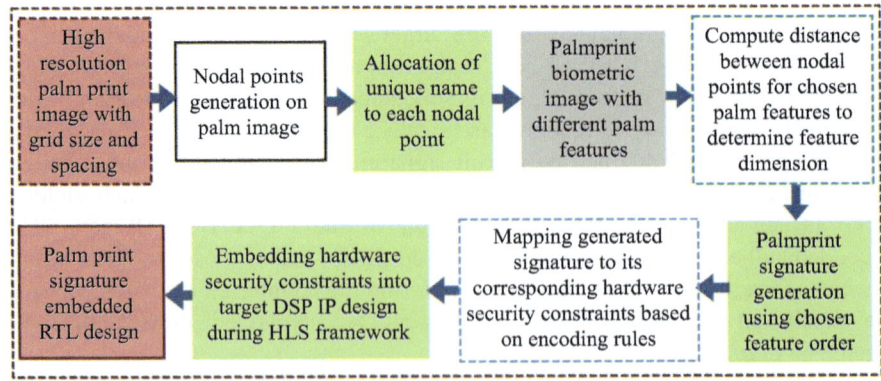

Figure 6.7   Flow chart of palmprint biometric approach

(ii)   The palmprint feature set and nodal points are now determined. Every feature and the nodal point is assigned a unique name for seamless mapping. Now, IP vendors can quickly identify particular palmprint characteristics with the help of a unique identifier/number. For example, the DTF feature has two nodal points, P9 and P22. A total of 25 nodal points are generated on the palmprint biometric image.

(iii) After generating nodal points and feature sets, the dimension corresponding to every feature set is computed with the help of obtained coordinates (a1, b1) and (a2, b2) of the respective nodal points of each generated feature. Manhattan distance is used to compute the dimension of a straight-line feature, and the Pythagoras theorem is used for the inclined line feature. The decimal dimension value is then converted into its binary equivalent. Table 6.4(a) shows the feature dimension and its equivalent binary representation corresponding to each feature. Out of 19 features generated, only 5 features have been used here for the demonstration. All 25 nodal points and 19 generated features have been highlighted in Figure 6.8.

(iv)   The concatenation of the palmprint features is chosen to produce the final secret palmprint signature. All binary equivalents are also concatenated, similar to the palmprint feature. Let the chosen order of concatenation be: "DL+DHL+WP+LP+DFF," where "+" indicates the concatenation operator. Based on IP vendor-selected concatenation order, the obtained digital template is: "*1000011011.10000010100011110111110000001.100110011001 1001101100111100.010110011001100110111111100101.110101110000101 0011101101.1100110011001001101.*" There are a total of 145 digits in the above palmprint signature template. However, we have only used 81 digits: the first 41 0s and 40 1s as the signature. We have not considered the binary points in our signature.

(v)   The obtained digital template is mapped to equivalent hardware security constraints using IP vendor-selected mapping rule given in Table 6.4(b). Thus, obtained 81 hardware security constraints corresponding to 41 0s and 40 1s are: $<M0, M2>$, $<M0, M4>$, $<M0, M6>$, $<M0, M8>$, $<M0, M10>$, $<M0, M12>$, $<M0, M14>$, $<M0, M16>$, $<M0, M18>$, $<M0, M20>$, $<M0, M22>$, $<M0, M24>$, $<M0, M26>$, $<M2, M4>$, $<M2, M6>$, $<M2, M8>$, $<M2, M10>$, $<M2, M12>$, $<M2, M14>$, $<M2, M16>$, $<M2, M18>$, $<M2, M20>$,

Table 6.4(a)   Feature measurement of chosen palmprint feature and their corresponding binary representation

Feature #	Feature name	Feature measurement	Binary representation
X1	DL	539.51	1000011011.1000001010001111011
X2	DHL	385.6	110000001.100110011001001101
X3	WP	636.35	1001111100.010110011001101101
X4	LP	485.84	111100101.1101011100001001
X5	DFF	219.8	11011011.1100110011001001101

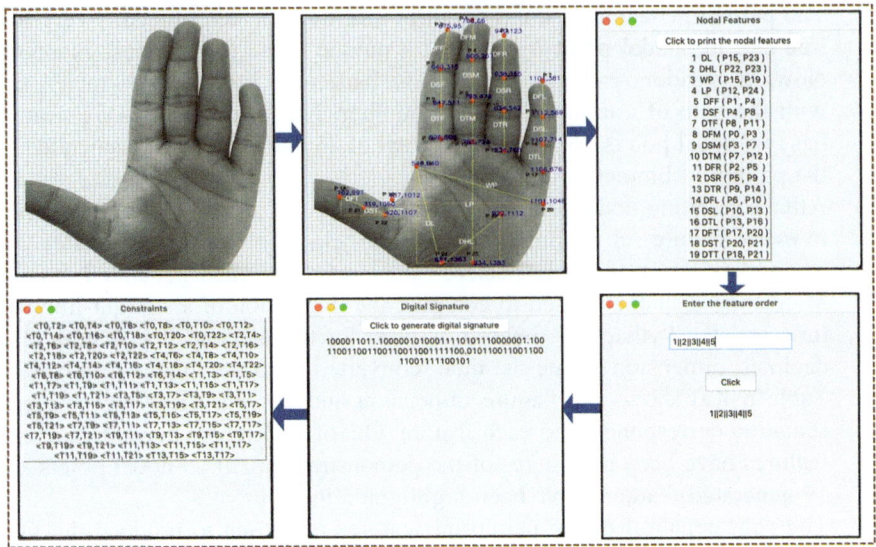

*Figure 6.8   Generation of palmprint biometric-based hardware security
constraints*

Table 6.4(b)   Mapping rule to generate hardware security constraints

Digit	Mapping rule
0	An edge between <even, even> storage variable node pair is added into CIG
1	An edge between <odd, odd> storage variable node pair is added into CIG

> <M2, M22>, <M2, M24>, <M2, M26>, <M4, M6>, <M4, M8>, <M4,
> M10>, <M4, M12>, <M4, M14>, <M4, M16>, <M4, M18>, <M4, M20>,
> <M4, M22>, <M4, M24>, <M4, M26>, <M6, M8>, <M6, M10>, <M6,
> M12>, <M6, M14>, <M6, M16> and <M1, M3>, <M1, M5>, <M1, M7>,
> <M1, M9>, <M1, M11>, <M1, M13>, <M1, M15>, <M1, M17>, <M1,
> M19>, <M1, M21>, <M1, M23>, <M1, M25>, <M3, M5>, <M3, M7>,
> <M3, M9>, <M3, M11>, <M3, M13>, <M3, M15>, <M3, M17>, <M3,
> M19>, <M3, M21>, <M3, M23>, <M3, M25>, <M5, M7>, <M5, M9>,
> <M5, M11>, <M5, M13>, <M5, M15>, <M5, M19>, <M5, M21>, <M5,
> M23>, <M5, M25>, <M7, M9>, <M7, M11>, <M7, M13>, <M7, M15>,
> <M7, M17>, <M7, M19>, <M7, M21>.

(vi)   Finally, the embedding of obtained hardware security constraints into the
schedule design is performed during the register allocation phase of the HLS
framework. Additional details of palmprint biometric methodology are discussed in Chapter 8.

6.4.2 Embedding of palmprint biometric signature on FIR filter

The embedding of palmprint-based hardware security constraints is performed in a similar fashion as explained in Section 6.3.3 (in the case of fingerprint biometric-based hardware security methodology).

6.4.3 Detection and validation of palmprint biometric for detective control against IP piracy and nullifying fraud claim of IP ownership

The detection and validation of palmprint biometrics for detective control against IP piracy and nullifying fraud claims of IP ownership require a pre-stored sample of an authentic IP vendor's palmprint image and rules (algorithm) for palmprint signature generation. Recapturing of palm image of the IP vendor during detection/validation process is not required. This is because the original palm image is securely pre-stored in a safe server of the IP vendor, which is subsequently used to generate the authentic palmprint signature and its respective hardware security constraints using the following security parameters: (a) grid size and spacing, (b) naming conventions of nodal points, (c) palm features coordinates, (d) concatenation order of features, and (e) constraints embedding rules (Sengupta *et al.*, 2021).

Palmprint signature template is regenerated using a pre-stored palmprint biometric image and the algorithm for palmprint signature generation. Subsequently, hardware security constraints are generated corresponding to regenerated palmprint biometric-based signature based on IP vendor-selected embedding rules. Further, storage variable register allocation information is extracted from the RTL datapath of the IP core (under test), and the generated hardware security constraints are matched with it. In the case of complete matching, the design is considered authentic; otherwise, the design may be a counterfeit (due to the absence of genuine IP vendor's palmprint signature). Further, it is impossible for an adversary to regenerate the exact signature template in an attempt to evade the counterfeit detection process, despite having access to a palmprint biometric image. He/she will additionally require several IP vendors selected aforesaid security parameters. Therefore, the dependency of the palmprint biometric-based hardware security approach on several such security factors above enhances the overall robustness of the technique (Sengupta *et al.*, 2021).

Moreover, it is impossible for the adversary to fraudulently claim the ownership of the IP, as they cannot match their signature with the embedded hardware security constraints of genuine IP vendor's palmprint biometric (extracted from RTL datapath of IP under test). However, a genuine IP vendor can easily claim a match between their palmprint signature and the register allocation information extracted from the RTL datapath of the IP under test (Sengupta *et al.*, 2021).

6.5   Digital signature using encrypted hash for IPP of DSP coprocessors

6.5.1   Summary of approach

The digital signature-based hardware security methodology (Sengupta *et al.*, 2019) accepts the following inputs: (a) transfer function of the DSP application or CDFG and (ii) IP vendor-specified resource constraints. Moreover, the CDFG of the DSP application can also be obtained using its respective transfer function. Further, the respective CDFG is scheduled using provided architectural constraints to generate a SDFG. Post-generation of SDFG, the storage variables (registers) are allocated in SDFG, and all operations are marked accordingly. Subsequently, a bitstream is generated based on the SDFG and IP vendor-selected encoding rule. If the parity of the operation number in SDFG and the control step is the same, then the encoded bit is "0" in the bitstream, otherwise "1." Further, the generated bitstream is fed as an input to the SHA-512 computation block. The output of the SHA-512 computation block is a bitstream digest of the target DSP application. One of the crucial operations of the SHA-512 computation block is word computation (W), which involves circular right shift (of 64-bit argument), left shift (of 64-bit argument), and addition modulo 264. Post-SHA-512 computation, the generated bitstream is divided and grouped into blocks of equal sizes (IP vendor-selected block size) and subsequently converted into their decimal equivalents. Further, RSA encryption is performed on each decimal value to yield encrypted decimal values using the IP vendor RSA private key. In the next step, all the obtained encrypted decimal values are converted back to their respective binary equivalents, which are further used to generate hardware security constraints based on the IP vendor-selected embedding rule. The generated hardware security constraints are embedded into the design of the DSP application during the register allocation phase of the HLS framework. Finally, a digital signature-embedded hardware DSP IP core is produced (Sengupta *et al.*, 2019).

6.5.2   Details of the approach

As discussed above, a scheduled data flow graph is generated using CDFG of the respective DSP application, the IP designer's-selected architectural constraints (functional units), and LIST scheduling algorithm. Subsequently, a bitstream corresponding to the DSP application is generated using the IP vendor/designer-selected encoding rule (explained in the encoding block of Figure 6.9). Next, the generated bitstream is fed as an input to the SHA-512 block to produce bitstream digest as an output. SHA-512 provides uniformity and high collision resistance to security. These additional properties enhance the security algorithm as they prevent the generation of the same bitstream digest for two bitstream inputs, which makes it challenging for an adversary to access the bitstream from its bitstream digest. The inclusion of round function computation for 80 rounds in SHA-512 supports it in producing well-distributed output, which decreases the

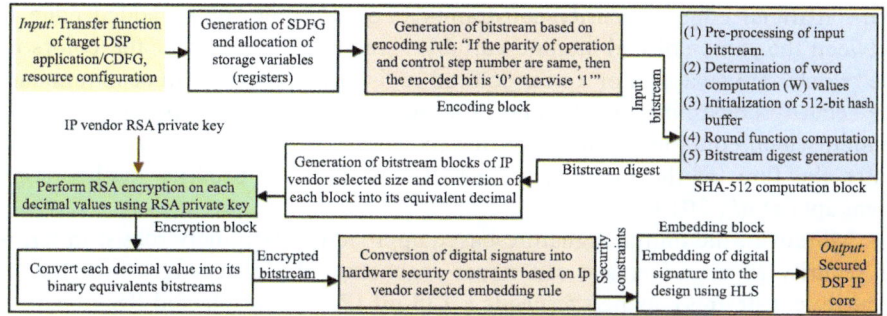

Figure 6.9   Digital signature-based hardware security approach

likelihood of getting the same bitstream digest value corresponding to two bitstream values (Sengupta *et al.*, 2019).

SHA-512 accepts input in multiple of 1,024. Therefore, the input bitstream is first padded with padding bits until it becomes a multiple of 1,024. Then, word computation values are determined using an updated bitstream. As discussed in the summary section, word computation involves circular right shift (of 64-bit argument), left shift (of 64-bit argument), and addition modulo 264. Now, the generated hash value is stored in an initialized 512-bit buffer. Finally, the round function is implemented for 80 iterations. The final generated hashed bitstream digest is the addition of input of the first iteration and output of 80 iterations. Further, the bitstream digest blocks are generated based on the IP vendor/designer-selected block size, and all block values are converted into their decimal equivalents. These decimal values are the input to the next RSA encryption block (Sengupta *et al.*, 2019).

RSA encryption (Rivest *et al.*, 1978) is an asymmetric type of encryption mechanism which imparts greater security strength to the security algorithm. The key component of the RSA encryption block is the IP designer-selected private key. The private key is used to perform encryption on each decimal block using: "$E = B^d$ mod n," where "E" is the encrypted/cipher data, "B" is the input decimal block values, "d" is the IP vendor/designer-selected private key value, and "n" = t*s (such that s and t are both prime and $s \neq t$). The inclusion of an RSA encryption block also enhances the robustness of the security as the private key value is unknown to the adversary. Subsequently, the encrypted decimal data is converted into their binary equivalents and concatenated into a single binary string. Further, the hardware security constraints are generated using determined encrypted binary string and IP vendor/designer-selected embedding rule.

An artificial edge between <even, even> storage variables (nodes) pair in CIG is added if the bit is "1." Otherwise, an artificial edge is added between <prime, prime> storage variable pairs.

Finally, all the generated covert security constraints are embedded into the register allocation design of DSP application using HLS framework. The target DSP application's register allocation (represented as CIG) is harnessed to embed extra artificial edges (security constraints). For example, if the added edge is between the same-colored storage variables (register), then either the colors of the register are swapped or a new colored register is allocated (to accommodate the security constraint). In the end, a secured DSP IP core is produced as the output, which contains security constraints as digital evidence. Figure 6.9 illustrates the flow of the digital signature-based hardware security methodology (Sengupta *et al.*, 2019).

Moreover, the digital signature-based approach is extremely robust in terms of security as an adversary can never regenerate the exact signature (list of security constraints), as he/she needs a 128-bit IP vendor-selected RSA encryption key, besides IP vendor-selected secret encoding rule, embedding rule, and block size. Any counterfeited IP core can easily be identified and isolated before integration into consumer electronic hardware electronic systems with the help of embedded digital signature during the counterfeit detection process. Further, it also helps in resolving a fraudulent claim of IP ownership problem (Sengupta *et al.*, 2019).

6.6   Qualitative comparison between fingerprint biometric for IPP vs. digital signature for IPP, digital signature for IPP vs. palmprint biometric for IPP, digital signature for IPP vs. palmprint biometric for IPP

This section presents a qualitative comparison between the discussed security approaches. Table 6.5 shows a qualitative comparison between a fingerprint-based hardware security approach for IPP (Sengupta and Rathor, 2020) and digital signature using an encrypted hash-based hardware security approach (Sengupta *et al.*, 2019). Next, Table 6.6 qualitatively compares the digital signature-based hardware security approach (Sengupta *et al.*, 2019) and the palmprint biometric-based hardware security approach for IPP (Chaurasia *et al.*, 2022; Sengupta *et al.*, 2021). Similarly, Table 6.7 provides a qualitative comparison between biometric-based hardware security methodologies, i.e., fingerprint-based hardware security approach (Sengupta and Rathor, 2020) and palmprint-based hardware security approach (Chaurasia *et al.*, 2022; Sengupta *et al.*, 2021). The qualitative comparison is performed based on the following parameters: (a) presence of unique natural identity, (b) security approach, (c) pre-processing, (d) requirement of external devices, (e) cost overhead due to embedded extra security constraints, (f) presence of digital evidence, (g) security strength, (h) implementational complexity, (i) counterfeit detection process, (j) dependency on external factors (such as the presence of dirt and grease on the optical scanner), (k) probability of coincidence (P_C), (l) tamper tolerance (TT), (m) digital template regeneration by adversary, (n) presence of encryption algorithm, and (n) security approach type (contactless or not contactless). A detailed analysis in terms of security strength is discussed in the next section.

Table 6.5   Qualitative comparison between fingerprint for IPP and digital signature for IPP

S. no.	Characteristics/ parameters	Fingerprint biometric for IPP (Sengupta and Rathor, 2020)	Digital signature for IPP (Sengupta *et al.*, 2019)
1.	Presence of unique natural identity	Yes; provides a unique natural fingerprint biometric identity of IP vendor	No; it does not provide any natural identity
2.	Security approach	Requires an IP vendor fingerprint image to generate minutiae points features	Depends on IP vendor selected RSA encryption private key and SHA-512 bit stream digest computation
3.	Pre-processing	It requires image pre-processing steps such as binarization, FFT, and thinning to perform image enhancement to extract accurate minutiae points	Not required
4.	Requirement of external devices	Requires an optical scanner	Not required
5.	Cost overhead due to embedded extra security constraints	Higher due to generation and embedding of larger security constraints	Lower than fingerprint biometric
6.	Presence of digital evidence	Provides higher digital evidence	Moderate
7.	Security strength	Provides more robust and higher security	Lower than fingerprint biometric
8.	Implementational complexity	More complex	Lower complexity
9.	Counterfeit detection process	Complex due to presence of pre-processing	Robust and seamless
10.	Dependency on external factors (such as dirt and grease)	No	No
11.	Probability of coincidence (P_C)	Lower P_C value indicating presence of higher digital evidence and stronger security	P_C value higher than fingerprint biometric
12.	Tamper tolerance capability (TT)	Higher TT indicating greater effort required to deploy brute force attack	Lower TT than fingerprint biometric approach
13.	Digital template regeneration by adversary	Not possible (as regeneration of digital template depends on minutiae points, feature set, feature order, grid size, concatenation order, etc.)	Possible in the case of compromised IP vendor selected encoding mechanism and RSA encryption key
14.	Presence of encryption algorithm	No	Yes; RSA and SHA-512 encryption algorithms

Table 6.6   Qualitative comparison between digital signature for IPP and palmprint biometric for IPP

S. no.	Characteristics/ parameters	Digital signature for IPP (Sengupta *et al.*, 2019)	Palmprint biometric for IPP (Chaurasia *et al.*, 2022; Sengupta *et al.*, 2021)
1.	Presence of unique natural identity	No; it does not provide any natural identity	Yes; provides a unique natural fingerprint biometric identity of IP vendor
2.	Security approach	Depends on IP vendor selected RSA encryption private key and SHA-512 bit stream digest computation	Requires an IP vendor palmprint image to generate nodal points and corresponding features
3.	Pre-processing	Not required	Not required
4.	Requirement of external devices	Not required	Requires a high-resolution camera for capturing palmprint image
5.	Cost overhead due to embedded extra security constraints	Lower than palmprint biometric	Higher due to generation and embedding of larger security constraints
6.	Presence of digital evidence	Moderate	Provides higher digital evidence
7.	Security strength	Lower than palmprint biometric	Provides more robust and higher security
8.	Implementational complexity	Lower complexity	More complex due to involvement of nodal points generation on palmprint image
9.	Counterfeit detection process	Robust and seamless	Complex than digital signature-based approach
10.	Dependency on external factors	No	No
11.	Probability of coincidence (P_C)	P_C value higher than palmprint biometric	Lower P_C value indicating presence of higher digital evidence and stronger security
12.	Tamper tolerance capability (TT)	Lower TT fingerprint biometric approach	Higher TT indicating greater effort required to deploy brute force attack
13.	Digital template regeneration by an adversary	Possible in the case of compromised IP vendor-selected encoding mechanism and RSA encryption key	Not possible (as regeneration of digital template depends on nodal points, feature set, feature order, grid size, etc.)
14.	Presence of encryption algorithm	Yes; RSA and SHA-512 encryption algorithms	No

Table 6.7 *Qualitative comparison between fingerprint biometric for IPP and palmprint biometric for IPP*

S. no.	Characteristics/ parameters	Fingerprint biometric for IPP (Sengupta and Rathor, 2020)	Palmprint biometric for IPP (Chaurasia *et al.*, 2022; Sengupta *et al.*, 2021)
1.	Presence of unique natural identity	Yes; provides a unique natural fingerprint biometric identity of IP vendor	Yes; provides a unique natural palmprint biometric identity of IP vendor
2.	Security approach	Requires an IP vendor fingerprint image to generate minutiae points features	Requires an IP vendor palm-print image to generate no-dal points and corresponding features
3.	Pre-processing	It requires image pre-processing steps such as binarization, FFT and thinning to perform image enhancement to extract accurate minutiae points	No-pre-processing is required
4.	Requirement of external devices	Requires an optical scanner to capture fingerprint image	Requires a high-resolution camera for capturing palm-print image
5.	Cost overhead due to embedded extra security constraints	Same as palmprint biometric	Higher due to generation and embedding of larger security constraints
6.	Presence of digital evidence	Provides higher digital evidence	Provides higher digital evidence
7.	Security strength	Provides stronger security	Provides stronger security
8.	Implementational complexity	More complex	Moderate
9.	Counterfeit detection process	Complex due to involvement of pre-processing	Less complex than fingerprint biometric
10.	Dependency on external factors	No	No
11.	Probability of coincidence (P_C)	P_C value higher than palmprint biometric-based security approach	Lower P_C value indicating presence of higher digital evidence and stronger security
12.	Tamper tolerance capability (TT)	TT lesser than palmprint biometric	Higher TT indicating greater effort required to deploy brute force attack
13.	Digital template regeneration by adversary	Not possible (as regeneration of digital template depends on minutiae points, feature order, feature set, grid size, concate-nation order, etc.)	Not possible (as regeneration of digital template depends on nodal points, feature order, feature set, grid size, etc.)
14.	Security approach type	Not contactless (depends on op-tical scanner)	Contactless

6.7   Analysis and discussions of results

In this section, the discussed security methodologies are analyzed based on two parameters: (a) probability of coincidence (P_C) and (b) tamper tolerance (TT).

(a) P_C: P_C has been used to measure the strength of security approaches. A lower P_C value indicates the presence of higher digital evidence and greater uniqueness in the generated signature, which helps in easy and smooth counterfeit detection. The formulation of P_C is given as follows:

$$P_C = (1 - 1/t)^x \tag{6.1}$$

where "t" = total colors (registers) in the CIG of DSP IP design and "x" = the number of edges (constraints) embedded.

Figures 6.10, 6.11, and 6.12 compare the value of P_C between (the fingerprint biometric and digital signature-based security approach), (the digital signature and

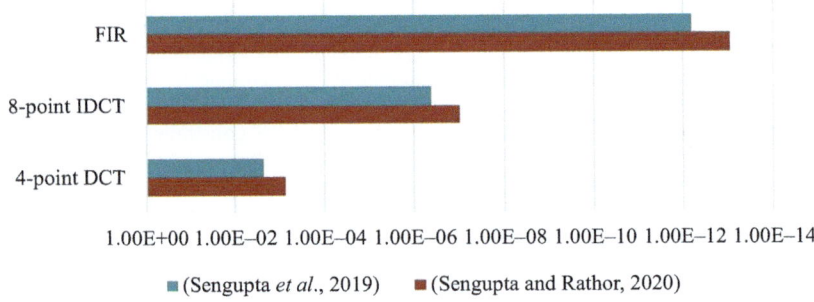

Figure 6.10   Comparison of P_C between fingerprint biometric and digital signature-based security approaches

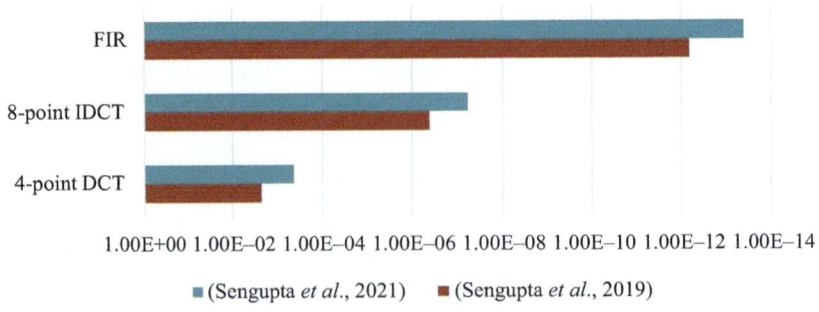

Figure 6.11   Comparison of P_C between palmprint biometric and digital signature-based security approaches

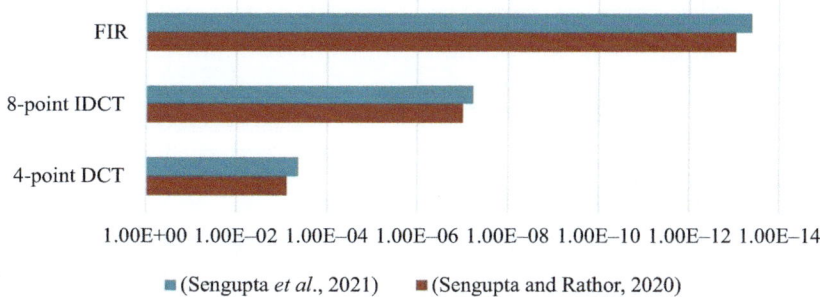

Figure 6.12 Comparison of P_C between palmprint biometric and fingerprint biometric-based security approaches

palmprint biometric-based security approach), and (the fingerprint and palmprint biometric-based security approach), respectively. As evident from Figure 6.10, the fingerprint biometric-based security approach is more robust than the digital signature-based approach. This is due to the generation and embedding of a huge amount of security constraints in the case of the fingerprint biometrics-based security approach. Next, the palmprint-based security approach surpasses the digital signature-based security approach in Figure 6.11 due to the generation of a greater number of security constraints. Similarly, Figure 6.12 illustrates the superiority of the palmprint biometric-based security approach over the fingerprint biometric approach for security. The greater the number of embedded security constraints, the lower the P_C value.

(b) TT: A higher value of tamper tolerance indicates greater security in terms of the effort required to perform a brute-force attack to regenerate the exact signature template. An adversary tries to regenerate the signature template (used to generate hardware security constraints while securing hardware IP core) to evade counterfeit detection. Therefore, the greater the TT value, the higher the effort required to regenerate the signature template by an adversary. The formulation of TT is given as follows:

$$TT = (x)^v \tag{6.2}$$

where "v" is the number of encoding variables used in the respective security approach and "x" is the number of edges (constraints) embedded.

Figure 6.13 illustrates a comparison of TT between digital signature-based and fingerprint-based security approaches, where the fingerprint-based security approach surpasses the digital signature-based security approach due to the generation of larger number of hardware security constraints. Further, from Figure 6.14, it is observed that the palmprint-based security approach provides more robust security than the digital signature-based security approach due to the generation of a larger number of security constraints and the involvement of greater encoding variables. Similarly, as evident from Figure 6.15, the palmprint-based

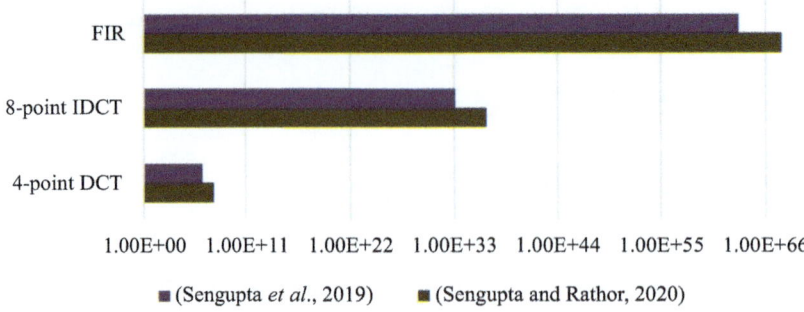

Figure 6.13   Comparison of TT between fingerprint biometric and digital signature-based security approaches

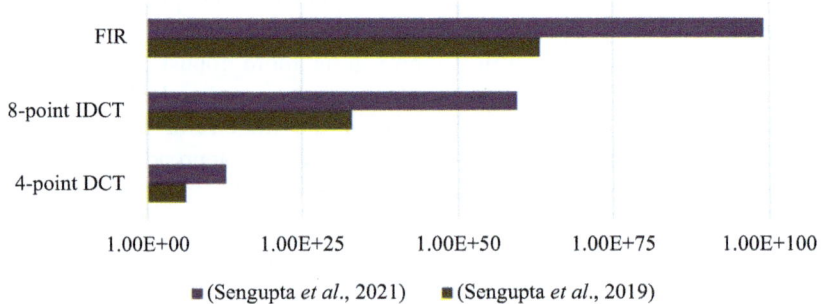

Figure 6.14   Comparison of TT between palmprint biometric and digital signature-based security approaches

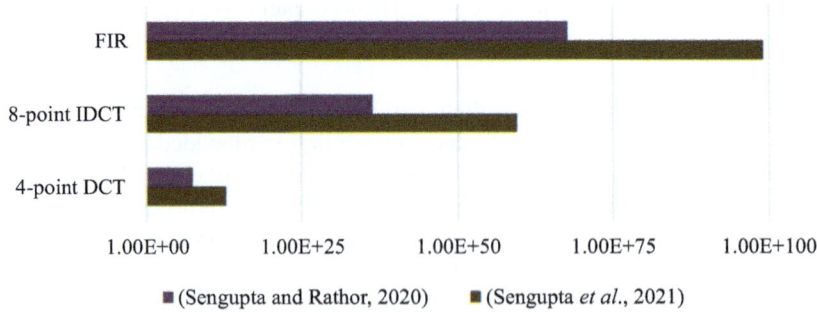

Figure 6.15   Comparison of TT between palmprint biometric and fingerprint biometric-based security approaches

security approach has a higher TT value than the fingerprint-based security approach due to generation of larger number of security constrictions and the involvement of greater encoding variables. The DSP benchmarks for performing analysis are adopted from (EXP BEN; Jain and Panda, 2007).

6.8 Conclusion

A discussion between three recent hardware security methodologies (i.e., fingerprint-based security methodology, palmprint-based security methodology, and digital signature encrypted hash-based security methodology) is presented in this chapter. Further, this chapter also presents a qualitative comparison and analysis between all three discussed security methodologies. The security aspect of the hardware IP cores is crucial as they perform essential and computation-intensive functions. As evident from the values of P_C and TT (shown in the analysis and discussion of results section), the palmprint-based security methodologies provide more robust security than fingerprint and digital signature-based security methodologies. And fingerprint-based security methodology is more robust than the digital signature-based security approach. Moreover, the discussed biometric-based security methodologies (i.e., fingerprint-based security methodology and palmprint-based security methodology) provide a unique natural biometric identity of IP vendor, which enhances the security of IP cores. The discussed methods offer robust and seamless detection of counterfeited IP cores from genuine ones and protect IP vendors and SoC integrators from piracy, fraudulent IP ownership claims, and end consumers from various safety concerns discussed in the introduction section.

6.9 Questions and exercise

1. Briefly explain the importance of DSP application in the context with electronic and computing devices.
2. What are the factors affecting the security of the DSP IP cores? Why it is necessary to secure DSP IP cores against different hardware threats?
3. Briefly describe the different IPP security approaches.
4. Defines the inputs of the fingerprint-based biometric approach for hardware security.
5. Explain the design flow of fingerprint biometric-based hardware security methodology.
6. What is importance of pre-processing phase in fingerprint biometric-based hardware security methodology?
7. Explain the generation process of minutiae points on a fingerprint and determination of its corresponding four parameters.
8. Describe the embedding of the generated fingerprint-based signature with an example of 8-point DCT IP core and determine its RAT corresponding after embedding the fingerprint-based signature.

9. What is the importance of colored interval graph while performing the embedding of fingerprint-based signature template?
10. Briefly explain the detection and validation of fingerprint biometrics for detective control against fraud claim of IP ownership and IP piracy.
11. Explain the design flow of palmprint biometric-based hardware security methodology in details.
12. What are the benefits of palmprint-based hardware security methodology over facial biometrics, hardware watermarking, hardware steganography, and digital signature based hardware security methodology?
13. What is the necessity of storing the original palmprint biometric image and where it is stored?
14. Explain digital signature-based hardware security approach in details.
15. Perform a qualitative comparison between (a) fingerprint biometric and digital signature for IPP and (b) digital signature and palmprint for IPP.
16. Draw a qualitative comparison between both biometric-based approaches i.e., fingerprint biometric and palmprint biometric-based security approaches.
17. What are the two security metrics used to compare and analyses the security strength of the discussed security approaches?
18. Perform a security analysis among all the three security approaches in context with number of generated hardware security constraints.

References

Chaurasia, R., A. Anshul, A. Sengupta, and S. Gupta (2022), "Palmprint biometric versus encrypted hash based digital signature for securing DSP cores used in CE systems," *IEEE Consumer Electronics Magazine*, vol. 11, no. 5, pp. 73–80.

Colombier, B. and L. Bossuet (2015), "Survey of hardware protection of design data for integrated circuits and intellectual properties," *IET Computers & Digital Techniques*, vol. 8, no. 6, pp. 274–287.

EXP BEN, University of California Santa Barbara Express Group. [Online]. Available: http://express.ece.ucsb.edu/benchmark/, accessed on March. 2022.

Islam, S.A., L.K. Sah, and S. Katkoori (2020), "High-level synthesis of key-obfuscated RTL IP with design lockout and camouflaging," *ACM Transactions on Design Automation of Electronic Systems,* vol. 26, no. 6, pp. 1–35.

Jain, R. and P. R. Panda (2007), "An efficient pipelined VLSI architecture for lifting based 2D-discrete wavelet transform," in: *Proceedings of IEEE International Symposium On Circuits Systems (ISCAS),* pp. 1377–1380.

Koushanfar, F., I. Hong, and M. Potkonjak (2005), "Behavioral synthesis techniques for intellectual property protection," *ACM Transactions on Design Automation of Electronic Systems,* vol. 10, no. 3, pp. 523–545.

Kumar, E. R., and N. P. Chandra (2019), "Embedding digital signature using encrypted hashing for protection of DSP Cores in CE," *IEEE Transactions on Consumer Electronics*, vol. 65, no. 3, pp. 398–407.

Le Gal, B. and L. Bossuet (2012), "Automatic low-cost IP watermarking technique based on output mark insertions," *Design Automation for Embedded Systems*, vol. 16, no. 2, pp. 71–92.

Pilato, C., S. Garg, K. Wu, R. Karri, and F. Regazzoni (2018), "Securing hardware accelerators: a new challenge for high-level synthesis," *IEEE Embedded Systems Letters*, vol. 10, no. 3, pp. 77–80.

Rivest, R., A. Shamir, and L. Adleman (1978), "A method for obtaining digital signatures and public-key cryptosystems (PDF)," *Communications of the ACM*, vol. 21, no. 2, pp. 120–126.

Rizzo, S., F. Bertini, and D. Montesi (2019), "Fine-grain watermarking for intellectual property protection," *EURASIP Journal on Information Security*, article number 10, https://doi.org/10.1186/s13635-019-0094-2.

Schneiderman, R. (2010), "DSPs evolving in consumer electronics applications," *IEEE Signal Processing Magazine*, vol. 27, no. 3, pp. 6–10.

Sengupta, A. and S. Bhadauria (2016), "Exploring low cost optimal watermark for reusable IP cores during high level synthesis," *IEEE Access*, vol. 4, pp. 2198–2215.

Sengupta, A. and M. Rathor (2019), "IP core steganography for protecting DSP kernels used in CE systems," *IEEE Transactions on Consumer Electronics*, vol. 65, no. 4, pp. 506–515.

Sengupta, A. and M. Rathor (2020), "Securing hardware accelerators for CE systems using biometric fingerprinting," *IEEE Transactions on Very Large Scale Integration (VLSI) Systems*, vol. 28, no. 9, pp. 1979–1992.

Sengupta, A., R. Chaurasia, and T. Reddy (2021), "Contact-less palmprint biometric for securing DSP coprocessors used in CE systems," *IEEE Transactions on Consumer Electronics*, vol. 67, no. 3, pp. 202–213.

Yasin, M., J. J. Rajendran, O. Sinanoglu, and R. Karri (2016), "On improving the security of logic locking," *IEEE Transactions on Computer-Aided Design of Integrated Circuits and Systems*, vol. 35, no. 9, pp. 1411–1424.

Chapter 7

Secured design flow using palmprint biometrics, steganography, and PSO for DSP coprocessors

Anirban Sengupta[1] and Aditya Anshul[1]

Reusable intellectual property (IP) cores are widely used in all modern digital integrated circuits (ICs) in the form of application-specific processors or hardware accelerators. Reusable IP cores integrated in system-on-chip (SoC) platforms of integrated circuits and systems provide a powerful blend of yielding superior design productivity with reduced design cycle time. However, leveraging advantages of IP core require low-cost security against threats such as piracy and fraudulent claim of ownership. This chapter discusses the following aspects: (a) security-aware end-to-end high-level synthesis (HLS)-based design methodology using low-cost steganography technique for protecting IP cores used in digital ICs; (b) security-aware end-to-end HLS-based design methodology using low-cost biometric signature for protecting IP cores used in digital ICs; (c) secured RTL designs of reusable IP core embedding secret steganography mark and biometric signature respectively offering low overhead, strong robustness, and greater security in terms of lower probability of coincidence and higher tamper tolerance. Further, the chapter also discusses the achieved robustness of $> 2\times$ (i.e., stronger digital evidence as evident from the lower probability of coincidence) and lower design cost ($\sim 6.17\%$) than state-of-the-art approaches.

7.1 Introduction

Reusable intellectual property cores are essential modules used in various digital IC-based computing devices such as smartphones, tablets, computers, digital cameras, smartwatches, and televisions. These IP cores execute various vital data-intensive and power-hungry applications involving massive computations like data compression–decompression, digital data filtering, and different complex mathematical calculations (Pilato *et al.*, 2018; Schneiderman, 2010). Finite impulse response (FIR) filters, discrete cosine transform (DCT), discrete wavelet transform (DWT), and inverse discrete cosine transform (IDCT) are some examples of the IP cores used in the

[1]Department of Computer Science and Engineering, Indian Institute of Technology Indore, India

above-mentioned digital IC devices. These IP cores are designed at the higher level of the design chain process using a HLS framework. Therefore, the overall design process of these IP cores involves using sophisticated knowledge from the perspective of the area, delay optimization, and CAD algorithms (Colombier and Bossuet, 2015).

The involvement of multiple offshore entities in the IP core's current design process makes it challenging to consider aforesaid optimization parameters. The complete system production process, starting from the procurement of IPs to the development and final product, involves various hardware threats like IP counterfeiting, cloning, and piracy (Rizzo *et al.*, 2019; Zhang, 2016). The involvement of these critical factors increases the overall design complexity of the IP cores. Therefore, it is crucial to secure and optimize the IP cores by making them capable of detecting piracy and counterfeiting. It is the responsibility of the IP designer and system-on-chip (SoC) integrator to protect against the hardware threats like IP counterfeiting and IP cloning.

This chapter discusses two security-aware design flows for designing low-cost secured register transfer level datapath of reusable IP cores using biometric signature and hardware steganography. It also includes schematic design implementations of secured IP cores for application-specific computing.

7.2 Emerging and contemporary approaches for IP core protection (IPP)

Figure 7.1 describes different state-of-the art/emerging approaches for securing/protecting IP cores.

7.3 Threat model and PSO-driven design space exploration

Threat model: Counterfeited or fake IPs create various problems for the end consumers as they may contain malicious logic/trojan and have not gone through rigorous testing and reliability analysis. Therefore, detecting counterfeited IPs is crucial to protect end consumers' safety, IP brand value, and revenue loss and isolate them before integrating into computing systems (Rizzo *et al.*, 2019). The discussed approaches can differentiate between authentic and fake IP cores (containing backdoor Trojan and other malicious logic) before integrating into SoC platforms (Sengupta and Bhadauria, 2016).

7.3.1 PSO-driven design space exploration in HLS

HLS is a process of converting a DSP application described in the form of the transfer function or data flow graph from its algorithmic level to its register transfer level during IC design process. Here the design is expressed in terms of its behavior, size, functionality, power, area, delay, etc. HLS process employs DSE module to explore an efficient hardware architecture to implement the RTL of the DSP application. The IC designs in the form of SoC often comprise of several IP cores of DSP and multimedia application. In the discussed low-cost (expressed mathematically in terms of design

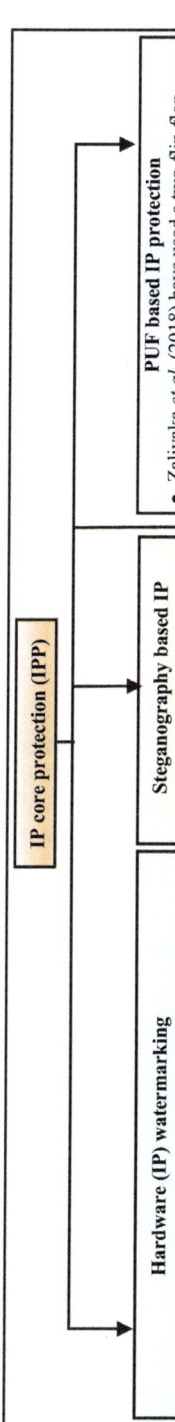

IP core protection (IPP)

Hardware (IP) watermarking

- From the perspective of hardware watermarking (Gal and Bossuet, 2012; Sengupta and Bhadauria, 2016; Koushanfar *et al.*, 2005), the embedded signature employs a number of signature variables, their combination/size, and mapping rules.
- Gal and Bossuet (2012) have exploited in-synthesis IP watermarking process to embed secret mark into dedicated hardware IP cores. The IP watermark in Gal and Bossuet (2012) is a set of mathematical relations between the IP input data, the initial values of the internal computation and the IP output. Each mathematical relation is called a sub-mark.
- Sengupta and Bhadauria (2016) have used four variable (i, I, T, !) signature encoding process to embed additional (artificial) edges in the colored interval graph (CIG) of the target application using HLS framework. Here each variable provides a unique mapping of the corresponding digit to the additional edge between storage variable pairs of the CIG. Further, Koushanfar *et al.* (2005) have used two variables (0, 1) signature encoding process to embed artificial edges in the register allocation graph of the DSP application using HLS framework.
- Watermark in Gal and Bossuet (2012) is not as robust as Sengupta and Bhadauria (2016) and Koushanfar *et al.* (2005) due to no involvement of signature encoding process. Though, Koushanfar *et al.* (2005) is robust in security, however, Sengupta and Bhadauria (2016) outperforms Koushanfar *et al.* (2005) in terms of better tamper tolerance ability.

Steganography based IP protection

- In hardware steganography (Sengupta and Rathor, 2019), stego-constraints (secret mark) are generated based on the encoding rules, threshold values, and confidential design data.
- The steganography technique (Sengupta and Rathor, 2019) is signature-free and provides more security through entropy threshold values. But it also becomes weak if encoding rules and secret value of chosen entropy threshold gets leaked.

PUF based IP protection

- Zalivaka *et al.* (2018) have used a two-flip-flop arbiter PUF, which produce trinary digit quadruple responses for metastability detection to increase the reliability of PUFs in FPGA implementations. A compressed quadbit is generated using combinations of first and last challenge bits, which depicts one of the five classes of trinary digit quadruple responses with greater reproducibility. This approach enables a smooth error correction along with a lower burden of complete challenge response pair storage used for authentication.
- Lao *et al.* (2017) have used a reliable, lightweight, and secure PUF-based authentication using two-level finite state machine (FSM). FSM in Lao *et al.* (2017) has been used to correct the erroneous behavior of PUF (due to environmental variations such as voltage, temperature, and age).

IP protection using machine learning techniques

- Xue, *et al.* (2020) have proposed an adversarial model from attacker's perspective. Further they have discussed about several parameters for hardware trojan implementation under the adversarial models.
- Similarly, Liakos *et al.* (2019) have discussed about detection of hardware trojans in ICs with the help of gate-level netlist features and classification approaches.
- Authors in Elnaggar and Chakrabarty (2018) have focused on both machine learning-based attack and defense mechanism for hardware along with classification of different hardware security threats using suitable machine learning algorithms.

Figure 7.1 Tree representation of emerging and contemporary approached in field of IPP

area and execution latency) security methodologies, the module of PSO-DSE is employed to explore a low-cost-secured hardware IP architecture corresponding to the target DSP application during HLS for digital IC design. The role of PSO-DSE in the discussed methodologies is to iteratively refine and prune the design search space comprising of secured architectural solutions (security constraints embedded). Due to its inherent capability of escaping local minima and achieving a lower cost (higher fitness) solution, PSO-DSE is therefore able to yield optimized, secure architectural solutions in terms of lower design cost. Thus, the integration of PSO-DSE mitigates the design overhead incurred (if any) due to embedding the security constraints (Mishra and Sengupta, 2014).

The area, delay, and design cost functions used in the PSO-DSE module of the discussed security-aware design flow are presented in (7.1), (7.2), and (7.3), respectively (Mishra and Sengupta, 2014):

$$Area = \sum_{i=1}^{2} (A(R_X) * (R_X)) \tag{7.1}$$

where $A(R_X)$ indicates the area of a resource type R_X and R_X shows the number of instances utilized for a particular resource type:

$$Delay(latency) = ((n_m * l_m) + (n_a * l_a)) \tag{7.2}$$

$$Design \cos t = X_1 * (Area/A_{Max}) + X_2 * (Delay/DMax) \tag{7.3}$$

where X_1 and X_2 are designer-defined weighing factors ($X_1 = X_2 = 0.5$) used for providing equal weightage during design cost function evaluation. A_{MAX} and D_{MAX} represent maximum design area (computed using allocating maximum functional resources available) and delay (computed using allocating minimum functional resources). Table 7.1 shows the nomenclature of the variables used in this chapter

Table 7.1   Nomenclatures used in the chapter

Symbol	Description
P_i	Resource architecture for ith particle position
P_i^+	New updated resource architecture of ith particle
P_{gb}	Global best particle position
P_{lbi}	Local best particle position for ith particle
Area (A)	Total area of the secured design
l_a	Delay of one adder
n_a	Total control step required by adder in scheduled data flow graph of IP core's application
V_{ix}	Velocity of ith particle in xth dimension
V_{ix}^+	New updated velocity of ith particle in xth dimension
Latency (D)	Total execution time of secured design (Delay)
C_f^{lbi}	Local best fitness cost of ith particle
C_f^i	Fitness cost of ith particle
l_m	Delay of one multiplier
n_m	Total control step required by multiplier in scheduled data flow graph of IP core's application

and (7.2) and (7.3) for the discussed methodologies. The IP vendor-selected swarm size is three, where the first particle position (configuration) is initialized with maximum functional resources corresponding to the IP core; the second particle position is initialized with the minimum number of functional resources corresponding to the IP core, and the third position is the average of first and second. Next, PSO is executed to obtain the global best design solution (resource configuration) corresponding to the IP core. At first, the initial cost value is computed for all the particle positions, and the local best corresponding to all particle positions (P_{lbi}) and global best (P_{gb}) is updated. The minimum cost particle is declared the global best particle (P_{gb}). Then, new particle positions are then computed based on adding computed velocity to the initial position (Mishra and Sengupta, 2014). Again, the cost is computed for all the updated particle positions, and local and global bests are updated if the computed cost is less than the previous one. At last, the mutation is performed to diversify the particle position. After mutation, the same cost computation process with local and global bests updating is followed until the terminating condition is satisfied (P_1 and P_2, mentioned in the result section) (Mishra and Sengupta, 2014).

7.3.2 Advantage of PSO over other search space algorithms

PSO has been preferred over various other design search spaces (such as genetic algorithm and bacterial foraging algorithm) because of the following reasons: (a) PSO includes a particular parameter to consider the previous velocity known as inertia weight. Such consideration of momentum of the prior iteration does not exist in the case of genetic algorithm-driven DSE (GA-DSE) (Sengupta and Bhadauria, 2016) and bacterial foraging-driven DSE BFO-DSE (Krishnan and Katkoori, 2006). Therefore, it gets stuck in local minima. (b) The value of inertia weight linearly decreases from 0.9 to 0.1, which provides a complete balance between exploration and exploitation (more significant steps at the beginning and smaller steps on nearing higher fitness solutions), which does not exist in case of GA-DSE, BFO-DSE. (c) PSO-DSE technique considers social and cognitive factor whose empirically obtained optimal values lie between 2 and 4. This enables achieving lower cost (higher fitness) solutions within very low exploration time (Mishra and Sengupta, 2014).

7.4   Palmprint biometric-based hardware security approach

7.4.1   Overview of the low-cost palmprint-based hardware security approach

Figure 7.2 depicts the security-aware low-cost design flow methodology using palmprint signature. The design flow integrates PSO-based design space exploration for generating a low-cost hardware architecture. Here, a unique palmprint biometric signature is generated using authentic vendor's palmprint (captured through a high-resolution camera). The generated signature is then converted into

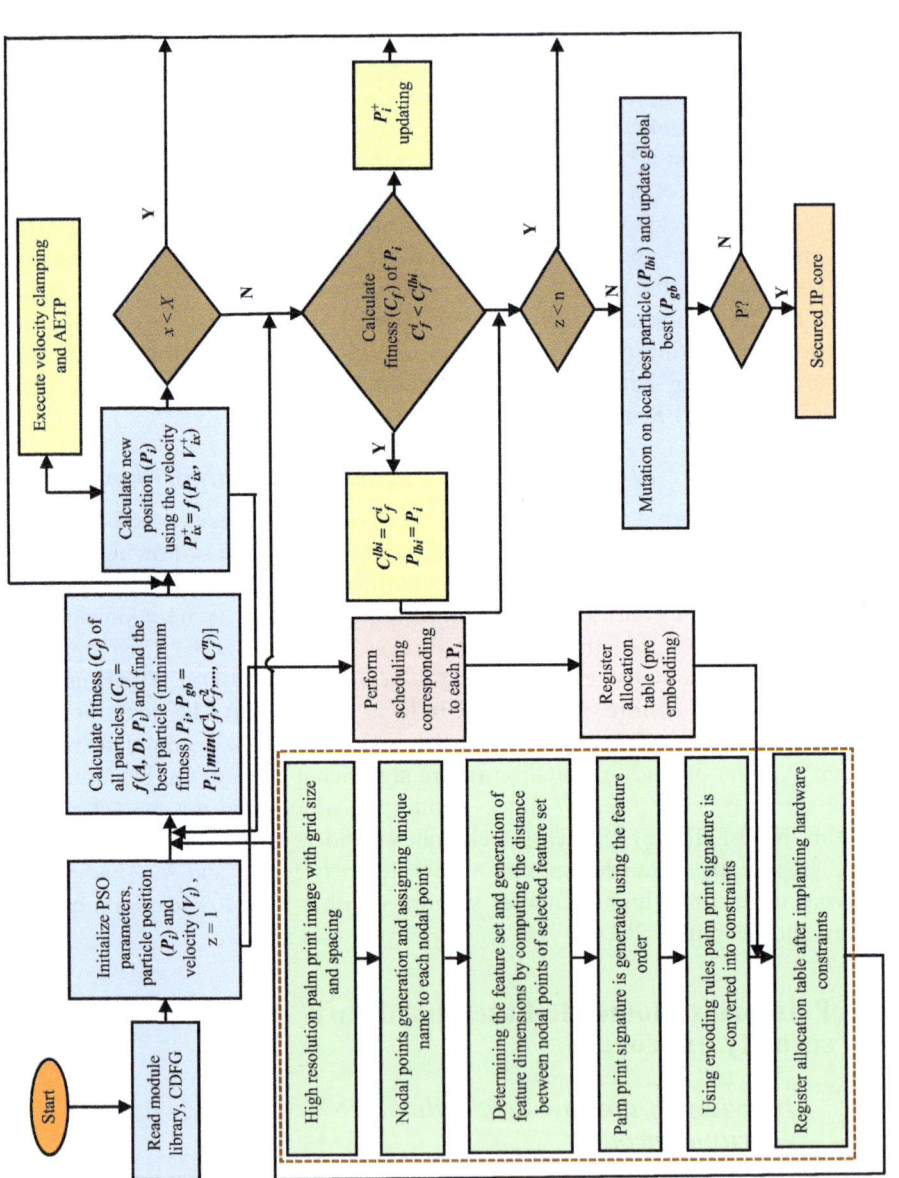

Figure 7.2   Security-aware design flow based on low-cost PSO-DSE-driven palmprint biometrics

Table 7.2(a)   Feature measurement of chosen palmprint feature and their corresponding binary representation

Feature #	Feature name	Dimension (magnitude)	Binary representation
Z1	DL	391.01	110000111.0000001010001111011
Z2	DHL	386.73	110000010.101110101110000101
Z3	WP	571.29	1000111011.01001010001111010111
Z4	LP	500	111110100
Z5	DFF	120	1111000

Table 7.2(b)   Mapping rule to generate hardware security constraints from palmprint signature

Digit	Mapping rule
0	Embed an edge between node pair <even, even> into CIG
1	Embed an edge between node pair <odd, odd> into CIG
.	Embed an edge between node pair <odd, prime> into CIG

their equivalent hardware security constraints based on mapping rule (Table 7.2(b)). It is then embedded into the RTL design of the target IP core using the HLS framework of the design process. First, as explained in Section 7.3.1, after initialization of swarm particles [combination of resource constraints (adders and multipliers)], the target IP core such as DCT, JPEG, and IDCT are scheduled with the help of LIST scheduling to yield scheduled dataflow graph (SDFG). Subsequently, design cost is computed for each swarm particle and global best (particle having lowest design cost) is selected. Further, palmprint-based generated signature is embedded into the design of target IP core and again the design cost and global best is computed. The process is repeated till the terminating condition is not reached (P_1 and P_2). Finally, a low-cost hardware IP architecture is obtained with an optimal number of resource constraints corresponding to palmprint-biometric signature-embedded secured IP core. Further, the obtained signature-embedded design (in the form of register allocation information and schedule) is converted into its RTL description and schematic implantation (Mishra and Sengupta, 2014; Sengupta *et al.*, 2021).

Advantages: Some of the key advantages of palmprint biometric-based security for IP cores over facial and fingerprint biometrics are as follows:

(a) Palmprint biometrics for IP core security is immune to aging than facial biometrics (Sengupta and Bhadauria, 2021).
(b) It is independent of various external factors like dirt, grease, etc. In contrast, the fingerprint biometric approach is not.
(c) Features extraction process for palmprint biometrics is easier than facial biometric (Sengupta *et al.*, 2021) and fingerprint biometric (Sengupta and Rathor,

2020). Minutiae points feature generation in the fingerprint biometric approach is more complex compared to palmprint biometric features generation.

(d) The number of overall features that can be generated is higher in the case of palmprint biometrics than facial and fingerprint biometrics.

(e) The palmprint-based technique is highly robust due to the vast number of features generated.

(f) The palmprint biometric approach is contactless, while the fingerprint-based (Sengupta and Rathor, 2020) approach is not.

Security properties: Moreover, palmprint biometric hardware security approach satisfies the following ISO/IEC standard 24745 properties: (a) *Unlikability*: using the discussed approach, it is possible to generate diverse palmprint biometric digital templates from the same palmprint image. These generated digital templates are exclusive of each other as they do not have any relation between them, and no cross-matching is possible. This is because distinct digital templates can be generated from the same palmprint image by varying the feature set, number of features, concatenation order, encoding algorithm, and the truncation size of the final signature. (b) *Revocability*: in the discussed approach, replacing the digital template of the palmprint image with a new one is possible from the same palmprint image by varying the number of selected features, order of concatenation of each feature, and encoding algorithm of the digital bits. This enables revoking the original template with a new one. (c) *Irreversibility*: recovering the original palmprint biometric data from the digital template is extremely challenging in the discussed approach. This is because to reverse the process from the digital template to arrive at the original palmprint biometric data, information such as the concatenation order of each selected feature, the numbers of chosen features, and details of the features embedded (feature name) are unknown to an adversary.

However, palmprint biometrics may have a few limitations as well (a) careful choice of palmprint features is necessary to ensure sturdy robustness; (b) careful choice of the encoding algorithm used for converting palmprint biometric signature into covert hardware security constraints; (c) requirement of a high-resolution camera to capture high-quality palmprint image signature generation and detection. The noise in the input palmprint image data can affect the feature dimension evaluation process. Therefore, the requirement of a high-resolution camera is mandatory, and (d) the requirement of designating the nodal points accurately on the palmprint grid image for accurately calculating the Manhattan distance between two nodal points during feature dimension estimation.

7.4.2 Details of the palmprint-based hardware security approach

Standard palmprint images [adopted from (CASIA Palmprint Database)] and palmprint images captured through high-resolution cameras have been used for embedding and demonstration purposes. We discuss the demonstration process as follows. At first, a high-resolution image of the palmprint is captured and placed under grid size and spacing to generate nodal points on the palmprint. After the

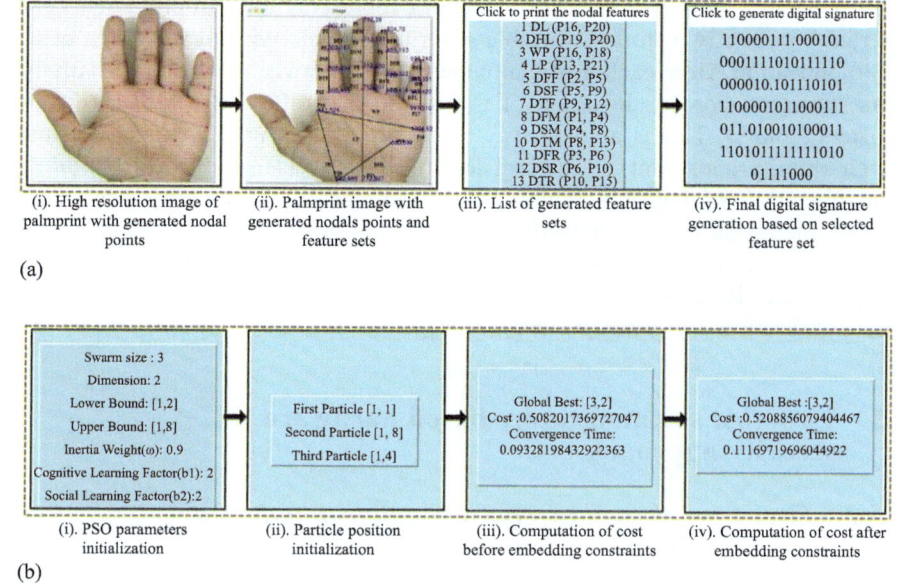

Figure 7.3   *(a) Signature generated using palmprint biometric security of IP cores. (b) Outputs generated using PSO-driven design space exploration.*

generation of nodal points, feature sets are determined. The biometric signature is generated based on the chosen feature set. Each nodal point on the palmprint is assigned with a unique name (identifier), as depicted in Figure 7.3(a) (i). Then, on the basis of nodal points, different features are generated, shown in Figure 7.3(a) (ii) and Table 7.2(a). After that, the dimension (magnitude of the selected feature) corresponding to each feature is computed with the help of obtained coordinates of each nodal point; Manhattan distance has been used to calculate the dimension (magnitude) of straight-line features and Pythagoras theorem for the remaining features (non-straight-line features). After that, all the decimal equivalents of the features are converted to their corresponding binary values. Finally, concatenation of different generated features is performed as per the required size of the signature to obtain the final palmprint-based signature. The selected feature set is highlighted in Table 7.2(a). The final generated template post-concatenation (of features Z1–Z5) is as follows: "110000111.00010100011110101111110000010.10111010111 00001011000111011.01001010001111010111101001111000." The generated digital template contains 104 digits. But for the sake of demonstration, we have only considered the first 47 ones and 43 zeros (81 bits). Decimal points are discarded here. After generating the palmprint signature, the signature template is then mapped to its equivalent hardware security constraints based on the mapping rule mentioned in Table 7.2(b). The equivalent hardware security constraints generated is embedded during the register allocation phase of HLS framework.

Each generated hardware security constraint represents an artificial edge being inserted between two storage variable pairs of the colored interval graph of the DSP application. Here each artificial edge inserted between two storage variable pairs represents forced allocation of the respective storage variables in distinct registers (colors). This is a cover way of strictly allocating IP vendor signature-driven security constraints in the register allocation phase of HLS framework by local alteration of registers (colors) to accommodate the security constraints. Finally, the constraints embedded register allocation phase of the design is converted into its respective RTL design of the target application (Note: The details of mapping rule, hardware security constraint generation, and embedding are discussed in Section 7.6; Sengupta *et al.*, 2021).

7.5 Low-cost steganography-based hardware security approach

7.5.1 Overview of the low-cost steganography-based hardware security approach

The section discusses a hardware security methodology based on a low-cost steganographic design flow to generate an optimized secure hardware architectural design solution. It also integrates PSO-based design space exploration with the steganographic design flow to explore an optimized (in terms of design area and latency) secured hardware architecture. The respective flow diagram is shown in Figure 7.4. Since embedding steganographic constraints into the IP design may incur design overhead, hence, it is vital to consider optimizing the possible overhead, if any. Therefore, it is imperative to determine an optimized architectural solution corresponding to the IP core hardware among numerous potential design solutions available in the design space before embedding steganographic security constraints into the reusable IP cores. Therefore, particle swarm optimization-driven design space exploration is integrated into the design flow to generate a low-cost design architecture corresponding to the target reusable IP cores. With the help of the embedded unique steganographic constraint, any counterfeited IP can be easily detected and isolated from original/genuine IP during the counterfeit detection process. Additionally, this will also secure the IP owner from any possible fraudulent claim of IP ownership (Mishra and Sengupta, 2014; Sengupta and Rathor, 2019).

7.5.2 Details of steganography-based hardware security approach

Steganography is a process of hiding secret information in some kind of file (such as an image or design file). In the discussed approach (Figure 7.4), steganography-based hardware secret security constraints are embedded in the register transfer level design file using the HLS framework of the IP design process to secure the IP against hardware threats like counterfeiting, piracy, and fraudulent claim of

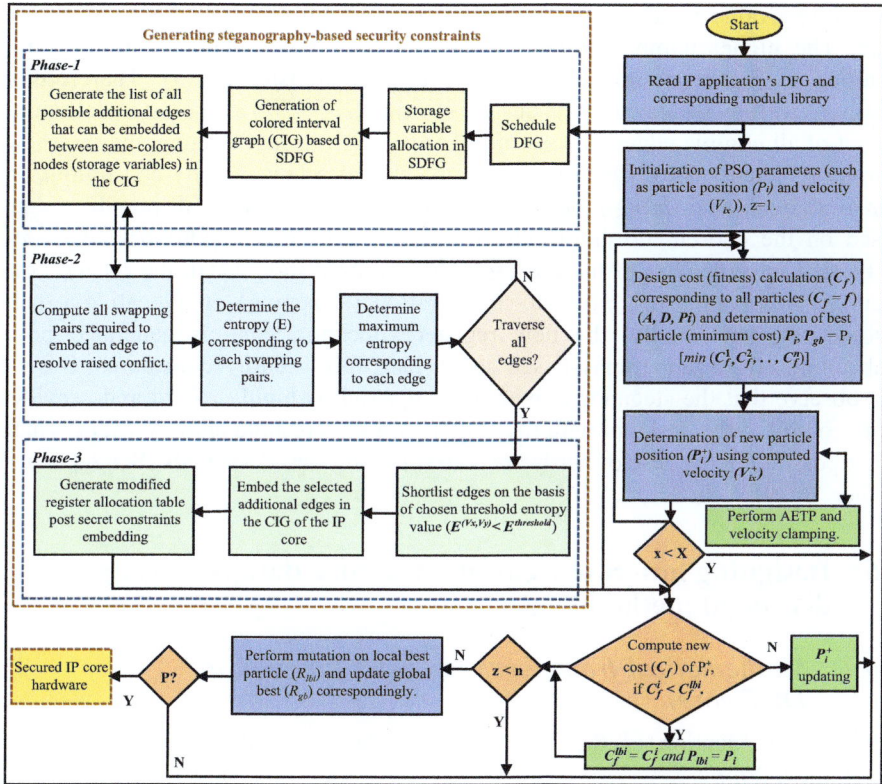

Figure 7.4   Security-aware design flow based on low-cost PSO-driven DSE steganography approach

ownership. The complete process of generating and embedding steganography-based hardware security constraints is divided into three phases: (a) *Phase 1:* first, the corresponding IP core application's DFG (data flow graph) is accepted as input. LIST scheduling generates corresponding scheduled DFG with explored low-cost resource architectural solution obtained using PSO-driven DSE (discussed in the next section). Based on SDFG, a register allocation table and colored interval graph (CIG) are created corresponding to IP hardware. All of the adjacent nodes (storage variables) of CIG must have different colors. The final output of *phase 1* is the set of all possible edges between the nodes of the same colors, which could be added in the CIG as an additional steganographic constraint. (b) *Phase 2:* now, embedding an edge between two same-colored nodes is not practically possible in the CIG (Ziener and Teich, 2008); thus, the conflict raised is resolved either by local alteration (swapping) of storage variables or by allocating the storage variable to a new colored register. Therefore, multiple color-swapping options are available to implant an additional edge between node pairs (V_X, V_Y). After that, corresponding edge entropy (E) is determined based on color swapping/transformation.

The entropy value equals the number of color transformations required to embed an edge. The higher number of color transformation yields higher entropy value resulting in higher design cost (more data path transformation). Therefore, a threshold entropy (selected by IC designer) value ($E^{threshold}$) is chosen to determine the set of all eligible edges (the set of edges having entropy less than the threshold entropy selected value), which will be embedded as secret security constraints. (c) *Phase 3* (*stego-embedding phase*): after determining the list of all possible edges based on the chosen threshold entropy value, all of these additional edges are embedded into the CIG of the corresponding IP hardware. Then, the related conflicts are resolved using either local alteration (swapping) or allocating a new colored register. Finally, the new register allocation table is generated post-embedding all steganography-based secret hardware security constraints. So, we can observe that the steganography-based approach is highly robust and provides more control to the designer with the help of chosen threshold entropy value, which is not possible in the case of hardware watermarking-based methods (Sengupta and Rathor, 2019).

7.6   Designing low-cost secured DCT core datapath using discussed methodologies

7.6.1   Mathematical framework (transfer function for DCT core)

The general equation of the DCT core can be formulated as (Obukhov and Kharlamov, 2008):

$$T(x) = z(x)\sqrt{\frac{2}{N}} \sum_{i=0}^{N-1} t(i) \cos\left[\frac{(2i+1)x\pi}{2N}\right] \tag{7.4}$$

For the above equation,

$$z(x) = \begin{cases} \dfrac{1}{\sqrt{2}}; f\ or\ x = 0 \\ 1; f\ or\ x \neq 0 \end{cases}; x = 0, 1, 2, 3, ..., N-1$$

Now, $t(i)$ is the input signal, $t(x)$ is the output signal, and N expresses the total number of data points:

$$T(0) = \frac{1}{\sqrt{2}}\sqrt{\frac{2}{4}} \sum_{i=i}^{i=3} t(i)\cos[0]$$

$$= \frac{1}{2}t(0) + \frac{1}{2}t(1) + \frac{1}{2}t(2) + \frac{1}{2}t(3)$$

$$= 0.5t(0) + 0.5t(1) + 0.5t(2) + 0.5t(3)$$

Similarly, the output signal values for $T(1)$ to $T(3)$ are computed:

$$T(1) = 0.65t(0) + 0.27t(1) - 0.27t(2) - 0.65t(3)$$

$$T(2) = 0.5t(0) + 0.5t(1) - 0.5t(2) - 0.5t(3)$$

$$T(3) = 0.27t(0) + 0.65t(1) - 0.65t(2) - 0.27t(3)$$

The final matrix expressing the output signals of 4-point DCT in matric multiplication form is given below:

$$\begin{bmatrix} T(1) \\ T(2) \\ T(3) \\ T(4) \end{bmatrix} = \begin{bmatrix} 0.5 & 0.5 & 0.5 & 0.5 \\ 0.65 & 0.27 & -0.27 & 0.65 \\ 0.5 & -0.5 & -0.5 & 0.5 \\ 0.27 & -0.65 & 0.65 & -0.27 \end{bmatrix} \begin{bmatrix} T(1) \\ T(2) \\ T(3) \\ T(4) \end{bmatrix}$$

The above matrix multiplication form expression can also be expressed as:

$$\begin{bmatrix} T(1) \\ T(2) \\ T(3) \\ T(4) \end{bmatrix} = \begin{bmatrix} C_1 & C_1 & C_1 & C_1 \\ C_2 & C_3 & -C_3 & -C_2 \\ C_1 & -C_1 & -C_1 & C_1 \\ C_3 & -C_2 & C_2 & -C_3 \end{bmatrix} \begin{bmatrix} T(1) \\ T(2) \\ T(3) \\ T(4) \end{bmatrix}$$

The general equation of 8-point DCT for computing the first output signal $(T(0))$ is expressed as (k_1 to k_4 are DCT coefficients):

$$T(0) = k_1 * t(0) + k_2 * t(1) + k_3 * t(2) + k_4 * t(3) \tag{7.5}$$

7.6.2 Designing DCT core datapath using low-cost palmprint biometric hardware security

The obtained DCT function in (7.5) is used to derive the data flow graph (DFG) shown in Figure 7.5(a). The discussed low-cost palmprint biometric hardware security is scalable in nature due to the possibility of generating a high number of security constraints corresponding to inclusion of larger number of features in the palm feature set. However, for the sake of brevity, this chapter presents the demonstration of discussed approach on a small application (DCT core). Figure 7.5 (a) and (b) illustrates the scheduled DFG of a 4-point DCT core (T0–T10 are storage variables used in different control steps). According to mapping rule explained in Table 7.2(b), an (even, even) node pair (storage variables is embedded into the CIG (such as T0, T2) corresponding to 0 and an (odd, odd) node pair is embedded into the CIG (such as T1, T3), and so on. The generated security constraints, using the encoding (mapping) rule explained in Table 7.2(b), are obtained as: <T1, T3>, <T1, T5>, <T1, T7>, ——————————————, <T13, T15>, <T13, T17>, <T13, T19> and <T0, T2>, <T0, T4>, <T0, T6>, ——————————————, <T6, T12>, <T6, T14>. Here T0, T1, T2, etc., are the storage variables shown in Figure 7.5(a) and (b). In Figure 7.5(a) and (b), storage variables (and their registers) in the scheduled DFG of 4-point DCT (before and after embedding the generated secret hardware security constraints) are depicted. The register allocation table of storage variables (before and after embedding the palmprint signature) of 4-point DCT has been shown in Tables 7.3 and 7.4, respectively. The secret security constraints corresponding to 34 0s and 47 1s have already been highlighted above. Subsequently, these generated extra hardware security constraints are embedded in the colored interval graph (CIG), representing the register allocation phase of the design process of the 4-point DCT. Figure 7.5(a) illustrates the CIG of 4-point DCT core before embedding hardware security constraints. Embedding these constraints (additional edges) between the nodes of similar color (in CIG) is not viable. Therefore, to avoid such conflicts, either local alteration (swapping) between the register allocation of storage variable is performed, or a new register (color) is used to meet the modified CIG of the target IP core. For example, storage variables T1 and T5 are allocated on the green-colored register in Table 7.3 (register allocation

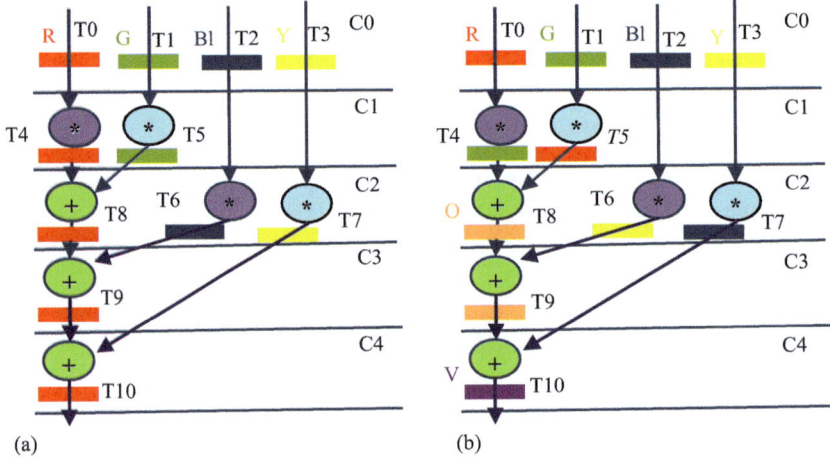

(a)   (b)

Figure 7.5   (a) Scheduled data flow graph of 4-point DCT with 1 adder and
2 multipliers before secret constraint embedding. (b) Scheduled data
flow graph of 4-point DCT with 1 adder and 2 multipliers after
embedding secret constraint.

Table 7.3   Register allocation of storage variable (T0–T10) of
DCT-4

	R	**G**	**Bl**	**Y**
C0	T0	T1	T2	T3
C1	T4	T5	T2	T3
C2	T8	–	T6	T7
C3	T9	–	–	T7
C4	T10	–	–	–

Table 7.4 *Register allocation of storage variable (T0–T10) of DCT-4 post signature embedding*

	R	G	Bl	Y	O	V
C0	T0	T1	T2	T3	–	–
C1	T5	T4	T2	T3	–	–
C2	–	–	T7	T6	T8	–
C3	–	–	–	T6	T9	–
C4	–	–	–	–	–	T10

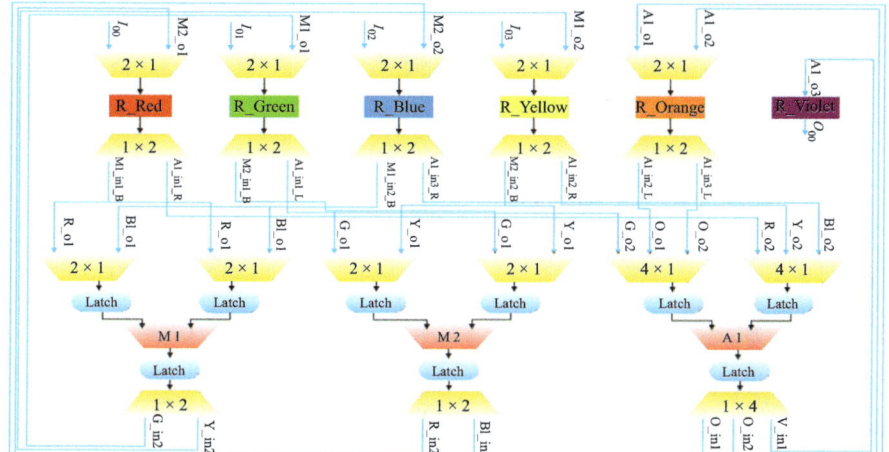

Figure 7.6 *Register transfer level (RTL) datapath of low-cost palmprint hardware security constraints secured 4-point DCT IP core*

table of storage variable of DCT-4). But, due to the embedding of additional hardware security constraint <T1, T5>, the position of T4 and T5 is swapped in Table 7.4. Similarly, storage variables T8, T9, and T10 are earlier allocated on the red-colored register in Table 7.3. But, due to <T0, T8>, <T5, T9>, and <T0, T10>, they are now allocated to new orange-, orange-, and violet-colored registers, respectively. The same procedure is repeated for all the specified security constraints. Figure 7.6 illustrates the register transfer level datapath of discrete cosine transforms IP core corresponding to embedded hardware security constraints obtained using palmprint biometric security approach. Table 7.5 shows the respective control signals for the secured DCT core datapath designed. The control signal table shows the corresponding control signals for the six registers used and three functional units used in the datapath design. Figure 7.7 illustrates the schematic design implementation of DCT core RTL datapath using palmprint biometric security approach.

Table 7.5　*Controller signals of secured low-cost palmprint hardware security constraints embedded RTL datapath of DCT*

Registers	Control signals	Functional units	Control signals	Functional units	Control signals
R_Red	2:1 Mux Selector_Red	**Multiplier (M1)**	2:1 Mux Selector_M1	**R_Violet**	Register_Violet Strobe_V
	Register_Red-Strobe_R		Input Latch Strobe_L	–	
	1:2 Demux Deselect_Red		Multiplier unit En_M1		
R_Green	2:1 Mux Selector_Green		Output Latch Strobe_L		
	Register_Green Strobe_G		1:2 Demux Deselect_M1		
	1:2 Demux Deselect_Green	**Multiplier (M2)**	2:1 Mux Selector_M2		
R_Blue	2:1 Mux Selector_Blue		Input Latch Strobe_L		
	Register_Blue Strobe_B		Multiplier unit En_M2		
	1:2 Demux Deselect_Blue		Output Latch Strobe_L		
R_Yellow	2:1 Mux Selector_Yellow		1:2 Demux Deselect_M2		
	Register_Yellow Strobe_Y	**Adder (A1)**	4:1 Mux Selector_A1		
	1:2 Demux Deselect_Yellow		Input Latch Strobe_L		
R_Orange	2:1 Mux Selector_Orange		Adder unit En_A1		
	Register_Orange Strobe_O		Output Latch Strobe_L		
	1:2 DemuxDeselect_Orange		1:4 Demux Deselect_A1		

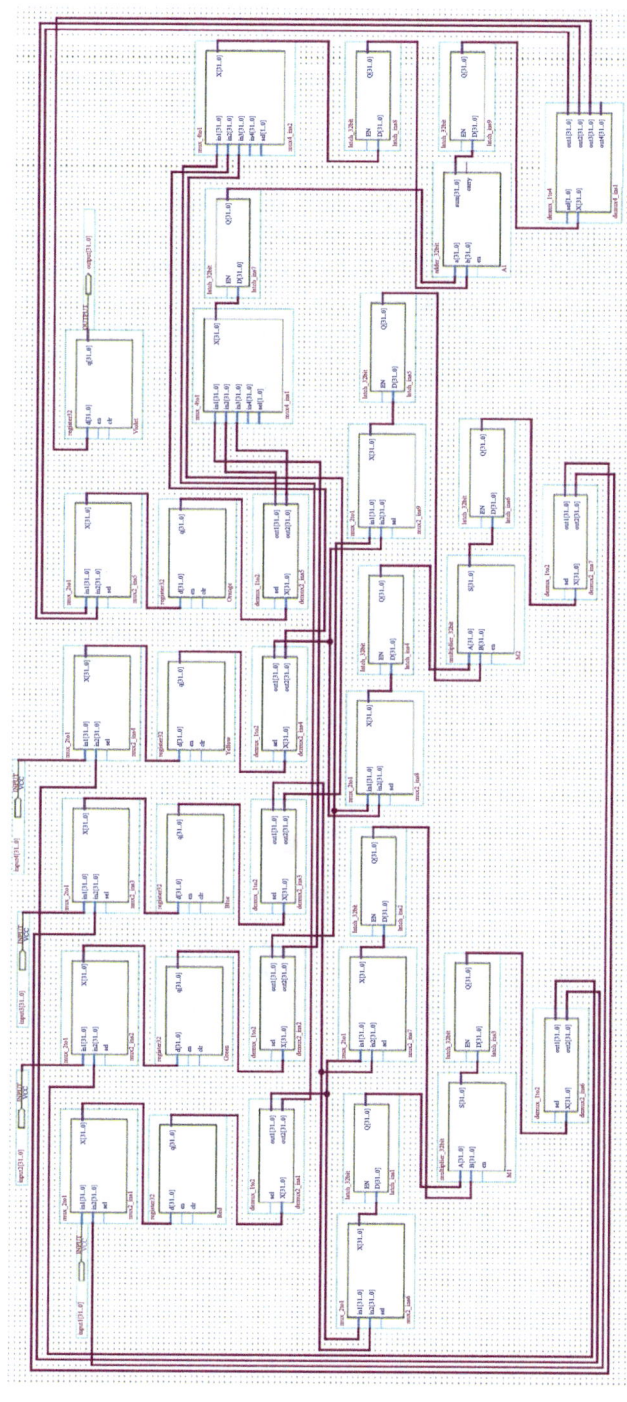

Figure 7.7   Schematic implementation of low-cost palmprint biometric-based hardware security approach

7.6.3   Designing DCT core datapath using low-cost steganographic-based hardware security

This sub-section presents the designing process of the secured DCT core using steganography-based security-aware design flow. The design process is initiated by generating the respective colored interval graph (CIG) as depicted in Figure 7.8(a) of 4-point DCT core which is used to implant additional edges chosen based on threshold entropy value (that acts as a security parameter). First, the entropy value for all possible additional edges between same-colored registers is computed. The concept of entropy in the context of hardware IP steganography has been discussed earlier in Section 7.5.2. A set of additional edges is selected based on chosen threshold entropy value. The set of all possible edges between same-colored registers for 4-point DCT is <T1, T5>, <T2, T6>, <T3, T7>, <T0, T4>, <T0, T9>, <T0, T10>, <T4, T8>, <T4, T9>, <T4, T10>, <T8, T9>, <T8, T10>, <T9, T10> (generated from register allocation in Table 7.6). There are two ways

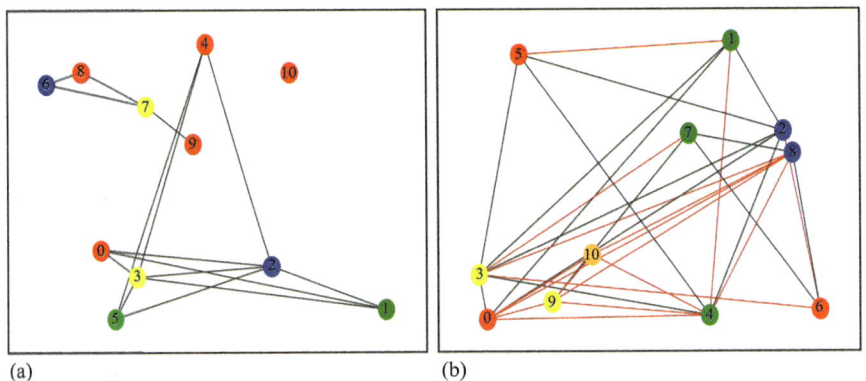

(a)   (b)

Figure 7.8   *(a) Colored interval graph (CIG) of 4-point DCT before embedding secret hardware constraints. (b) Colored interval graph (CIG) of 4-point DCT after embedding secret hardware steganographic constraints.*

Table 7.6   *Register allocation of storage variable (T0–T10) of DCT-4*

	R	G	Bl	Y
C0	T0	T1	T2	T3
C1	T4	T5	T2	T3
C2	T8	–	T6	T7
C3	T9	–	–	T7
C4	T10	–	–	–

to resolve the conflict while implanting an additional edge: (i) local alteration (swapping) in the colors of corresponding storage variables (this will increase entropy by two due to two color transformations), and (ii) allocation of storage variable on a new colored register (this will increase entropy by 1). Let us take an example of embedding the edge <T1, T5>: ((([CS:0] (T1 ⇔ T0), E (entropy) = 2, (R ⇔ G)), ([CS:1] (T5 ⇔ T4), E (entropy)=2, (R ⇔))). So, the maximum entropy for embedding edge <T1, T5> is 2. Here, while embedding edge <T1, T5>, (i) the colors of storage variables T1 and T0 are swapped in control step (CS)-0 between the *Red (R) register* and the *Green (G) register* to resolve the conflict. The resultant *Entropy (E)* value is two due to two color transformations. This is the same in control step (CS)-1 between T5 and T4. Therefore, the maximum entropy value is 2 (maximum of both) for embedding the edge <T1, T5>. And, for <T2, T6>: ((([CS:0, 1] (T2 ⇔ T0, T4), E=3, (R ⇔ BL)), ([CS:0, 1] (T2 ⇔ T1, T5), E=3, (G ⇔ BL)), ([CS:0, 1] (T2 T2 ⇔ T3 T3), E=2, (BL ⇔ Y)), ([CS:2] (T6 ⇔ T8), E=2, (BL ⇔ R)), ([CS:2,3] (T6 ⇔ T7), E=2, (BL ⇔ Y)), ([CS:2] (T6 ⇔ G), E=1)). So, the maximum entropy value for implanting edge <T2, T6> is 4. Similarly, we will compute the entropy for the remaining additional edges. Refer to Table 7.7. Now, the shortlisted edges for embedding (on the basis of designer-selected threshold entropy i.e., $E^{threshold}$ = 3) are <T1, T5>, <T2, T6>, <T3, T7>, <T0, T4>, <T0, T9>, <T0, T10>, <T4, T8>, <T4, T9>, <T4, T10>, <T8, T9>, <T8, T10>, <T9, T10>. That is, all edges that have an entropy threshold below or equal to three are shortlisted. After embedding the additional edges (constraints), the final register allocation table is shown in Table 7.8. Figure 7.9 illustrates the designed secured RTL datapath of the DCT core corresponding to embedded hardware security constraints obtained using the steganography-based approach. Table 7.9 describes the control signals for the steganographic hardware security constraints embedded RTL datapath of the DCT core. Figure 7.10 illustrates the schematic design implementation of DCT core RTL datapath using steganography security approach.

Table 7.7   Shortlisted list of possible edges and their corresponding maximum entropy value for 4-point DCT

Possible edge	Maximum entropy	Possible edge	Maximum entropy
<T1, T5>	2	<T0, T9>	3
<T2, T6>	3	<T0, T10>	3
<T3, T7>	3	<T4, T8>	3
<T0, T4>	2	<T4, T9>	3
<T0, T8>	3	<T8, T10>	3
<T4, T10>	3	<T9, T10>	3
<T8, T9>	2	–	–

Table 7.8   *Register allocation of storage variable (T0–T10) of DCT-4 post signature embedding*

	R	G	Bl	Y	O
C0	T0	T1	T2	T3	–
C1	T5	T4	T2	T3	–
C2	T6	T7	T8	–	–
C3	–	T7	–	T9	–
C4	–	–	–	–	T10

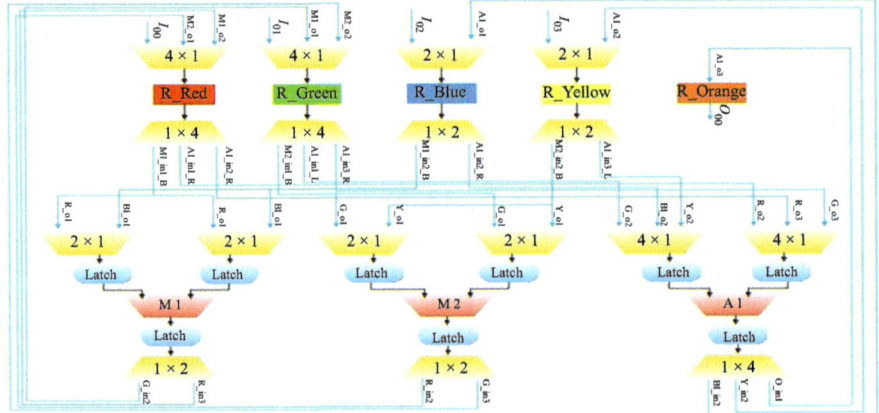

Figure 7.9   *Secured RTL datapath of 4-point DCT IP core embedding low-cost steganographic hardware security constraints*

7.7   Analysis and discussions

The discussed approaches and its validation during experimental analysis have been performed on Intel(R) Core (TM) i7-10510U CPU @ 1.80 GHz 2.30 GHz processor with an 8GB RAM configuration system. The parameters used for experimental evaluation of low-cost palmprint biometric security approach are palmprint biometric of original IP vendor, biometric signature strength, palmprint features, and the number of features in the palmprint feature set. The value of threshold entropy used in the experiment is three for the low-cot steganography-based approach. The common PSO-DSE parameters for both approaches are swarm size $(S) = 3$; acceleration coefficient (b_1 and b_2) = 2; inertia weight (ω) = linearly decreasing between 0.9 and 0.1; random numbers (r_1 and r_2) = 1, stopping

Table 7.9   *Controller signals of secured low-cost steganographic constraints embedded RTL datapath of DCT*

Registers	Control signals	Functional units	Control signals
R_Red	4:1 Mux Selector_Red Register_Red Strobe_R 1:4 Demux Deselect_Red	**Multiplier (M1)**	2:1 Mux Selector_M1 Input Latch Strobe_L Multiplier unit En_M1
R_Green	4:1 Mux Selector_Green Register_Green Strobe_G 1:4 Demux Deselect_Green		Output Latch Strobe_L 1:2 Demux Deselect_M1
		Multiplier (M2)	2:1 Mux Selector_M2
R_Blue	2:1 Mux Selector_Blue Register_Blue Strobe_B 1:2 Demux Deselect_Blue		Input Latch Strobe_L Multiplier unit En_M2 Output Latch Strobe_L
R_Yellow	2:1 Mux Selector_Yellow Register_Yellow Strobe_Y 1:2 Demux Deselect_Yellow	**Adder (A1)**	1:2 Demux Deselect_M2 4:1 Mux Selector_A1 Input Latch Strobe_L
–	–		Adder unit En_A1
	–		Output Latch Strobe_L
	–		1:4 Demux Deselect_A1
R_Orange	Register_Orange Strobe_O	–	–

criterion $= P_1$ (no improvement in cost till ten consecutive iterations) or P_2 (exhaustion of total iteration, $P = 50$). To evaluate the area and the latency (delay) of the IP core benchmarks (adopted from (EXP BEN), Jain and Panda, 2007) of the IP design, a 15-nm technology scale based on the Nan Gate library (15 nm open cell library) has been used.

IP core application is available in the form of DFG and module library. The control data flow graph (CDFG) of the respective IP core hardware is generated with the help of functional description of the respective IP core hardware, which is then provided as the input to the experiment for the calculation of design area, latency, and cost. The conversion of hardware's functional description into its corresponding CDFG involves parsing the transfer function of respective IP

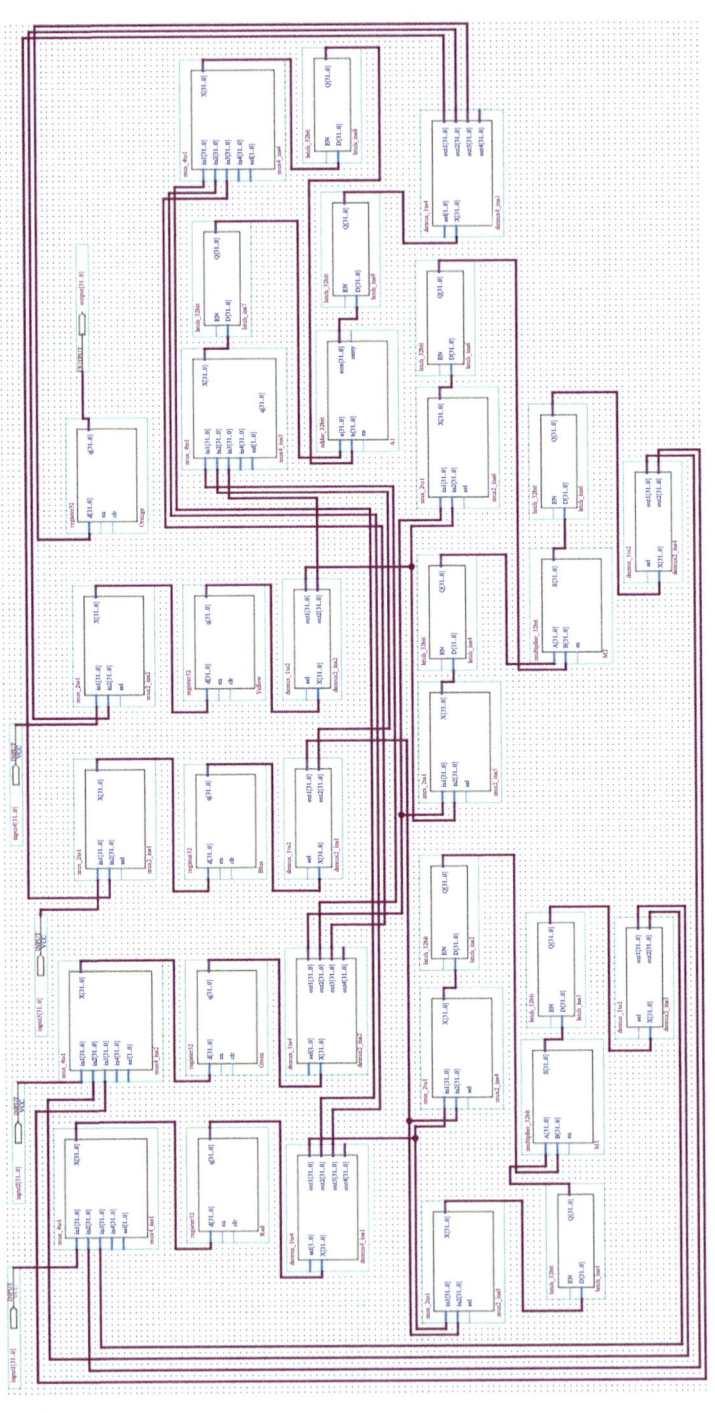

Figure 7.10   Schematic implementation of low-cost steganography-based hardware security approach

hardware into matrix multiplication format (comprising of input data samples, output samples, and DCT-coefficients). Further, mathematical function computing nth-output data samples is formed from the matrix multiplication format. Finally, the obtained mathematical function is depicted with a connected graph (where mathematical operations are the nodes of the graph and data dependency between the nodes is represented with the help of edges). CDFG is converted into SDFG using LIST scheduling.

7.7.1 Design cost analysis of low-cost palmprint biometric-based security approach

Table 7.10 describes the obtained design area, latency (delay), and cost of the IP core before and after implanting the low-cost palmprint biometric-based hardware security constraints. It also contains the final explored low-cost resource architectural solutions corresponding to respective IP core acquired through PSO-driven DSE. For example, the design area corresponding to FIR IP core before and after implanted palmprint-biometric hardware security constraints is 213.91 μm^2 and 219.41 μm^2, respectively. Table 7.11 shows that with an average cost overhead of 0.28 %, the low-cost palmprint-based hardware security approach can be implemented to provide security against piracy and counterfeiting and generate low-cost resource architecture against respective IP core. A comparison of the low-cost palmprint-based hardware security approach in terms of design cost is illustrated in Figure 7.11 with Sengupta and Bhadauria (2016). Table 7.12 illustrates that low-cost palmprint-based hardware security approach is more optimized as compared to Sengupta and Bhadauria (2016) as it shows an average cost reduction of 10.43% due to the integration of PSO-DSE. Additionally, the reductions in the design cost obtained through the low-cost palmprint-based hardware security approach compared to Koushanfar *et al.* (2005) have been reported in Figure 7.11 and Table 7.12, respectively. The average implementation run-time for low-cost palmprint-based approach is 95.75 ms.

7.7.2 Design cost analysis of low-cost steganography-based security approach

Similarly, Table 7.13 describes the obtained design area, latency (delay), and cost of the IP core before and after implanting the low-cost steganography-based hardware security constraints. It also contains the final explored low-cost resource architectural solutions corresponding to respective IP core, acquired through PSO-driven DSE. For example, the design area corresponding to FIR IP core before and after implanted palmprint-biometric hardware security constraints are 213.91 μm^2 and 213.91 μm^2, respectively. Table 7.11 shows that with an average cost overhead of 0.05%, the low-cost biometric approach can be implemented to provide security against piracy and counterfeiting and generate low-cost resource architecture against respective IP core. A comparison of the low-cost biometric approach in terms of cost is illustrated in Figure 7.8 with watermarking approach (Sengupta and Bhadauria, 2016). Table 7.12 illustrates that our low-cost approach is more

Table 7.10 *Area, latency, cost, and resource configuration of low-cost palmprint-based hardware security approach*

Benchmarks	Explored resource configuration	Design area (µm²) before signature embedding	Design area (µm²) after signature embedding	Latency (ps) before signature embedding	Latency (ps) after signature embedding	Design cost before signature embedding	Design cost after signature embedding
FIR	3(+), 2(*)	213.91	219.41	1,391.09	1,391.09	0.410	0.412
8-point IDCT	1(+), 2(*)	176.16	178.52	1,324.86	1,324.86	0.443	0.445
JPEG-sample	2(+), 1(*)	122.68	122.68	1,656.07	1,656.07	0.588	0.588
4-point DCT	1(+), 2(*)	173.02	174.59	662.43	662.43	0.561	0.562

Table 7.11   Cost overhead of discussed methodologies before and after embedding hardware security constraints

Benchmarks	Cost overhead of low-cost palmprint-based approach	Cost overhead of low-cost steganography-based approach
FIR	0.49%	0%
8-point IDCT	0.45%	0%
JPEG-sample	0%	0%
4-point DCT	0.18%	0.18%

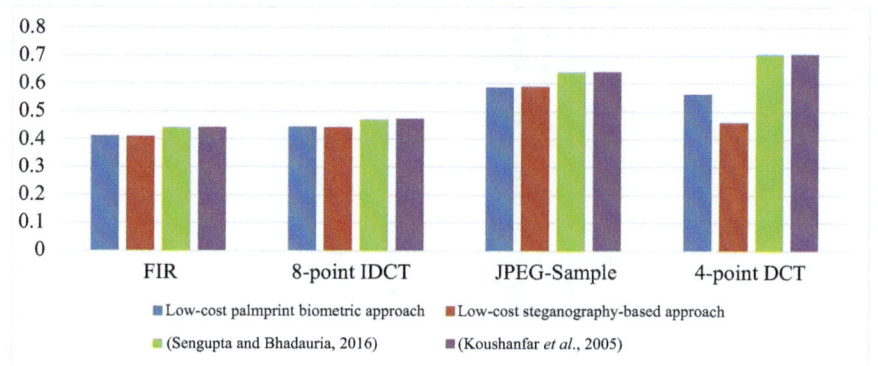

Figure 7.11   Comparison of design cost between the discussed approaches (Koushanfar et al., 2005; Sengupta and Bhadauria, 2016)

Table 7.12   Reduction in cost of discussed methodologies compared to Sengupta and Bhadauria (2016) and Koushanfar et al. (2005)

Benchmarks	Design cost reduction of low-cost palmprint-based approach compared to Sengupta and Bhadauria (2016)	Design cost reduction of low-cost steganography-based approach compared to Sengupta and Bhadauria (2016)	Design cost reduction of low-cost palmprint-based approach compared to Koushanfar et al. (2005)	Design cost reduction of low-cost steganography-based approach compared to Koushanfar et al. (2005)
FIR	7.21%	7.66%	7.21%	7.66%
8-Point IDCT	5.72%	6.14%	5.72%	6.14%
JPEG-sample	8.40%	8.40%	8.40%	8.40%
4-Point DCT	20.40%	20.40%	20.40%	20.40%

Table 7.13 *Area, latency, cost, and resource configuration of low-cost steganography-based hardware security approach*

Benchmarks	Explored resource configuration	Design area (µm²) before signature embedding	Design area (µm²) after signature embedding	Latency (ps) before signature embedding	Latency (ps) after signature embedding	Design cost before signature embedding	Design cost after signature embedding
FIR	3(+), 2(*)	213.91	213.91	1,391.09	1,391.09	0.410	0.410
8-point IDCT	1(+), 2(*)	176.16	176.16	1,324.86	1,324.86	0.443	0.443
JPEG-sample	2(+), 1(*)	122.68	122.68	1,656.07	1,656.07	0.589	0.589
4-point DCT	1(+), 2(*)	173.02	173.80	662.43	662.43	0.561	0.562

optimized as compared to the hardware watermarking (Sengupta and Bhadauria, 2016) approach as it shows an average cost reduction of 10.65% due to the integration of PSO-DSE. Additionally, the reductions in the design cost obtained through the low-cost biometric approach compared to Koushanfar *et al.* (2005) have been reported in Figure 7.11 and Table 7.12, respectively. The average implementation run-time for this low-cost biometric approach is 160.75 ms.

7.7.3 Security analysis of low-cost palmprint biometric-based security approach

A comparison of the security strength (robustness) measured through the probability of coincidence (P_C) between the low-cost palmprint biometric security approach and Sengupta and Bhadauria (2016) and Koushanfar *et al.* (2005) are shown in Table 7.14 and Figure 7.12, respectively. The probability of coincidence (P_C) has been used to measure the strength of low-cost biometric method. It represents the probability of obtaining the same palmprint signature in a non-palmprint embedded (baseline unsecured) design. The formulation of PC is given as:

$$P_C = (1 - 1/t)^x \qquad (7.6)$$

where "t" is the number of colors in the CIG of IP design before embedding and "x" is the number of edges (constraints) embedded. The lower the P_C, the better will be the security. As evident, Table 7.12 and Figure 7.12 depict that the low-cost palmprint-based hardware security approach achieves desirable lower Pc compared to Sengupta and Bhadauria (2016) and (Koushanfar *et al.* (2005) due to the higher strength of embedded palmprint biometric (owing to more palmprint features, resulting in more hardware security constraints), thereby providing stronger digital evidence and higher robustness. On an average, the low-cost palmprint-based hardware security approach reduces P_C below half the probability of coincidence

Table 7.14   The P_C (indicating the strength of digital evidence) of the discussed approaches compared to Sengupta and Bhadauria (2016) and Koushanfar et al. (2005)

Security approach	FIR	8-Point IDCT	JPEG P_C	4-Point DCT
	#E_C (Effective constraints embedded)			
Low-cost palmprint-based hardware security	2.0E−5 81	2.0E−5 81	3.42E−65 81	7.56E−11 81
Low-cost steganography-based hardware security	2.0E−1 12	3.2E−3 43	4.95E−19 23	2.38E−2 13
Hardware watermarking approach (Sengupta and Bhadauria, 2016)	2.6E−1 10	1.3E−1 15	1.15E−12 15	5.6E−2 10
Hardware watermarking approach (Koushanfar *et al.*, 2005)	2.6E−1 10	1.3E−1 15	1.15E−12 15	5.6E−2 10

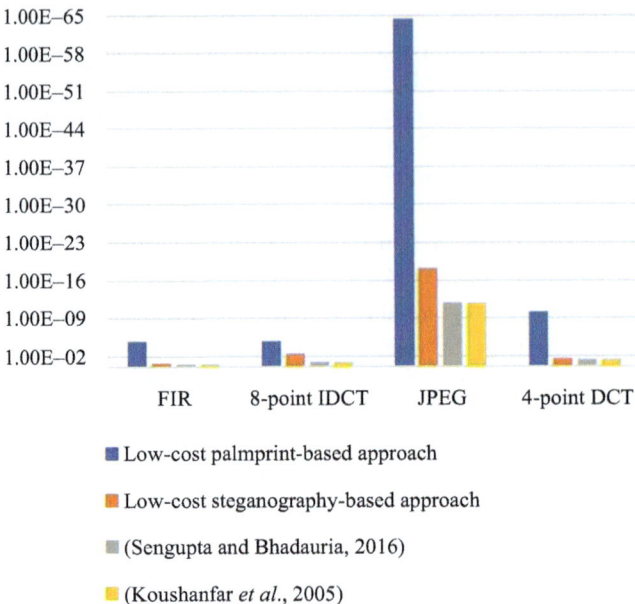

1.00E–65
1.00E–58
1.00E–51
1.00E–44
1.00E–37
1.00E–30
1.00E–23
1.00E–16
1.00E–09
1.00E–02

FIR   8-point IDCT   JPEG   4-point DCT

■ Low-cost palmprint-based approach

■ Low-cost steganography-based approach

■ (Sengupta and Bhadauria, 2016)

■ (Koushanfar *et al.*, 2005)

*Figure 7.12   Comparison of probability of coincidence between the discussed
approaches and Sengupta and Bhadauria (2016) and Koushanfar
et al. (2005)*

compared to Sengupta and Bhadauria (2016) and Koushanfar *et al.* (2005), thereby
enhancing digital evidence security by more than 2×.

The palmprint biometric signature is highly robust and difficult to guess from an
adversarial perspective. The effort required to guess the exact palmprint biometric
signature embedded can be quantified as: $e = 1 \div 2^s$, where "s" stands for the number
of embedded palmprint signature bits. For example, if the value of "s" embedded is
81 bits, then the value of e is 4.1359031e−25. Further, the strength of the signature
space (X) is given as: 2^s, which is $X = 2.4178516e+24$, corresponding to $s = 81$ bits.
This is an estimation of the entropy of low-cost biometric approach in terms of the
hardness of the adversarial guessing and effort. In low-cost biometric approach, the
order in which the palmprint features are concatenated to obtain the palmprint signature is unknown to an adversary. In low-cost biometric approach, 14 palmprint
features can be concatenated to generate a robust palmprint signature. This results in
14! (Factorial 14) permutations. Therefore, the total effort (T_E) required from an
adversarial perspective is $1/n! * 1/2^s$, where n is the number of the palmprint features
used for the signature bit generation. For example, when $n = 14$ and $s = 81$, then
$T_E = 4.7441894e−36$. The larger the value of n and s, the higher the effort required
from an adversarial perspective. Therefore, the higher is the low-cost biometric
approach's entropy.

Table 7.15 indicates the comparison of the tamper tolerance (TT) ability of
low-cost palmprint-based hardware security approach with Sengupta and

Table 7.15 *Comparison of tamper tolerance ability between the discussed approaches and Sengupta and Bhadauria (2016) and Koushanfar et al. (2005)*

Security approach	FIR	8-Point IDCT	JPEG TT	4-Point DCT
		#E$_C$ (Effective constraints embedded)		
Low-cost palmprint-based hardware security	4.43E+38 81	4.43E+38 81	4.43E+38 81	4.43E+38 81
Hardware watermarking approach (Sengupta and Bhadauria, 2016)	1.04E+6 10	1.07E+9 15	1.07E+9 15	1.04E+6 10
Hardware watermarking approach (Koushanfar et al., 2005)	1.02E+3 10	3.2E+4 15	3.2E+4 15	1.02E+3 10

Bhadauria (2016) and Koushanfar *et al.* (2005) evaluated using the following function:

$$TT = v^w \qquad (7.7)$$

where "v" represents the signature encoded variables, and "w" indicates the number of security constraints embedded into the design. As evident from the table, the low-cost palmprint signature-based approach achieves higher TT than that of Sengupta and Bhadauria (2016) and Koushanfar *et al.* (2005) due to the larger number of signatures encoding variables and signature strength. This indicates higher robustness obtained using the low-cost palmprint signature-based approach. However, TT ability for low-cost steganographic approach has not been reported due to the unavailability of a signature encoding mechanism.

7.7.4 Security analysis of low-cost steganography-based security approach

This sub-section compares low-cost steganography security approach and Sengupta and Bhadauria (2016) and Koushanfar *et al.* (2005) in the context of robustness and security strength. Both low-cost steganography approach and Sengupta and Bhadauria (2016) and Koushanfar *et al.* (2005) specifically deal with the hardware threats such as IP hardware cloning, counterfeiting, piracy, and fraudulent claim of ownership. Here also, the probability of coincidence (P_C) has been used as a security metric to measure the strength of low-cost steganography and Sengupta and Bhadauria (2016) and Koushanfar *et al.* (2005) approaches.

From Table 7.14 and Figure 7.12, low-cost steganography approach shows a lower probability of coincidence than that of Sengupta and Bhadauria (2016) and Koushanfar *et al.* (2005) due to generating a significant number of constraints, leading to higher security and more robustness.

7.8   Conclusion

This chapter discusses two low-cost hardware security methodologies that generate optimized secured designs corresponding to IP core using PSO-DSE-driven palm-print biometric and steganography-based approach. Further, the chapter discusses designs of secured RTL datapath of DCT core using low-cost palmprint signature and steganographic constraints. The discussed approach is more robust and provides stronger digital evidence than Sengupta and Bhadauria (2016) and Koushanfar *et al.* (2005), as evidenced by the lower probability of coincidence and higher tamper tolerance. Securing the computing systems from threats like IP cloning and counterfeiting is critical as pirated IPs are unsafe and could jeopardize the safety of computing systems. With the help of the discussed low-cost security approaches, any pirated IP can be easily detected and isolated before integration into a computing system.

7.9   Questions and exercise

1. Explain the taxonomy of different emerging approaches in the field of protection of hardware IP cores.
2. What are the benefits of integration of design space exploration along with the security aspect of the hardware IP core?
3. Describe the design flow showing the integration of particle swarm optimization with the palmprint biometric and steganography-based hardware security methodologies.
4. What the benefits of PSO over other design space exploration algorithm such as genetic and bacterial foraging?
5. Explain the formulation of cost function for obtaining low-cost optimized output corresponding to IP vendor/designer specified high-level specifications.
6. Describe the summary of the low-cost palmprint biometric-based hardware security methodology and low-cost steganography-based hardware security methodology.
7. What are the advantages of using palmprint biometrics over fingerprint and facial biometrics methodologies for hardware security?
8. Briefly explain the different ISO/IEC standard 24745 properties in context with biometric-based hardware security approaches.
9. Describe the details of low-cost palmprint-based hardware security methodology.
10. Explain the different phases of low-cost steganography-based hardware security approach in details.
11. What distance metrics are used to compute the distance between two nodal points (feature dimension) of a palmprint biometric image?
12. Describe the process of generation of hardware security constraints from obtained signature template using low-cost palmprint-based security methodology.

13. Explain the process of generation of steganographic secret security constraints.
14. Determine the general equation for 8-point DCT for computing first output signal using DCT transfer function.
15. What do you means by entropy corresponding to a specific hardware security constraints?
16. What is the importance of threshold entropy in context to low-cost steganographic hardware security approach?
17. Demonstrates the embedding of generated hardware security constraints (either generated using low-cost palmprint-based security methodology or low-cost steganography-based security approach) into the design of a DSP application (take 6-order FIR DSP application for demonstration).
18. Explain the importance of colored interval graph during embedding of hardware security constraints.
19. Draw a comparison between the both low-cost security approaches based on the presence of digital evidence into the design and security strength of the security methodologies.
20. What is the design cost overhead as a result of embedding additional security constraints (in the form of additional edges) in the design (GIG) of the respective DSP application?
21. What is the total effort required from an adversary perspective in order to break the palmprint biometric-based hardware security methodology?
22. How the discussed both low-cost security methodologies helps in providing protection from IP piracy and fraudulent claim of IP ownership?

References

15 nm open cell library. [Online], Available: https://si2.org/open-celllibrary/, last accessed on January 2022.

CASIA Palmprint Database, NIST, Chinese Academy of Sciences, http://biometrics.idealtest.org/dbDetailForUser.do?id=5#/, accessed on January 2022.

Colombier, B. and L. Bossuet (2015), "Survey of hardware protection of design data for integrated circuits and intellectual properties," *IET Computers & Digital Techniques*, vol. 8, no. 6, pp. 274–287.

EXP BEN, University of California Santa Barbara Express Group. [Online]. Available: http://express.ece.ucsb.edu/benchmark/, accessed on March, 2022.

Jain, R. and P. R. Panda (2007), "An efficient pipelined VLSI architecture for lifting based 2D-discrete wavelet transform," in: *Proceedings of IEEE International Symposium on Circuits Systems (ISCAS)*, pp. 1377–1380.

Koushanfar, F., I. Hong, and M. Potkonjak 2005), "Behavioral synthesis techniques for intellectual property protection," *ACM Transactions on Design Automation of Electronic Systems*, vol. 10, no. 3, pp. 523–545.

Krishnan, V. and S. Katkoori (2006), "A genetic algorithm for the design space exploration of datapaths during high-level synthesis," *IEEE Transactions on Evolutionary Computation*, vol. 10, no. 3, pp. 213–229.

Le Gal, B. and L. Bossuet (2012), "Automatic low-cost IP watermarking technique based on output mark insertions," *Design Automation for Embedded Systems*, vol. 16, no. 2, pp. 71–92.

Mishra, V. P. and A. Sengupta (2014), "MO-PSE: adaptive multi objective particle swarm optimization based design space exploration in architectural synthesis for application specific processor design," *Advances in Engineering Software*, vol. 67, pp. 111–124.

Obukhov, A. and A. Kharlamov (2008), "Discrete cosine transform for 8x8 blocks with CUDA," *Nvidia White Paper Document*.

Pilato, C., S. Garg, K. Wu, R. Karri, and F. Regazzoni (2018), "Securing hardware accelerators: a new challenge for high-level synthesis," *IEEE Embedded Systems Letters*, vol. 10, no. 3, pp. 77–80.

Rizzo, S., F. Bertini, and D. Montesi (2019), "Fine-grain watermarking for intellectual property protection," *EURASIP Journal on Information Security*, 2019, vol. 10, p. 804.

Schneiderman, R. (2010), "DSPs evolving in consumer electronics applications [special reports]," *IEEE Signal Processing Magazine*, vol. 27, no. 3, pp. 6–10.

Sengupta, A. and S. Bhadauria (2016), "Exploring low cost optimal watermark for reusable IP cores during high level synthesis," *IEEE Access*, vol. 4, pp. 2198–2215.

Sengupta, A. and M. Rathor (2019), "IP core steganography for protecting DSP kernels used in CE systems," *IEEE Transactions on Consumer Electronics*, vol. 65, no. 4, pp. 506–515.

Sengupta, A. and M. Rathor (2020), "Securing hardware accelerators for CE systems using biometric fingerprinting," *IEEE Transactions on Very Large Scale Integration (VLSI) Systems*, vol. 28, no. 9, pp. 1979–1992.

Sengupta, A. and M. Rathor (2021), "Facial biometric for securing hardware accelerators," *IEEE Transactions on Very Large Scale Integration (VLSI) Systems*, vol. 29, no. 1, pp. 112–123.

Sengupta, A., R. Chaurasia, and T. Reddy (2021), "Contact-less palmprint biometric for securing DSP coprocessors used in CE systems," *IEEE Transactions on Consumer Electronics*, vol. 67, no. 3, pp. 202–213.

Zhang, J. (2016), "A practical logic obfuscation technique for hardware security," *Transactions on Very Large Scale Integration (VLSI) Systems*, vol. 24, no. 3, pp. 1193–1197.

Ziener, D. and J. Teich (2008), "Power signature watermarking of IP cores for FPGAs," *Journal of Signal Processing Systems*, vol. 51, no. 1, pp. 123–136.

Chapter 8

Methodology for exploration of security–design cost trade-off for signature-based security algorithms

Anirban Sengupta[1] and Rahul Chaurasia[1]

The chapter describes a methodology for the exploration of security–design cost trade-off for signature-based security algorithms for digital signal processing (DSP) hardware design architecture (Chaurasia and Sengupta, 2022d). In this methodology, security–design cost analysis has been discussed for three different hardware security algorithms based on facial biometrics, encrypted-hash, and hardware watermarking. Further, to generate an optimal hardware design architectural solution, particle swarm optimization (PSO)-based design space exploration (DSE) has been performed. Thus, signature-based security algorithms are integrated with the PSO-DSE framework for exploring the hardware architecture trade-offs of security–design cost. Therefore, this methodology enables the intellectual property (IP) core designer and/or consumer electronics (CE) integrator to decide the choice of their DSP hardware IP architecture such that it meets the end objective of robust security (against fake/pirated IPs) and lower design cost.

The organization of the chapter is as follows: Section 8.1 discusses about the introduction of the chapter; Section 8.2 discusses the motivation for performing security–design cost trade-off; Section 8.3 presents the discussion and analysis of similar existing work; Section 8.4 explains the details of the methodology for exploration of security–design cost trade-off for secure optimal hardware design architecture; Section 8.5 provides analysis and discussion and Section 8.6 concludes the chapter.

8.1 Introduction

Complex, data-intensive, and computationally intensive tasks require higher energy, resources, more space and time, etc. In large computing devices or systems, during such scenarios, hardware accelerators are used to achieve higher performance and efficacy by accelerating the process. During the acceleration process of an application, certain computing tasks are offloaded into specialized hardware

[1]Department of Computer Science and Engineering, Indian Institute of Technology Indore, India

components. In general, hardware acceleration is employed to improve application performance. For example, image and video processing tasks may be offloaded onto a graphics card to enable faster, higher-quality playback of videos and games. On the other hand, in tethering systems, operations involving tethering are offloaded onto a WiFi chip when acting as a WiFi hotspot. This, therefore, enables system performance by freeing up the processor. Further, several applications like cryptographic applications are performed using cryptographic IP cores while fingerprint, face recognition, and handprint biometric require DSP and image processing IP cores. Further artificial intelligence (AI) applications require AI cores, sound processing via sound card, computer networking via network processor and network interface controller and digital signal processing via digital signal co-processor, etc.

The DSP IP cores are the integral part of consumer electronic systems used for facilitating applications such as image, audio and video processing with higher efficacy in terms of enhanced operating speed, efficient performance of computationally intensive tasks, etc., at lower design cost. Hence, due to this efficacy, they have become an important and integral part of modern electronic/automated devices such as smart watches, laptops, hearing aids, mobile phones, tablets, audio–video receivers and digital camera and many other electronic devices. Moreover, DSP cores such as infinite impulse response (IIR) filter, discrete cosine transform (DCT), finite impulse response (FIR) filter, discrete wavelet transform (DWT), and auto-regression filter (ARF) are widely used to facilitate image compression–decompression, digital data filtering, sound processing, signal coding, gait analysis, and so on (Mahdiany *et al.*, 2001; Schneiderman, 2010; Sengupta and Rathor, 2020).

Due to rapid growth in modern technology, to match demand and supply ratio and time to market, the design process of IP cores has been globalized. This is because an IP core before its integration into an integrated circuit (ICs), may take several years of research, development, and design. Further, exploring an optimal design architecture is also a tiresome task. In the present technology era, the demand for hardware IP core designs that are secure and low cost (optimal) have become very significant and imperative. Thus, the exercise of designing optimal IP design architecture can play a major role in generating low-cost computing and CE systems. Further, since DSP applications are computationally intensive, therefore their optimal hardware can be designed using high-level synthesis (HLS) process integrated with design space exploration process such as particle swarm optimization (PSO) (Mishra and Sengupta, 2014). On the other hand, besides the optimality issue, security threats arising due to the involvement of offshore design houses in modern design supply chain render a third-party IP (3PIP) core highly untrustworthy. The major security challenges involved during system design of an end product includes IP counterfeiting, IP cloning, and false claim of IP ownership (Chaurasia and Sengupta, 2022a; Sengupta and Rathor, 2020, 2021). Thus, it is important to ensure the generation of a design which is both secure as well as low-cost. Major factors that encourage performing security–design cost trade-off are as follows:

1. Rapid growth in modern technology.
2. Demand for hardware IP core designs that are secure and low-cost.

3. Orthogonal issues: optimized design architecture yielding lower design cost as well as enhanced security.
4. To ensure concurrent optimization and robust security.

The approach presented by Chaurasia and Sengupta (2022d) is capable of obtaining an optimal secured design solutions for DSP hardware IP, used in electronics systems, based on security–design cost trade-off using PSO for different signature-based security algorithms.

This chapter mainly focuses on exploring a low-cost solution for secure DSP hardware architecture using HLS. It considers the security of DSP hardware corresponding to signature-based security algorithms for generating secure architecture design. This methodology integrates a signature generation toolbox comprising of signature-based security algorithms. However, the methodology is applicable to any signature-based security algorithm. Further, a low-cost solution is ensured by performing the design space exploration using PSO framework, which ensures the optimal value of security–design cost fitness function. Therefore, it results into an optimal and secure–design architecture which offers lesser design cost along with their robust security. Thus, satisfying the end objective corresponding to IP designer and CE integrator by generating a low-cost secure design and thereby also ensuring the security and integrity of end consumer (Chaurasia and Sengupta, 2022d).

8.2   Why perform security–design cost trade-off?

Ensuring concurrent optimization and robust security of the IP core designs are major goals for any IP core designer. However, choosing one security approach over the other is a tedious task from a designer's perspective. Additionally, to ensure scalability of the design as a function of the signature strength (capable to operate with large signature size) and design size is equally important. However, the selection process is influenced by several crucial parameters such as tamper tolerance ability, strength of IP ownership proof, vulnerability and replicability of the security mechanism, robustness of the counterfeit detection, design cost overhead (area and latency), and implementation complexity. Further, there is trade-off involved between design cost optimization and hardware security as enhancing one may lead to influencing the other. Therefore, it is viable to design an optimal as well as secured IP with low design cost.

8.3   Summary of "Signature based Security Algorithms for Hardware IPs" in the literature

In the literature, fingerprint biometric (Sengupta and Rathor, 2020), facial biometric signature (Chaurasia and Sengupta, 2022a, 2022b; Sengupta and Chaurasia, 2022a; Sengupta and Rathor, 2021), IP watermarking (Le Gal and Bossuet, 2012; Hong and Potkonjak, 1999; Koushanfar *et al.*, 2005; Rathor *et al.*, 2023; Roy and Sengupta, 2019; Sengupta *et al.*, 2018; Sengupta and Bhadauria, 2016), encrypted hash (Sengupta *et al.*, 2019), contact-less palmprint biometric (Sengupta *et al.*, 2021), chromosomal DNA impression (Chaurasia and Sengupta, 2022c; Sengupta and Chaurasia, 2022b) based hardware security approaches have been presented to ensure the security of reusable IP core (s) used in CE systems against IP piracy/counterfeiting.

(a) ***Signature-based fingerprint biometric:*** In the fingerprint biometric approach (Sengupta and Rathor, 2020), first, the fingerprint biometric image corresponding to an authentic IP vendor is preprocessed. The preprocessing step comprises of image enhancement using fast Fourier transform (FFT), binarization, and thinning processes to enable accurate and precise minutiae extraction. Subsequently, the feature set comprising the details of ridge angle, co-ordinate points, and crossing number corresponding to each minutiae point is determined. These features of each minutiae point are then transformed into binary form to generate binarized digital template. Finally, by concatenating the binarized signature corresponding to IP vendor-selected minutiae points, the fingerprint template is generated. This binarized digital template is then encoded into hardware security constraints through IP vendor-specified encoding rules. These hardware security constraints post-embedding into the design (during register allocation phase of HLS process) offer detective control against IP piracy.

(b) ***Signature-based facial biometric:*** Sengupta and Rathor (2021) presented an approach for securing DSP hardware against IP piracy in HLS using IP vendor facial biometric. Sengupta and Chaurasia (2022a) exploited IP vendor facial biometric-driven digital signature to secure a custom reusable IP core corresponding to convolutional neural network (CNN) convolutional layer, against IP piracy. Chaurasia and Sengupta (2022a) exploited facial biometric driven-digital signature to protect Trojan-secured DSP designs against IP piracy. In this approach, facial biometric of system on chip (SoC) integrator is implanted into the Trojan-secured design. The embedded facial signature into Trojan-secured design safeguards SoC integrator by enabling definitive detective control against potential threat of IP piracy from an adversary that may present in foundry house. Chaurasia and Sengupta (2022b) presented facial biometric-based hardware security approach for ensuring symmetrical protection of IP rights corresponding to IP buyer and seller. In this approach, first the facial biometric signature of IP buyer is implanted into the target design during register allocation phase of HLS process. Subsequently, the facial biometric signature of IP seller is implanted into the design (during register allocation phase of HLS), obtained post-embedding facial biometric of IP buyer. This generates a unique one to one mapped IP core between IP buyer and seller. Thus, ensuring the protection of IP rights of both IP seller and buyer symmetrically. The embedded facial signature corresponding to IP seller safeguards his rights against false IP ownership claim by an adversary (may be present in SoC design house or fabrication house). Additionally, it also enables the detection of pirated IP designs during piracy detection process. The IPs containing the original facial signature corresponding to IP seller are considered as the authentic versions. Further, the facial signature corresponding to an IP buyer safeguards his right against an adversary (that may be present in the untrustworthy design houses) illegally attempting to resell the same custom IP without the knowledge of original IP buyer.

(c) ***Signature-based watermarking:*** Hong and Potkonjak (1999) and Koushanfar *et al.* (2005) presented watermarking-based hardware security approaches with two variable signatures. However, these approaches do not consider the optimization of design area and embedding cost. Gal and Bossuet (2012) presented an in-synthesis IP watermarking approach which uses marking scheme based on mathematical relationships between numeric values as inputs and outputs at specified time. It considers the optimization for low-cost watermark by modifying the controller. The embedded watermark ensures the security while satisfying the design area and latency constraints. Sengupta and Bhadauria (2016) presented a multivariable watermark approach for securing the DSP cores during HLS. It considers optimization of hardware area and embedded design cost only. Sengupta *et al.* (2017) presented an HLS-based triple phase watermarking approach for securing the reusable IP cores against IP piracy. The watermark in this approach comprises of multi-variable (seven variables) signature through three independent phases of HLS such as scheduling phase, hardware allocation phase, and register allocation phase. Roy and Anirban Sengupta (2019) presented a multi-level watermarking-based hardware security approach. In this approach, watermark is inserted during multiple higher design abstraction levels such as architectural level and register transfer level. The embedded watermark offers robust security of DSP cores against IP piracy threat. Rathor *et al.* (2022) presented a quadruple phase watermarking approach for protecting DSP designs against IP piracy and ownership infringement. In this approach, partitioning of graph, encoding tree, and multifold mapping are used to generate a robust hardware watermarking signature. The generated watermark signature has been embedded during four phases of behavioral synthesis namely scheduling phase, register binding phase, functional unit (FU) binding, and interconnect binding phase.

(d) ***Signature-based encrypted-hashing:*** Sengupta *et al.* (2019) presented encrypted hash-based hardware security approach to secure DSP IP cores against IP piracy. This approach comprises of dual encoding phase, secure hashing algorithm, and Rivest Shamir Adleman (RSA)-based encryption to generate encrypted digital signature to be embedded into the design. This encrypted digital signature in the form of encoded hardware security constraints is then inserted into the design during register allocation module of HLS.

(e) ***Signature-based palmprint biometric:*** Sengupta *et al.* (2021) presented a biometric-based approach using contact-less palmprint for securing DSP hardware designs against IP piracy and ownership threat. In this approach, first the palmprint of IP vendor is captured through high-resolution camera and subjected to a specific grid size to designate palm nodal points accurately. Next, palm image with IP vendor-selected palm feature set is generated. Subsequently, each of the palm features is exploited to derive its corresponding binarized signature. Thus, by concatenating the palm feature based on the order decided by IP vendor, palmprint biometric signature is generated. The palmprint signature is then embedded into the design in the form of secret hardware security constraints for enabling piracy detection and nullifying false ownership threat.

(f) ***Signature-based DNA biometric:*** Sengupta and Chaurasia (2022b) and Chaurasia and Sengupta (2022c) presented a hardware security methodology with double line of security using structural obfuscation and chromosomal deoxyribonucleic acid (DNA) biometric of IP vendor. In this approach, first the target DSP design is transformed using structural obfuscation to hinder potential threat of register transfer level (RTL) alteration by an adversary. Subsequently, the DNA signature is extracted from the DNA sequence of the IP vendor which comprises of four chemical elements (adenine, guanine, cytosine, and thymine). To do so, first the chemical components of the DNA sequence corresponding to the IP vendor are encoded using IP vendor-specified encoding rule. Post-encoding each chemical component is converted into its binarized form and is concatenated to form the binary-encoded DNA impression. Subsequently, the generated binarized DNA signature is fed into Feistel cipher encryption process to generate encrypted DNA biometric signature. This encrypted signature is then converted in the form of encoded hardware security constraints, which is then embedded into the structurally obfuscated design using HLS. This acts as the invisible digital proof to guard against IP piracy and false ownership claim.

The discussed methodology in this chapter provides a mechanism to perform trade-off between security and design cost of DSP cores using PSO-based design space exploration with respect to various security algorithms (Sengupta and Bhadauria, 2016; Sengupta and Rathor, 2021; Sengupta *et al.*, 2019). The trade-off mechanism discussed enables generation of an optimal secured DSP RTL hardware for securing CE systems against piracy. The optimal DSP RTL hardware solution is optimized in terms of robust security and lower design cost (design area, delay). This therefore enables the design of secured consumer electronics hardware that are secured against counterfeited/forged IP cores. Finally, this also helps the end consumer in using safe/secure CE products that are authentic by nature and do not contain any reliability hazards. Major signature-based hardware security algorithms are watermarking-based, encrypted hash-based, facial biometric-based, and fingerprint-based biometric algorithms. Furthermore, most approaches (e.g., hardware watermarking, encrypted hashing, and facial and fingerprint biometric) do not focus on performing trade-off between security and design cost and therefore does not generate optimal security solutions for DSP hardware IP. Thus, these approaches are incapable of generating RTL solution for DSP architectures that satisfy both the objective of robust security and the lower design cost in terms of design area and delay.

8.4 Methodology for exploration of security–design cost trade-off for signature-based security

Now, let us discuss the security–design cost trade-off methodology.

8.4.1 Summary

In this chapter, an approach for the exploration of security–design cost trade-off for signature-based security algorithms for DSP hardware has been discussed. A stochastic multi-objective particle swarm optimization algorithm has been employed for the same (Mishra and Sengupta, 2014). The primary inputs to the approach are signature generation toolbox, input DSP application (in form of C-code/transfer function), module library, and particle swarm optimization (PSO) input parameters such as population size, acceleration coefficient, inertia weight, and terminating criteria. The output of the approach is an optimal security–design cost solution for the DSP application based on the security algorithm selected from the signature generation toolbox by the designer. The major blocks of the approach are the following: PSO-based design space exploration, DSP application input block, HLS scheduling, allocation and binding, signature-embedding (SE) block, and security–design cost trade-off fitness function block. PSO-based design space exploration block is responsible for performing the exploration of a low-cost resource configuration by considering the parameters of security and design cost. DSP application input block is responsible to transform the behavioral description of DSP application into data flow graph. HLS-based scheduling block is responsible for scheduling the DFG of input DSP design based on PSO-driven resource configuration. Post scheduling, hardware is allocated and their binding is performed. Next, signature-embedding block is responsible for embedding the signature corresponding to IP designer-selected security algorithm. The security–design cost trade-off fitness function block results in an optimal security–design cost register transfer level architectural solution as an output based on the different signature-based security algorithms used for DSP applications. Moreover, the approach scales well as a function of the signature strength (capable to operate with a large signature size) and design size. This helps in analyzing the impact of choosing a particular security approach with respect to design optimality (in terms of design cost and robustness).

As shown in Figure 8.1, the input block of the discussed approach consists of PSO input parameters for performing design space exploration. The output of design space exploration is the hardware resource configuration. In each iteration, based on the output hardware resource configuration, the input DFG of the DSP application is scheduled. Subsequently, hardware is allocated to the operations of scheduled design and their binding is performed. The output of the HLS scheduling, allocation, and binding block from the HLS framework is then fed as an input into the signature-embedding (SE) block. Further, two other inputs of the SE block are: (a) signature-driven security algorithm selected from signature generation toolbox and (b) signature strength, both are selected by the IP designer. The SE block is responsible for generating the signature embedded design for the

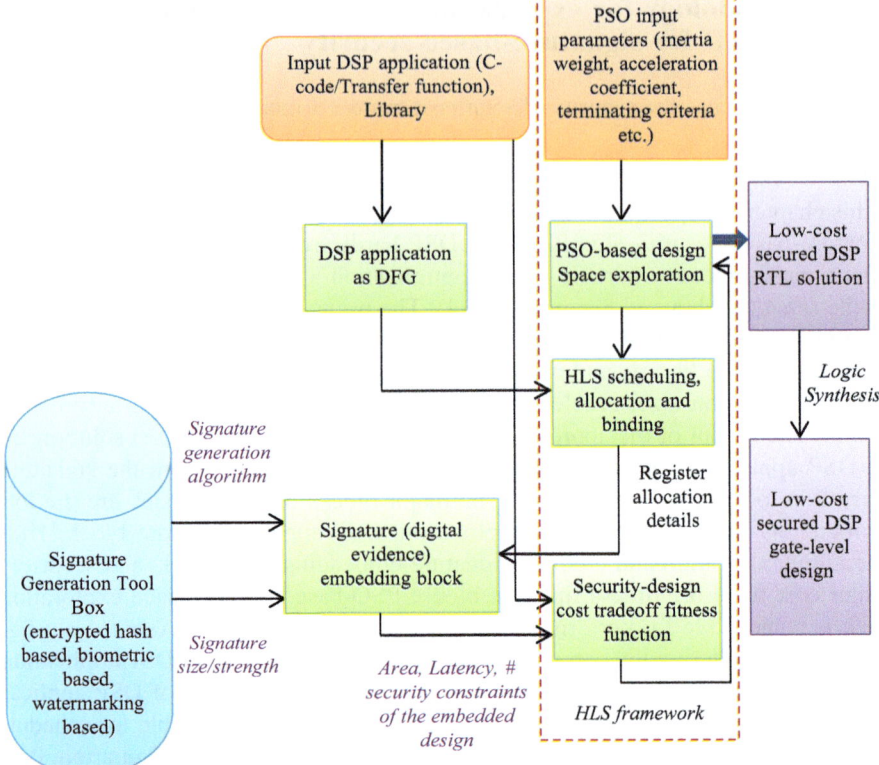

Figure 8.1   Details of the methodology for performing security–design cost trade-off

security–design cost trade-off fitness function. Security–design cost trade-off fitness function accepts the following two inputs: signature embedded design as the output of SE block and module library file. This therefore provides an optimal and secured RTL architectural solution corresponding to the particular DSP application and security algorithm, selected by the IP designer. Further, the initial RTL solution obtained in the first iteration of security–design cost trade-off fitness function was fed into the PSO-based DSE block, based on which the next iteration executes. The execution continues until a global best solution (Sgb) is explored by the algorithm. The PSO-based DSE provides an optimal RTL architectural solution for a particular DSP application after the convergence of the initial solution to the global minima.

8.4.2   Details

Now, we discuss the details of the methodology presented in Chaurasia and Sengupta (2022d).

8.4.2.1 Signature generation toolbox

As shown in Figure 8.1, a signature generation tool-box comprises of the signature-based security algorithms. An IP designer may select any of the signature-based security algorithm (with specific signature strength) for embedding into the target design. In the discussed methodology, the security algorithms that are mainly considered for analyzing the security–design cost trade-off are: IP watermarking based (Sengupta and Bhadauria, 2016), encrypted hash based (Sengupta *et al.*, 2019), and facial biometric based (Sengupta and Rathor, 2021). Further, an IP designer can also select any of the DSP application to obtain its corresponding secure and low-cost architectural solution. The signature (digital evidence) can be generated with respect to any DSP application and any security algorithm selected by an IP designer. This, therefore, enables an IP designer to obtain an optimal and secured architectural solution for DSP applications corresponding to security algorithm chosen by IP designer. In the watermarking approach, signature is generated based on auxiliary multi-variable combination of IP designer-chosen signature length. Subsequently, multi-variable signature is encoded to generate its corresponding secret hardware security constraints. These security constraints act as digital evidence against IP piracy, post embedding into the target design. In encrypted hash-based algorithm, it uses multi-level encoding, secure hash algorithm (SHA)-512, RSA algorithm, and 128-bit private key for generating the digital signature. This encrypted digital signature is then encoded into secret hardware security constraints. These encoded security constraints act as digital evidence against IP piracy, post embedding into the target design. Furthermore, in the signature generation process using biometric approach, it exploits facial features (always unique in the form of nodal points) of an individual (IP vendor). Signature can be generated by the combination of different facial features (more the number of features, more the signature strength) and by the different ordering possibilities. The details of precise co-ordinates of nodal points, type of feature selected, ordering of the features, position of the bits (0 and 1), and grid size are unknown to an adversary. This makes facial biometric approach more secured as compared to watermarking and encrypted hash-based approach. It is more robust to guard a true IP designer from the fraudulent claim of IP ownership.

8.4.2.2 Signature-based security algorithms

The details of different security algorithms for securing DSP IP core design are discussed below:

- *Watermarking-based hardware security*: In the watermarking-based hardware security methodology (Sengupta and Bhadauria, 2016), first, the watermark signature is formed based on the input variables (i, I, T, !). The strength of the signature is decided by the IP vendor. In the next phase, the conversion of the watermark signature into hardware security constraints is performed using a possible encoding rule:
  - *Signature bit "i"*: representing storage variable pair indices ($V_s x$ and $V_s y$) of the prime number required to be allocated to distinct registers, where x and y can be of any integer value.

- ○ *Signature bit "I"*: representing storage variable pair indices ($V_s x$ and $V_s y$) of even number required to be allocated to distinct registers, where x and y can be of any integer value.
- ○ *Signature bit "T"*: representing storage variable pair indices ($V_s x$ and $V_s y$) of odd–even number required to be allocated to distinct registers, where x and y can be of any integer value.
- ○ *Signature bit "!"*: representing storage variable pair indices between $V_s x$ and $V_s y$, required to be allocated to distinct registers, where the first variable "x" is 0 and the second variable "y" can be of any integer value (except the generated pairs, depending on the size of the DSP application).

- Storage variable pairs ($V_s x$ and $V_s y$) are formed using storage variables denoted "Vn" corresponding to DFG of the design. Subsequently, in the last phase of watermarking scheme, embedding of the signature constraints is performed, as shown in Figure 8.2.
- Further, corresponding to other types of watermarking methodologies, for example: the single-phase watermarking that embeds the security constraints in register allocation phase and triple-phase watermarking which embeds the security constraints in scheduling, FU vendor, and register allocation phases.
- **Encrypted hash-based hardware security**: Furthermore, in the encrypted hash-based security algorithm (Sengupta *et al.*, 2019) the scheduled DFG of the corresponding DSP application is encoded into a bit stream based on the following encoding rule:
  - ○ If the operation number and the control step number assigned to the operation are of the same parity then, it is encoded as bit = "0"
  - ○ and if the operation number and the control step number assigned to the operation are of different parity then, it is encoded as bit="1."
- In the next phase, SHA-512 algorithm is employed for obtaining bitstream digest value corresponding to input bitstream of the previous phase. Subsequently, bitstream is bifurcated into blocks and their conversion into decimal has been performed. Next, RSA encryption has been performed on each decimal input value. This therefore results in the encrypted digital signature after performing the RSA encryption using 128-bit private key chosen by the IP vendor. The generated encrypted digital signature is then converted into binarized bitstream. In the next phase, encrypted digital signature is transformed into secret hardware security constraints using the following encoding rule:
- *Signature bit "0"*: representing storage variable pair indices (*Jx* and *Jy*) of prime number required to be allocated to distinct registers, where *Jx* and *Jy* can be of any integer value.
- *Signature bit "1"*: representing storage variable pair indices (*Jx* and *Jy*) of even number required to be allocated to distinct registers, where *Jx* and *Jy* can be of any integer value.
- **Facial biometric-based hardware security**: In this methodology (Sengupta and Rathor, 2021), first, the facial biometric of IP vendor has been captured. Subsequently, the captured facial image is subjected to specific grid size. In the

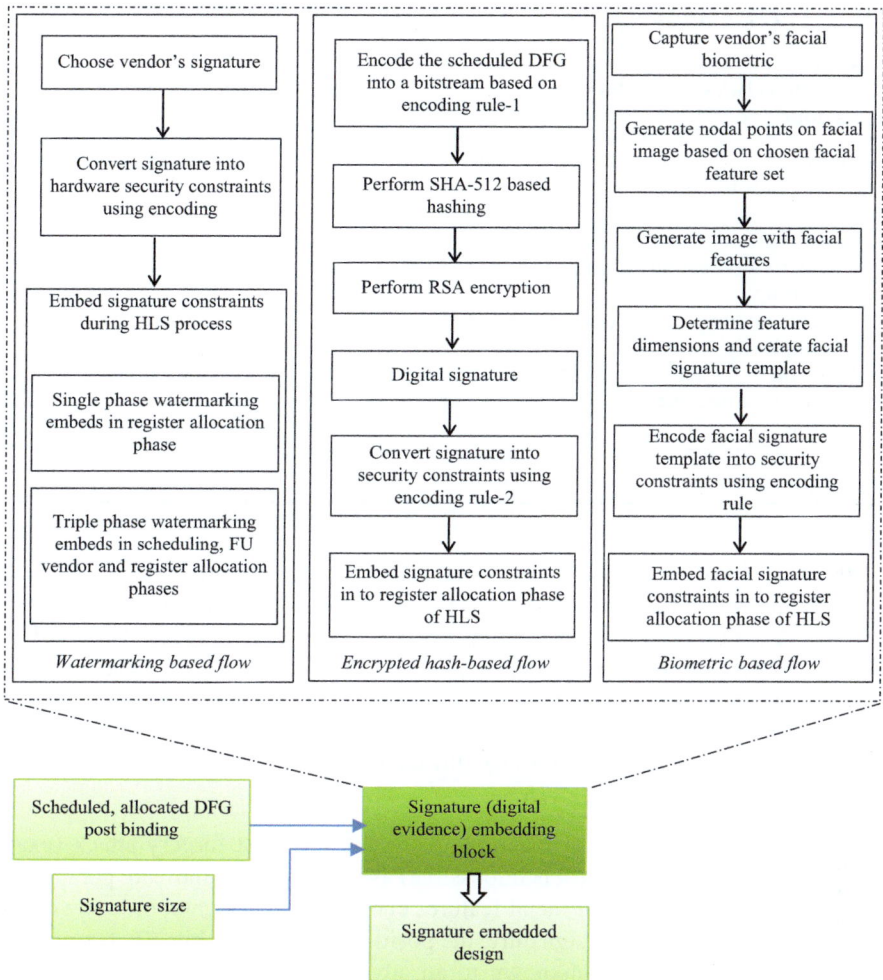

Figure 8.2   Details of the process of embedding the security constraints during HLS process of different signature-based security algorithms

next phase, nodal points are generated on captured facial image based on the IP vendor-chosen facial feature set. Depending on the security strength to be achieved, an IP vendor may select or discard the facial feature(s) from feature set. Subsequently in the next phase, facial image with the IP vendor-selected facial features has been produced. Next, feature dimensions corresponding to each feature have been determined using Manhattan distance. The dimensions corresponding to each feature are then transformed into binarized form. Thus, facial signature template has been generated by concatenating the binarized signature (in the concatenation order decided by IP vendor). Subsequently, in the next phase, facial signature template has been encoded into hardware

Table 8.1   Qualitative comparison between the security approaches

S. no.	Characteristics/ parameters	Biometric	Watermarking	Digital signature
1.	Security mechanism	Natural biometric features (minutiae points or facial nodal points).	Signature and encoding rules.	RSA encryption, SHA-512
2.	Counterfeit detection control	Strong	Less	Less
3.	Implementation complexity	Less	More	More
4.	Proof of IP ownership by a genuine owner	Seamless	Difficult	Arduous
5.	Vulnerability and replicability	Almost Impossible	Yes	Yes

security constraints using IP vendor-specified encoding rule. Following encoding rule has been employed for encoding the facial signature template:

o   *Signature bit "0"*: representing storage variable pair indices ($V_s x$ and $V_s y$) of even number required to be allocated to distinct registers, where x and y can be of any integer value.

o   *Signature bit "1"*: representing storage variable pair indices ($V_s x$ and $V_s y$) of odd number required to be allocated to distinct registers, where x and y can be of any integer value.

•   Facial biometric-based approach is more robust against forgery attack (exact regeneration of secret mark is impossible) as the employed intricate parameters such as grid size, types of facial features chosen by IP designer, ordering of the features for deriving the signature, the position of signature bits (0s,1s), and the encoding rules, all are unknown to an adversary. Furthermore, a qualitative comparison among the above security approaches is shown in Table 8.1.

8.4.2.3   PSO-based design space exploration for performing trade-off with respect to different signature based security algorithms

To explore an optimal design space solution corresponding to different DSP applications, multi-objective PSO as shown in Figure 8.3, has been employed (Mishra and Sengupta, 2014). The primary inputs to the PSO-DSE are inertia weight (τ), acceleration coefficient (c1, c2), terminating criteria (T), and population size (Sp). Besides the once mentioned, the secondary input is the global best resource configuration "Sgb." Sgb is obtained corresponding to minimum security–design cost value (maximum fitness value) in current iteration and based on which solution for the next iteration is evaluated. This process continues until the design solution converges to global minima (as shown in Figure 8.1 earlier). Thus, the output of the PSO block is the low-cost resource configuration.

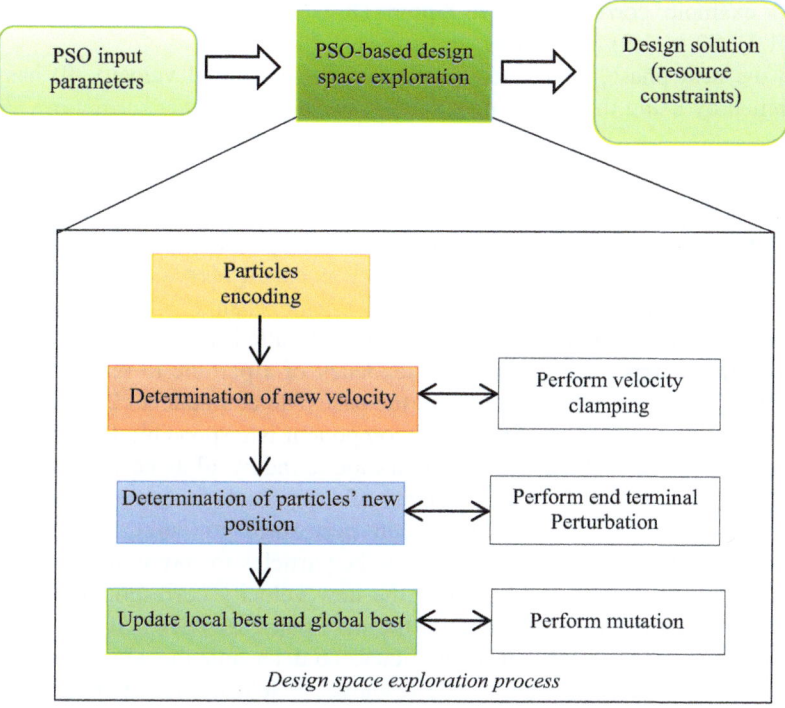

Figure 8.3   Details of PSO-based design space exploration

Now, we discuss the different phases of PSO-DSE process. The first phase of PSO-DSE is responsible for performing the encoding of the particles chosen by the IP designer. Here each particle corresponds to different hardware resources for example adder(s), multiplier(s), etc. The first particle's position is initialized by minimum resources: $S1 = (P_1^{min}, P_2^{min})$, where P_1 and P_2 are the hardware resource types, adder(s) and multiplier(s), respectively (available in the library file). Therefore, P_1^{min}, P_2^{min} represent the minimum resource configuration corresponding to respective DSP application. The second particle's position is initialized by maximum resources: $S2 = (P_1^{max}, P_2^{max})$, where P_1^{max}, P_2^{max} represent the maximum possible resource configuration corresponding to respective DSP application. Further, third particle's position (S3) is initialized by average of maximum and minimum resource values. The rest of the particle's position (S4 ... Sn) is initialized by the resources, extracted from the following equation:

$$Sid = (a + b)/2 \pm \triangledown \tag{8.1}$$

where "a" and "b" are the minimum and maximum resource value and "$\triangledown$" is any random number between "a" and "b." Further, Sid represents resource configuration of *i*th particle in *d*th dimension.

For example, corresponding to 8-point DCT application, particle positions are S1 = (1,1), S2 = (1,8), S3 = (1,4), and so on.

In the next phase, new velocity of the particles (initial velocity=0) has been determined by using the following equation, as shown below:

$$v_{id}^{+} = \tau.v_{id} + c1 \times 1(Slbi - Sid) + c2 \times 2(Sgb - Sid) \qquad (8.2)$$

where "v_{id}" and "v_{id}^{+}" signifies the velocity of i particle in d dimension corresponding to previous and next iteration, respectively, and $\times 1$, $\times 2$ are random numbers between [0,1]. The component $\tau.v_{id}$ is called inertia component. Inertia component is responsible for preventing a drastic change in the direction of a particle. The other component $c1 \times 1.(Slbi-Sid)$ is called the cognitive component. The cognitive component signifies the tendency of a particle to return to its individual best resource configuration from the past. The component $c2 \times 2.(Sgb-Sid)$ is called the social component. The social component is responsible for directing the particle towards the best resource configuration found by all its neighbors including itself. However, in case if the new velocity of any particle outreaches the boundary space, then velocity clamping has been performed to control the excessive exploration drift. This, therefore, manages the particles to stay in the design space by taking the step size sensibly. Thus, the new velocity corresponding to all the particles, is determined. Subsequently, the next phase of PSO-DSE is responsible for determining new position of the particles. To determine the new position of the particles, the previous position of the resources is added with their new velocity (computed in the previous phase). However, this may cause a particle to outreach the boundary space. Therefore, if the new position of the particle outreaches the boundary space, then "end terminal perturbation" has been performed to keep the particle in its design space boundary. Subsequently, the next phase is responsible for performing the updation of local best and Sgb of the particle. In this phase, local best solution of each particle has been updated if the solution with minimum security–design cost is found. Correspondingly, based on the previous updation, Sgb is also updated in each iteration. Furthermore, to diversify the solution and perform better exploration of the design space as well as to avoid getting stuck in the local minima, mutation is performed on each particle position. Next, the fitness cost corresponding to each mutated particle is computed and local and global best resource configurations are also updated. Following process continues until terminating criteria T is met. The termination criteria are: (a) if the solution executes for a certain number of times or (b) solution converges and does not get updated for next 10 iterations. Thus, the optimal architectural solution (global best resource constraints) is explored using the PSO-DSE by converging the initial solution to the global minima.

8.4.2.4   Embedding signature (digital evidence) on DSP design: a case study on 8-point DCT

So far, we have discussed the process of generating low-cost architectural solution using PSO-based design space exploration. Now, we discuss the process of embedding the signature into DSP design. To do so, a case study on 8-point DCT design has been discussed in this chapter. As shown in Figure 8.1, the inputs of the signature (digital evidence) embedding block are the signature generation algorithm and signature strength (size) chosen by the IP designer and the scheduled and allocated/binded DFG of the DSP application. The output of the SE block is the signature-embedded design.

The DFG corresponding to algorithmic description of 8-point DCT design is shown in Figure 8.4(a), where k1–k8 represents generic values of DCT coefficients and $x[0]$ to $x[7]$ indicate input values. $X[n]$ indicates nth output where "n" ranges from 0 to 7. *Note*: more details corresponding to 8-point DCT framework (derivation of generic equation, coefficient matrix and conversion of DCT function into DFG) are available in Salivahanan and Vallavaraj (2001) and Sengupta and Mohanty (2019). Now, to obtain the scheduled design, scheduling of the DFG of respective DSP application has been performed based on the resource constraints generated as the output of the PSO-DSE. List scheduling algorithm has been used for scheduling the DGF of corresponding DSP application. The scheduled DFG of 8-point DCT application (based on one adder and four multipliers) is shown in Figure 8.4(b). The scheduled design is used for inferring the following details: the number of control steps (C_T) using multiplier and the control steps using the adder only. This information is useful in determining the design latency (D_T). Subsequently, allocation and binding of the hardware resources (registers) has been performed for each operation, as shown in Figure 8.4(b). Post generating the scheduled and allocated hardware design, its corresponding register allocation information is extracted. The register allocation table corresponding to 8-point DCT design (pre-embedding the signature) is shown in Table 8.2. The register allocation table comprises of the following details: the number of registers required for accommodating the storage variables of the design, number of control steps required to schedule the design, and the position of storage variables based on their dependency information corresponding to the functional behavior of the design. As evident from Table 8.2, the number of required registers corresponding to 8-point DCT is eight (R^1 to R^8), where each register is designated using a different color. Further, the number of required control steps is nine (C_T0 to C_T8).

In the case of hardware watermarking approach, for performing the embedding of generated watermark signature into the target design, the details of embedding are discussed below.

Assuming that the IP designer-chosen watermark signature based on the variables (i, I, T, !) is 16 bit long (for the sake of brevity). However, the discussed approach is easily scalable as a function of the signature and design size (IP designer can also select a signature of larger size). Let us consider the 16-bit watermark signature as follows:

!, i, I, i, !, T, i, !, i, !, I, i, !, i, I, I

Now, we discuss the generation of secret hardware security constraints corresponding to the above signature. The security constraints corresponding to above

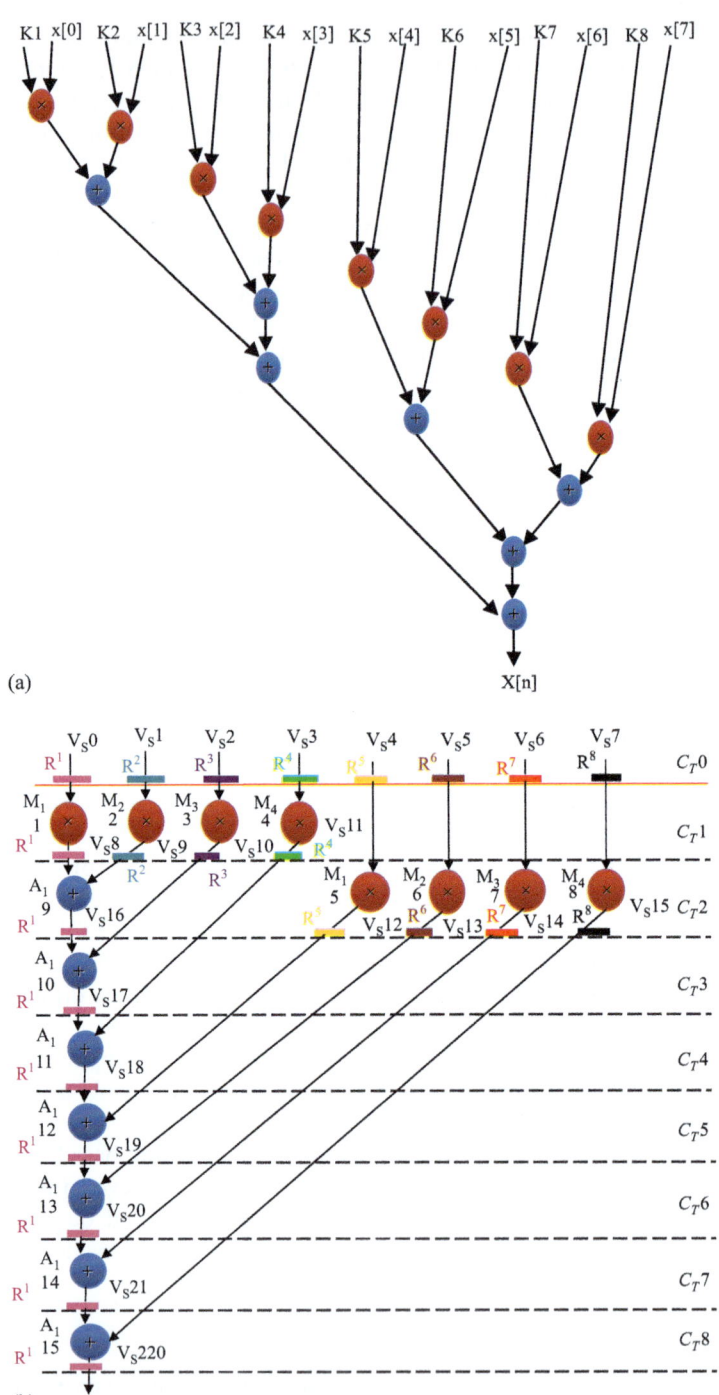

Figure 8.4   *(a) DFG of 8-point DCT IP core design. (b) Scheduled DFG of 8-point DCT IP core design using one adder (A) and four multipliers (M).*

Table 8.2 Register allocation of 8-point DCT application (pre-embedding signature)

C_T	R^1	R^2	R^3	R^4	R^5	R^6	R^7	R^8
C_T0	V_S0	V_S1	V_S2	V_S3	V_S4	V_S5	V_S6	V_S7
C_T1	V_S8	V_S9	V_S10	V_S11	V_S4	V_S5	V_S6	V_S7
C_T2	V_S16	–	V_S10	V_S11	V_S12	V_S13	V_S14	V_S15
C_T3	V_S17	–	–	V_S11	V_S12	V_S13	V_S14	V_S15
C_T4	V_S18		–	–	V_S12	V_S13	V_S14	V_S15
C_T5	V_S19	–	–	–	–	V_S13	V_S14	V_S15
C_T6	V_S20	–	–	–	–	–	V_S14	V_S15
C_T7	V_S21	–	–	–	–	–	–	V_S15
C_T8	V_S22	–	–	–	–	–	–	–

Table 8.3 Hardware security constraints corresponding to watermark signature

For "!"	For "I"	For "I"	For "T"	–
< V_S0, V_S1>	< V_S2, V_S3>	< V_S2, V_S4>	< V_S1, V_S2>	–
< V_S0, V_S2>	< V_S2, V_S5>	< V_S2, V_S6>	–	–
< V_S0, V_S3>	< V_S2, V_S7>	< V_S2, V_S8>	–	–
< V_S0, V_S4>	< V_S2, V_S11>	< V_S2, V_S10>	–	–
< V_S0, V_S5>	< V_S2, V_S13>	–	–	–
–	< V_S2, V_S17>	–	–	–
–	–	–	–	–

watermark signature are derived using designer-specific encoding rule (as discussed earlier in Section 8.4.2.2). Therefore, the resulting security constraints corresponding to chosen watermark signature are shown in Table 8.3. Next, these generated hardware security constraints are covertly embedded into the register allocation phase of HLS process. To do so, local alteration of the registers for re-allocation of the storage variables into the register allocation table is performed based on the rule: "*both storage variables of any security constraint pair cannot be allocated into the same register.*" Thus, the register allocation table post embedding the watermark signature-driven secret hardware security constraints into 8-point DCT design is shown in Table 8.4, where the storage variables marked in brown color are indicating their locally altered position (through swapping) after embedding the watermark constraints. These embedded secret watermark constraints act as secret digital evidence for enabling the detective control (security) against IP piracy and nullifying fraudulent ownership claim. The scheduled 8-point DCT design with implanted watermark signature is shown in Figure 8.5, where the dotted boundary (in red color) represents modification in the design due to implanted watermark signature. *Note: the modification is only during register assignments corresponding to storage variables.* The original functionality of the target design remains intact post embedding watermark signature. This

Table 8.4   Register allocation of 8-point DCT application (post embedding watermark signature)

C_T	R^1	R^2	R^3	R^4	R^5	R^6	R^7	R^8
C_T0	V_s0	V_s1	V_s2	V_s3	V_s4	V_s5	V_s6	V_s7
C_T1	V_s8	V_s9	V_s11	V_s10	V_s4	V_s5	V_s6	V_s7
C_T2	V_s16	–	V_s11	V_s10	V_s12	V_s13	V_s14	V_s15
C_T3	V_s17	–	V_s11	–	V_s12	V_s13	V_s14	V_s15
C_T4	V_s18	–	–	–	V_s12	V_s13	V_s14	V_s15
C_T5	V_s19	–	–	–	–	V_s13	V_s14	V_s15
C_T6	V_s20	–	–	–	–	–	V_s14	V_s15
C_T7	V_s21	–	–	–	–	–	–	V_s15
C_T8	V_s22	–	–	–	–	–	–	–

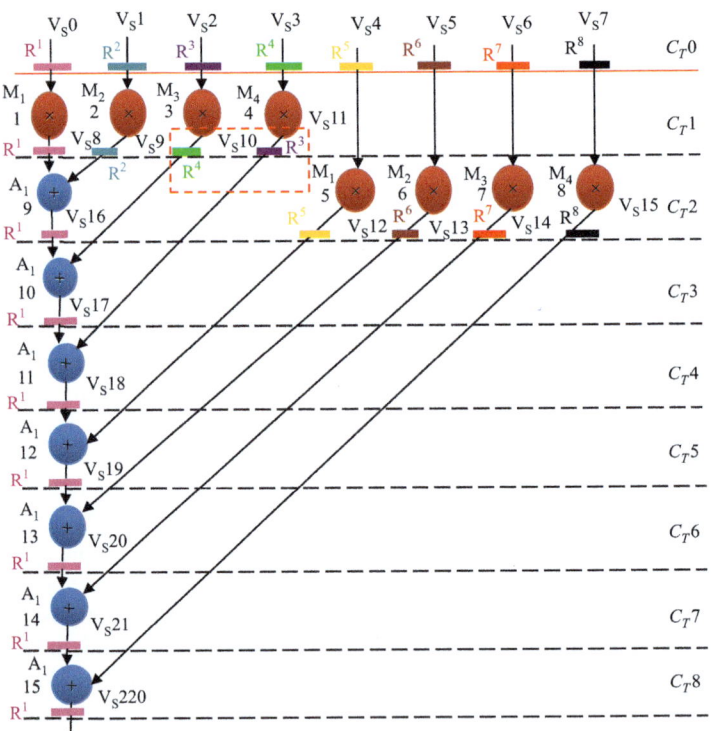

Figure 8.5   Scheduled DFG of 8-point DCT IP core design post embedding watermark signature

modification during the register allocation stage of the design process gets reflected subsequently at the lower abstraction level such as RT-level (in terms of change in interconnectivity of data selector units such as mux and demux). Thus, it leads to security impression (in terms of digital evidence) from RT-level (obtained using

HLS) to gate level/netlist level (obtained using RTL synthesis) and layout level (obtained using gate level and physical level synthesis) of an IP design process.

Further, in the case of encrypted digital signature-based approach, for performing the embedding of generated digital signature into the target design, the details of embedding are discussed below.

Assuming that the IP designer-chosen encrypted digital signature is 16-bit long (for the sake of brevity). However, the discussed approach is easily scalable as a function of the signature and design size (IP designer can also select a signature of larger size). Let us consider the 16-bit encrypted digital signature as follows:

1,1,0,1,0,1,0,1,0,1,0,1,1,0,0,1

Now we discuss the generation of secret hardware security constraints corresponding to the above signature. The security constraints corresponding to above-encrypted digital signature are derived using designer-specific encoding rule (as discussed earlier in Section 8.4.2.2). Therefore, the resulting hardware security constraints corresponding to chosen digital signature are shown in Table 8.5. Next, these generated hardware security constraints are embedded into design during the register allocation phase of HLS. To do so, local alteration of the registers for reallocation of storage variables into the register allocation table is performed based on the rule: "*both variables of any security constraint pair cannot accommodate into same register*." Thus, the register allocation table of 8-point DCT design post embedding the encrypted digital signature-driven secret hardware security constraints is shown in Table 8.6, where the variables marked in brown color are indicating their locally altered position (through swapping). On the other hand, the storage variables marked in green color (Sv16, Sv18) are representing the altered position (updated position), after embedding of the digital signature-based hardware security constraints. These embedded digital signature-driven secret hardware security constraints act as digital evidence for enabling the detective control against IP piracy. The scheduled 8-point DCT design with implanted encrypted digital signature is shown in Figure 8.6, where the dotted boundary (in red color) represents modification in the design due to implanting encrypted digital signature.

Similarly, in the case of facial biometric-based security algorithm, for performing the embedding of generated facial signature into the target design, the details of embedding are discussed below.

Table 8.5 *Security constraints corresponding to encrypted hash-based digital signature*

For "0"		For "1"	
$< V_S2, V_S3 >$	–	$< V_S0, V_S2 >$	$< V_S0, V_S16 >$
$< V_S2, V_S5 >$	–	$< V_S0, V_S4 >$	$< V_S0, V_S18 >$
$< V_S2, V_S7 >$	–	$< V_S0, V_S6 >$	–
$< V_S2, V_S11 >$	–	$< V_S0, V_S8 >$	–
$< V_S2, V_S13 >$	–	$< V_S0, V_S10 >$	–
$< V_S2, V_S17 >$	–	$< V_S0, V_S12 >$	–
$< V_S2, V_S19 >$	–	$< V_S0, V_S14 >$	–

Assuming that the IP designer-chosen facial biometric signature is 16-bit long (for the sake of brevity). Let us consider, 16-bit facial biometric signature as follows:

1,0,1,1,1,1,1,1,0,0,0,0,0,1,0,1

Table 8.6   Register allocation of 8-point DCT application (post embedding encrypted hash-based digital signature)

C_T	R^1	R^2	R^3	R^4	R^5	R^6	R^7	R^8
C_T0	V_s0	V_s1	V_s2	V_s3	V_s4	V_s5	V_s6	V_s7
C_T1	V_s9	V_s8	V_s10	V_s11	V_s4	V_s5	V_s6	V_s7
C_T2	–	V_s16	V_s10	V_s11	V_s12	V_s13	V_s14	V_s15
C_T3	V_s17	–	–	V_s11	V_s12	V_s13	V_s14	V_s15
C_T4	–	V_s18	–	–	V_s12	V_s13	V_s14	V_s15
C_T5	V_s19	–	–	–	–	V_s13	V_s14	V_s15
C_T6	V_s20	–	–	–	–	–	V_s14	V_s15
C_T7	V_s21	–	–	–	–	–	–	V_s15
C_T8	V_s22	–	–	–	–	–	–	–

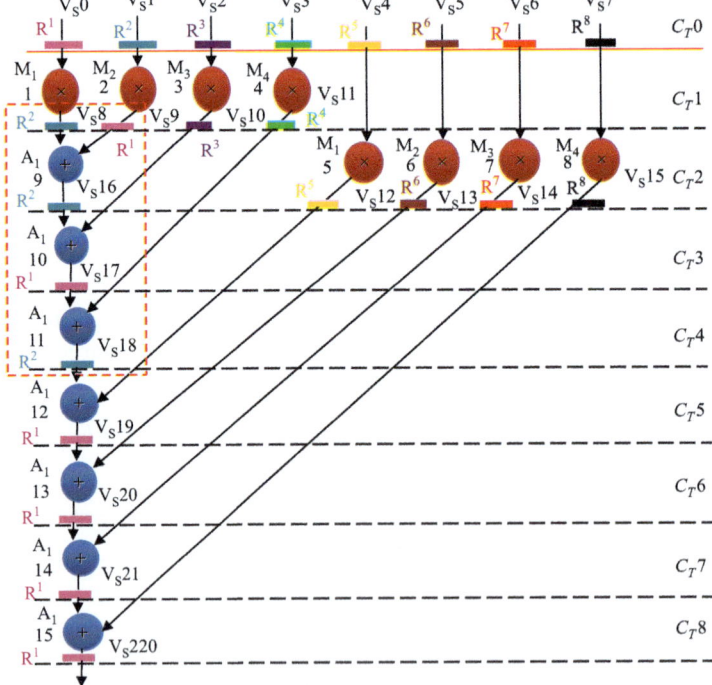

Figure 8.6   Scheduled DFG of 8-point DCT IP core design post embedding encrypted digital signature

Table 8.7   Security constraints corresponding to facial biometric signature

For signature bit "0"		For signature bit "1"	
$< V_S0, V_S2>$	–	$< V_S1, V_S3>$	$< V_S1, V_S17>$
$< V_S0, V_S4>$	–	$< V_S1, V_S5>$	$< V_S1, V_S19>$
$< V_S0, V_S6>$	–	$< V_S1, V_S7>$	–
$< V_S0, V_S8>$	–	$< V_S1, V_S9>$	–
$< V_S0, V_S10>$	–	$< V_S1, V_S11>$	–
$< V_S0, V_S12>$	–	$< V_S1, V_S13>$	–
$< V_S0, V_S14>$	–	$< V_S1, V_S15>$	–

Table 8.8   Register allocation of 8-point DCT application (post embedding facial biometric signature)

C_T	R^1	R^2	R^3	R^4	R^5	R^6	R^7	R^8
C_T0	V_S0	V_S1	V_S2	V_S3	V_S4	V_S5	V_S6	V_S7
C_T1	V_S9	V_S8	V_S10	V_S11	V_S4	V_S5	V_S6	V_S7
C_T2	V_S16	–	V_S10	V_S11	V_S12	V_S13	V_S14	V_S15
C_T3	V_S17	–	–	V_S11	V_S12	V_S13	V_S14	V_S15
C_T4	V_S18	–	–	–	V_S12	V_S13	V_S14	V_S15
C_T5	V_S19	–	–	–	–	V_S13	V_S14	V_S15
C_T6	V_S20	–	–	–	–	–	V_S14	V_S15
C_T7	V_S21	–	–	–	–	–	–	V_S15
C_T8	V_S22	–	–	–	–	–	–	–

Therefore, the generated security constraints corresponding to chosen facial biometric signature are shown in Table 8.7. Next, these generated hardware security constraints are embedded into target design based on the same rule stated earlier. Thus, the register allocation table of 8-point DCT design post embedding the facial biometric signature-driven secret hardware security constraints is shown in Table 8.8, where the variables marked in brown color are indicating their locally altered position (through swapping). These embedded facial security constraints act as digital evidence for enabling the detective control against IP piracy. Further, as the facial security constraints also associate the unique facial identity of IP vendor. This therefore enables the definitive proof of IP ownership for an original IP vendor. The scheduled 8-point DCT design with implanted facial biometric signature is shown in Figure 8.7, where the dotted boundary (in red color) represents modification in the design due to implanted facial signature.

8.4.2.5   Security–design cost trade-off fitness function

So far, we have discussed the process of embedding the generating signature corresponding to security algorithm. Now, we discuss the details of security–design cost trade-off fitness function. As shown in Figure 8.1, the inputs to the security–design cost fitness function are: the signature embedded design and the library.

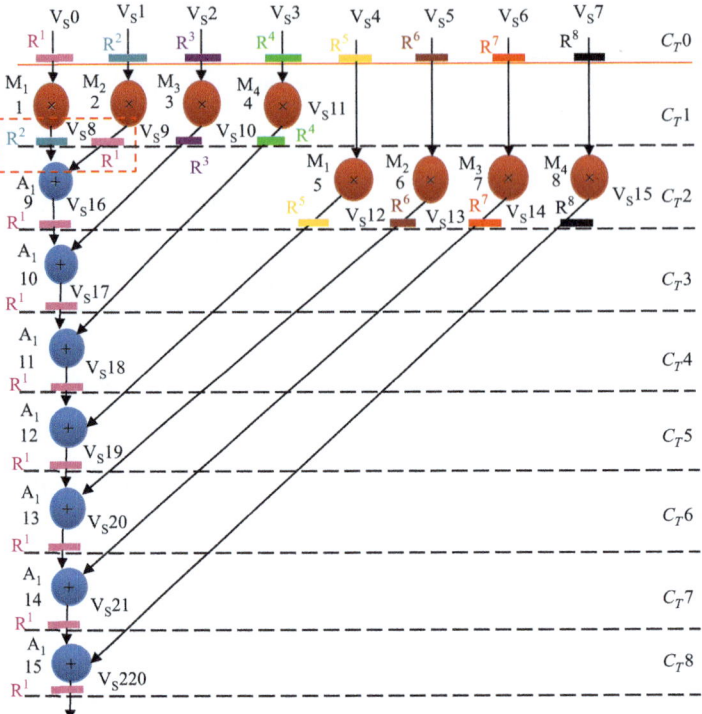

Figure 8.7   Scheduled DFG of 8-point DCT IP core design post embedding facial biometric signature

Signature-embedded design is used to obtain the following information such as design area, latency, and security constraints. Further, based on the embedded security constraints corresponding to a particular signature-based security algorithm, security metric in terms of embedded constraints size of the corresponding signature $\left(S_m^n\right)$ can be determined as follows:

$$S_m^n = X/Z \tag{8.3}$$

where "X" represents the number of embedded security constraints corresponding to signature-based security methodology chosen by IP vendor and "Z" represents total possible security constraints corresponding to security methodology. This security metrics signifies the robustness in terms of the proof of digital evidence against piracy and false IP ownership. This is because the more is the number of embedded secret security constraints the more is the robust security. Thus, safeguarding the design against the threat of piracy and fraudulent claim of IP authorship. Furthermore, library file (a 15-nm open-cell library) comprises of the following information such as count of hardware resources (adders and multipliers), area of the adder, multiplier and register unit and delay (time consumed) of the adder and multiplier unit. Based on that, the area of the design ("D_A") and latency ("D_T") has been computed using the following (8.4) and (8.5) as shown below:

$$D_A = n * (\text{area of adder}) + m * (\text{area of multiplier}) + P * (\text{area of register})$$
(8.4)

where "n" indicates the number of adders and "m" indicates the number of multipliers and "P" indicates the total number of registers required during embedding of the generated security constraints (based on the signature).

$$D_T = (\#C_T \text{ using multiplier} * \text{delay of 1 multiplier})$$

$$+ (\#C_T \text{ using adder only} * \text{delay of 1 adder})$$
(8.5)

where C_T represents the control steps required for scheduling the DFG of DSP application. The design cost (Zc) of a particular DSP application can be determined using the following design cost metric, as shown below:

$$Zc(Sid) = W_a(A_D/A_{MAX}) + W_T(T_D/T_{MAX})$$
(8.6)

where "A_{MAX}" represents the maximum design area (can be computed using maximum available hardware resources (P_1^{max}, P_2^{max})). "T_D" represents the maximum latency (can be computed based on the most serial execution using minimum possible hardware resources (P_1^{min}, P_2^{min})). W_a and W_T are the weighing factors of normalized design area and latency in the cost function (indicating the priority provided by the IP vendor to the design area and latency during cost evaluation). The assumed values by the IP vendor corresponding to "W_a and W_T" are 0.5.

So far, we discussed the computation of security metric (S_m^n) and design cost metric (Zc). Now, we discuss the details of security–design cost fitness function.

The security–design cost trade-off fitness value can be determined by using the following equation as shown below:

$$Security - cost \ trade\text{-}off \ \text{fitness} \ function = W_s(S_m^n) + W_d(Zc)$$
(8.7)

where W_S and W_D are the weighing factors of normalized security and design cost in the fitness function (indicating the priority provided by the IP designer to the security and design cost during trade-off), "S_m^n" refers to the security and "Zc" refers to the design cost. Thus, post performing the computation of fitness value of each particle, the global best resource configuration can be determined. The particle with the minimum function value is declared as the fittest (Sgb) among all other particles in each iteration. This process is followed in each iteration until the process gets converged or the terminating criterion is met. This, therefore, at the end results into an optimal security–design cost architectural solution for respective DSP application corresponding to signature-based security algorithms using PSO-DSE.

8.5   Analysis and discussion

This section analyses security–design cost trade-off for the signature-based security algorithms corresponding to DSP hardware using PSO. It enables an IP designer to choose an optimal DSP hardware's architectural solution. Furthermore, it also guides an IP designer to achieve maximum security strength and minimal design cost overhead in parallel (Chaurasia and Sengupta, 2022d).

8.5.1   Security analysis

The discussed methodology analyzes the security of the signature-based security algorithms for DSP applications. The security (in terms of strength of ownership proof) is assessed using the following probability of coincidence (Xp) metric (Koushanfar *et al.*, 2005; Sengupta and Rathor, 2019):

$$Xp = \left(1 - \frac{1}{x}\right)^{y} \tag{8.8}$$

where "x" signifies the number of registers/colors, before implanting secret hardware security constraints in the design during the register allocation phase of HLS and "y" refers the number of hardware security constraints (storage variable pairs) to be implanted in the design covertly. The "Xp" metric specifies the probability of coincidently detecting security constraints of an authentic design in an unsecured version; hence, it is intended for it to be low as much as possible. The Xp values achieved for the respective security algorithms (IP watermarking, encrypted hash-based digital signature, and IP facial biometric) for 8-point DCT and ARF are shown in Tables 8.9 and 8.10, respectively. Similarly, "Xp" metric values have been obtained for DWT, FIR and 4-point DCT applications (Express Benchmark Suite, 2022). The "Xp" values for DWT, FIR, and 4-point DCT application are shown in Tables 8.11–8.13, respectively. It can be observed that the "Xp" for the

Table 8.9   *Comparison of Xp for different signature-based security algorithms based on their encoding rules for 8-point DCT application (number of registers (x)=8)*

Max. constraints	Fingerprint biometric (Sengupta and Rathor, 2021)		IP watermarking (Sengupta and Bhadauria, 2016)		Digital signature (Sengupta *et al.*, 2019)	
	Embedded security constraints	*Xp*	Embedded security constraints	*Xp*	Embedded security constraints	*Xp*
32	32	1.39E−02	32	1.39E−02	32	1.39E−02
64	64	1.94E−04	64	1.94E−04	55	6.46E−04
128	110	4.17E−07	102	1.21E−06	83	1.53E−05

Table 8.10 *Comparison of Xp for different signature-based security algorithms based on their encoding rules for ARF application (number of registers (x)=8)*

Max. constraints	Fingerprint biometric (Sengupta and Rathor, 2021)		Hardware watermarking (Sengupta and Bhadauria, 2016)		Digital signature (Sengupta *et al.*, 2019)	
	Embedded security constraints	*Xp*	Embedded security constraints	*Xp*	Embedded security constraints	*Xp*
32	32	1.39E−02	32	1.39E−02	32	1.39E−02
64	64	1.94E−04	64	1.94E−04	64	1.94E−04
128	128	3.77E−08	127	4.31E−08	110	4.17E−07

Table 8.11 *Comparison of Xp for different signature-based security algorithms based on their encoding rules for DWT application (number of registers (x)=5)*

Max. constraints	Fingerprint biometric (Sengupta and Rathor, 2021)		Hardware watermarking (Sengupta and Bhadauria, 2016)		Digital signature (Sengupta *et al.*, 2019)	
	Embedded security constraints	*Xp*	Embedded security constraints	*Xp*	Embedded security constraints	*Xp*
32	32	7.92E−04	32	7.92E−04	32	7.92E−04
64	64	6.27E−07	64	6.27E−07	55	4.67E−06
128	110	2.00E−11	101	1.60E−10	83	9.05E−09

Table 8.12 *Comparison of Xp for different signature-based security algorithms based on their encoding rules for FIR application (number of registers (x)=8)*

Max. constraints	Fingerprint biometric (Sengupta and Rathor, 2021)		Hardware watermarking (Sengupta and Bhadauria, 2016)		Digital signature (Sengupta *et al.*, 2019)	
	Embedded security constraints	*Xp*	Embedded security constraints	*Xp*	Embedded security constraints	*Xp*
32	32	1.39E−02	32	1.39E−02	32	1.39E−02
64	64	1.94E−04	64	1.94E−04	64	1.94E−04
128	128	3.77E−08	122	8.41E−08	100	1.58E−06

Table 8.13 Comparison of Xp for different signature-based security algorithms based on their encoding rules for 4-point DCT application (number of registers (x)=4)

Max. constraints	Fingerprint biometric (Sengupta and Rathor, 2021)		Hardware watermarking (Sengupta and Bhadauria, 2016)		Digital signature (Sengupta et al., 2019)	
	Embedded security constraints	Xp	Embedded security constraints	Xp	Embedded security constraints	Xp
32	24	1.00E−03	27	4.23E−04	21	2.37E−03
64	25	7.52E−04	37	2.38E−05	21	2.37E−03
128	25	7.52E−04	41	7.54E−06	21	2.37E−03

facial biometric-based security algorithm is lesser than the "*Xp*" for watermarking (Sengupta and Bhadauria, 2016) and encrypted hash-based security (Sengupta *et al.*, 2019) in 8-point DCT, ARF, DWT, FIR, and 4-point DCT applications. This is because lesser "*Xp*" is achieved if more number of constraints can be embedded using that security algorithm. Facial biometric approach (Sengupta and Rathor, 2021) results into more security constraints as it generates the signature based on unique and non-replicable facial features as well as uses a greater number of features from the features set to generate larger size security constraints. The number of embedded constraints has been generated based on the signature strength.

Security against tampering attack has been evaluated using the tamper tolerance ability (*TA*). A larger signature size proportionately increases the signature space, thereby enhancing the resistance for an attacker to find the exact security signature impression implanted into the design. The tamper tolerance ability is measured using the metric (Koushanfar *et al.*, 2005; Sengupta and Bhadauria, 2016):

$$TA = (v)^e \qquad (8.9)$$

where "*v*" signifies the number of signature variables used in the security algorithm, which is two ("0" & "1") for both encrypted hash (Sengupta *et al.*, 2019) and facial biometric (Sengupta and Rathor, 2021) and four ("i," "I," "T," "!") for watermarking approach (Sengupta and Bhadauria, 2016) and "*e*" denotes the chosen signature strength. Since the number of encoding variables in watermarking approach is comparatively higher, thus the tolerance of watermarking approach against tampering attack is sturdier than facial biometric and encrypted hash-based security. The comparison of *TA* ability of the watermarking, encrypted hash, and facial biometric approach, based on different signature sizes is shown in Figure 8.8. As evident, the "*TA*" of the watermarking approach is higher than facial biometric and encrypted hash-based digital signature approaches.

8.5.2 Analyzing the impact of signature strength on fitness value and register count for DSP applications

The impact of signature strength on fitness value and register count for DSP applications based on different security algorithms such as watermarking (Sengupta and Bhadauria, 2016), encrypted hash (Sengupta *et al.*, 2019), and facial biometric (Sengupta and Rathor, 2021) is analyzed using the security–design cost trade-off function (as shown in (8.7)). The corresponding results for 8-point DCT and ARF are shown in Figures 8.9 and 8.10, respectively. Similarly, for other DSP, applications such as DWT, FIR, and 4-point DCT are shown in Figures 8.11–8.13, respectively. As evident, the bigger signature size results in more security constraints (higher security strength) than the smaller signature size; hence, more possibility of design overhead (in the form of register count on embedding all the effective security constraints).

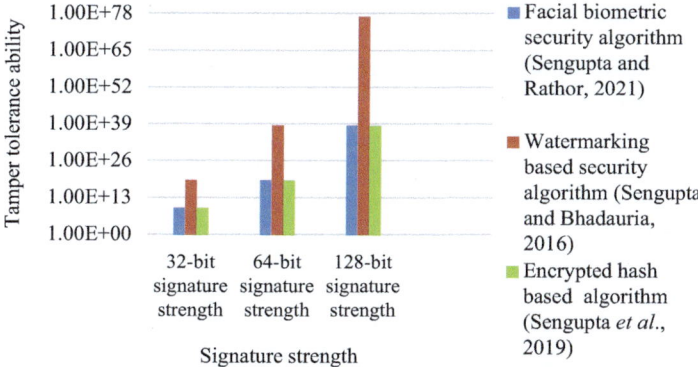

Figure 8.8 *Comparison of tamper tolerance ability of security algorithms for different signature strengths*

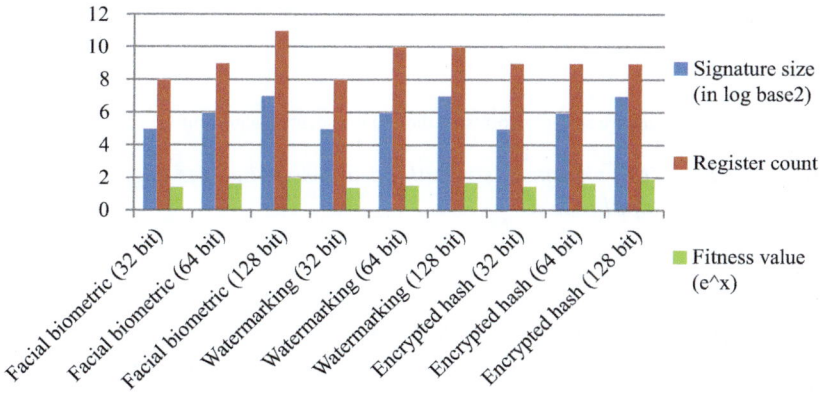

Figure 8.9 *Impact of signature strength on fitness value and register count in 8-point DCT application*

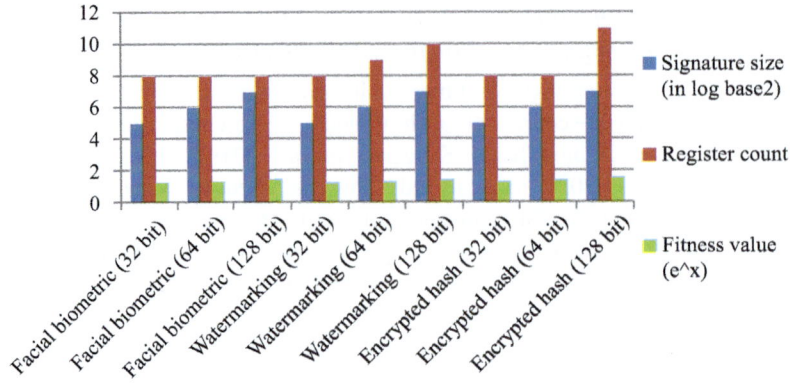

Signature strength corresponding to the security algorithm

Figure 8.10   Impact of signature strength on fitness value and register count in ARF application

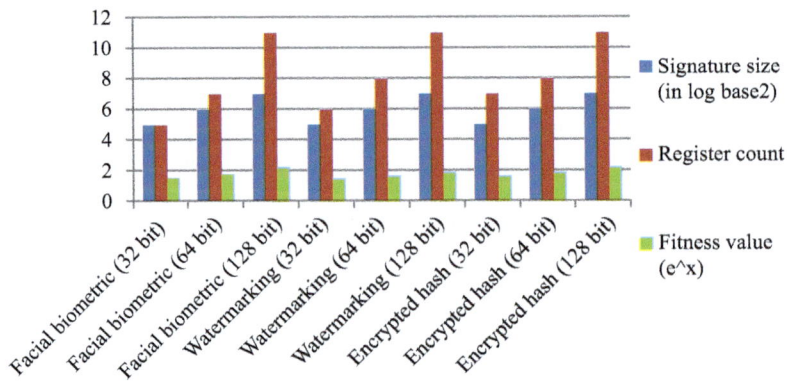

Signature strength corresponding to the security algorithm

Figure 8.11   Impact of signature strength on fitness value and register count in DWT application

8.5.3   Analyzing the security algorithms in terms of hardware cost, embedded security constraints, and exploration time

The details of the security constraints, fitness function, design area, delay, Sgb, and average exploration time obtained using PSO-DSE for the signature-based security algorithms for 8-point DCT and ARF are shown in Tables 8.14 and 8.15,

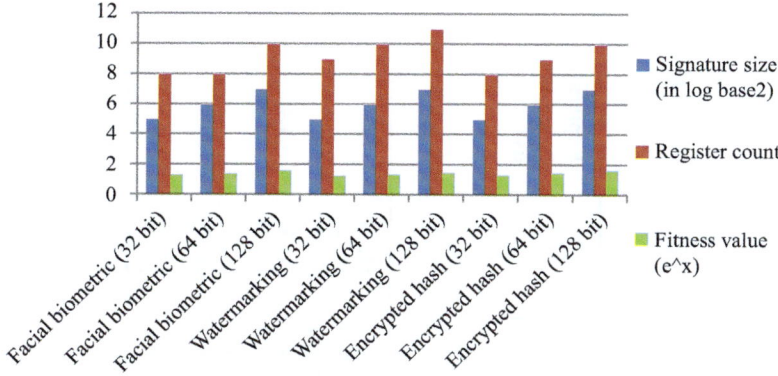

Figure 8.12 *Impact of signature strength on fitness value and register count in FIR application*

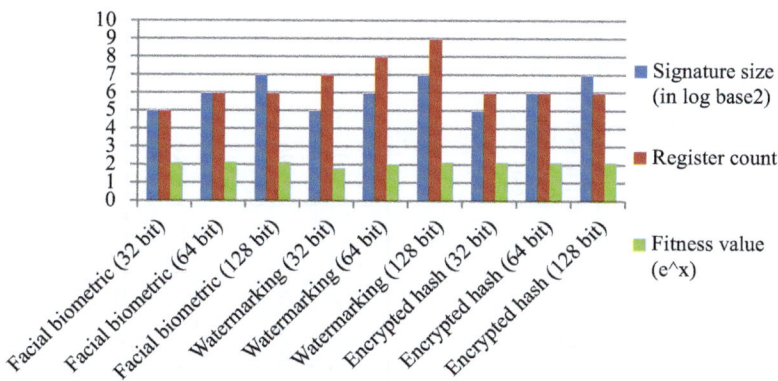

Figure 8.13 *Impact of signature strength on fitness value and register count in 4-point DCT application*

respectively. Similarly, the results corresponding to other DSP applications such as DWT, FIR, and 4-point DCT are shown in Tables 8.16–8.18, respectively. The global best resource configuration (hardware solution) reported by the security–design cost trade-off approach for 8-point DCT and ARF are (1A, 4M) and (2A, 4M), respectively. The PSO-DSE process during security–design cost trade-off always converges to the Sgb. Further, the details of hardware units obtained during trade-off exploration (security–design cost) are reported in Table 8.19.

Table 8.14 *Details of the security constraints, fitness value, Sgb, and average exploration time of security–design cost trade-off methodology for 8-point DCT w.r.t. various security algorithms*

Application framework	Signature size (in bits)	Embedded constraints corresponding to the encoding rule of the algorithm	Security algorithm	Fitness value (Security–design cost)	Design area "Ad" (in μm²)	Design latency "Ld" (in ms.)	Sgb	Exploration time (Avg. in μs)
8-DCT	32	32	Facial biometric	0.3682	327.156	927.3992	[1, 4]	173.7
	64	64	Facial biometric	0.5008	327.9424	927.3992	[1, 4]	
	128	110	Facial biometric	0.6915	329.5152	927.3992	[1, 4]	
8-DCT	32	32	Watermarking	0.3296	327.156	927.3992	[1, 4]	164.4
	64	64	Watermarking	0.4238	328.7288	927.3992	[1, 4]	
	128	102	Watermarking	0.5349	328.7288	927.3992	[1, 4]	
8-DCT	32	32	Encrypted hash	0.4065	327.9424	927.3992	[1, 4]	150
	64	55	Encrypted hash	0.5289	327.9424	927.3992	[1, 4]	
	128	83	Encrypted hash	0.6778	327.9424	927.3992	[1, 4]	

Table 8.15 *Details of the security constraints, fitness value, Sgb and average exploration time of security–design cost trade-off methodology for ARF framework w.r.t. various security algorithms*

Application framework	Signature size (in bits)	Embedded constraints corresponding to the encoding rule of the algorithm	Security algorithm	Fitness value (security–design cost)	Design area "Ad" (in μm^2)	Design latency "Ld" (in ms)	Sgb	Exploration time (avg. in μs)
ARF	32	32	Facial biometric	0.2512	346.0304	1,391.0988	[2, 4]	168.8
	64	64	Facial biometric	0.2980	346.0304	1,391.0988	[2, 4]	
	128	128	Facial biometric	0.3916	346.0304	1,391.0988	[2, 4]	
ARF	32	32	Watermarking	0.2466	346.0304	1,391.0988	[2, 4]	157.2
	64	64	Watermarking	0.2890	346.8168	1,391.0988	[2, 4]	
	128	127	Watermarking	0.3721	347.6032	1,391.0988	[2, 4]	
ARF	32	32	Encrypted hash	0.2814	346.0304	1,391.0988	[2, 4]	153.4
	64	64	Encrypted hash	0.3583	346.0304	1,391.0988	[2, 4]	
	128	110	Encrypted hash	0.4697	348.3896	1,391.0988	[2, 4]	

Table 8.16 Details of the security constraints, fitness value, Sgb, and average exploration time of the security–design cost trade-off methodology for DWT framework w.r.t. various security algorithms

Application framework	Signature size (in bits)	Embedded constraints corresponding to the encoding rule of the algorithm	Security algorithm	Fitness value (security–design cost)	Design area "Ad" (in μm^2)	Design latency "Ld" (in ms)	Sgb	Exploration time (avg. in μs)
DWT	32	32	Facial biometric	0.4492	98.3040	2384.7408	[1, 1]	154.9
	64	64	Facial biometric	0.5955	99.8768	2384.7408	[1, 1]	
	128	110	Facial biometric	0.8063	103.0224	2384.7408	[1, 1]	
DWT	32	32	Watermarking	0.4115	99.0904	2384.7408	[1, 1]	168.7
	64	64	Watermarking	0.5198	100.6632	2384.7408	[1, 1]	
	128	101	Watermarking	0.6452	103.0224	2384.7408	[1, 1]	
DWT	32	32	Encrypted hash	0.4974	99.8768	2384.7408	[1, 1]	180.7
	64	55	Encrypted hash	0.6364	100.6632	2384.7408	[1, 1]	
	128	83	Encrypted hash	0.8063	103.0224	2384.7408	[1, 1]	

Table 8.17 *Details of the security constraints, fitness value, global best solution, and average exploration time of the security–design cost trade-off methodology for FIR framework w.r.t. various security algorithms*

Application framework	Signature size (in bits)	Embedded constraints corresponding to the encoding rule of the algorithm	Security algorithm	Fitness value (security–design cost)	Design area "Ad" (in μm^2)	Design latency "Ld" (in ms)	Sgb	Exploration time (avg. in µs)
FIR	32	32	Facial biometric	0.2932	383.7792	993.642	[4, 4]	153.1
	64	64	Facial biometric	0.3644	383.7792	993.642	[4, 4]	
	128	128	Facial biometric	0.5071	385.3520	993.642	[4, 4]	
FIR	32	32	Watermarking	0.2757	384.5656	993.642	[4, 4]	163.6
	64	64	Watermarking	0.3293	385.3520	993.642	[4, 4]	
	128	122	Watermarking	0.4262	386.1384	993.642	[4, 4]	
FIR	32	32	Encrypted hash	0.3191	383.7792	993.642	[4, 4]	161.8
	64	64	Encrypted hash	0.4163	384.5656	993.642	[4, 4]	
	128	100	Encrypted hash	0.5257	385.3520	993.642	[4, 4]	

Table 8.18 Details of the security constraints, fitness value, global best solution, and average exploration time of security–design cost trade-off methodology for 4-point DCT framework w.r.t. various security algorithms

Application framework	Signature size (in bits)	Embedded constraints corresponding to the encoding rule of the algorithm	Security algorithm	Fitness value (security– design cost)	Design area "Ad" (in μm²)	Design latency "Ld" (in ms)	Sgb	Exploration time (avg. in μs)
4-point DCT	32	24	Facial biometric	0.7611	98.3040	1126.1276	[1, 2]	150.1
	64	25	Facial biometric	0.7817	99.0904	1126.1276	[1, 2]	
	128	25	Facial biometric	0.7817	99.0904	1126.1276	[1, 2]	
4-point DCT	32	27	Watermarking	0.6116	99.8768	1126.1276	[1, 2]	155.0
	64	37	Watermarking	0.7342	100.6632	1126.1276	[1, 2]	
	128	41	Watermarking	0.7835	101.4496	1126.1276	[1, 2]	
4-point DCT	32	21	Encrypted hash	0.7817	99.0904	1126.1276	[1, 2]	157.3
	64	21	Encrypted hash	0.7817	99.0904	1126.1276	[1, 2]	
	128	21	Encrypted hash	0.7817	99.0904	1126.1276	[1, 2]	

Table 8.19   The details of DSP hardware units obtained by using trade-off exploration (security–design cost)

Application framework	Security algorithm	Post embedding register count based on signature size (bits)			#Adder unit(s)	#Multiplier unit(s)	#Multiplexer units	#Demultiplexer units
		32	64	128				
4-point DCT	Facial biometric	5	6	6	1	2	6	3
	Watermarking	7	8	9	1	2	6	3
	Encrypted hash	6	6	6	1	2	6	3
8-point DCT	Facial biometric	8	9	11	1	4	10	5
	Watermarking	8	10	10	1	4	10	5
	Encrypted hash	9	9	9	1	4	10	5
FIR	Facial biometric	8	8	10	4	4	16	8
	Watermarking	9	10	11	4	4	16	8
	Encrypted hash	8	9	10	4	4	16	8
DWT	Facial biometric	5	7	11	1	1	4	2
	Watermarking	6	8	11	1	1	4	2
	Encrypted hash	7	8	11	1	1	4	2
ARF	Facial biometric	8	8	8	2	4	12	6
	Watermarking	8	9	10	2	4	12	6
	Encrypted hash	8	8	11	2	4	12	6

8.6   Conclusion

This chapter discussed an approach for the exploration of security–design cost trade-off corresponding to signature-based security algorithms and DSP hardware IPs. It provides an optimal design architectural solution for secured IP cores such as 8-point DCT, ARF, DWT, FIR, 4-point DCT, etc. used in CE systems, by employing PSO-based design space exploration. The security–design cost trade-off methodology considers three different signature-based hardware security algorithms based on facial biometrics, encrypted-hash, and watermarking for integration with the PSO-DSE framework for exploring the hardware architectural trade-offs of security–design cost. Experimental results in terms of security, design cost (area, delay), exploration time, and other vital parameters validate the exploration of low-cost secure design architectural solution that offers an IP designer and SoC integrator to employ low-cost and secured IP cores for integration in modern electronic/automated system designs. Furthermore, it also ensures an IP designer to achieve minimal design cost overhead and maximum-security strength concurrently.

At the end of this chapter, a reader gains the knowledge about the following:

- The significance of security–design cost trade-off for generating the low-cost architectural solution corresponding to signature-based security algorithms and DSP designs that are used in CE and computing systems.
- Mechanism of different signature-based hardware security algorithms.
- Significance of PSO-based design space exploration for generating low-cost architectural solution.
- Comparative analysis of signature-based hardware security algorithms in terms of robust security and design cost.
- Low-cost architectural solution for different DSP applications.
- A case study on 8-point DCT application in terms of generating scheduled design with embedded signature (digital evidence) corresponding to security algorithms such as hardware watermarking, encrypted hashing, and facial biometric.

8.7   Questions and exercise

1. Discuss the significance of performing security–design cost trade-off for different DSP applications.
2. What are different signature-based hardware security algorithms?
3. Why are the security and design cost called orthogonal in nature?
4. Discuss the hardware security methodologies using watermarking, encrypted hashing, and facial biometric approaches in details.
5. Discuss the trade-off for different signature-based security algorithms in terms of security robustness and design cost.
6. Discuss the SE process corresponding to 8-point DCT application using encrypted hash-based security algorithm.

7. Discuss the SE process corresponding to 8-point DCT application using watermarking-based security algorithm.
8. Discuss the SE process corresponding to 8-point DCT application using facial biometric-based security algorithm.
9. Discuss the qualitative comparison between security algorithms based on different security parameters.
10. Which security methodology achieves more robust hardware security and why?
11. Discuss the process of design space exploration using PSO.
12. Discuss the significance of performing velocity clamping and end terminal perturbation in PSO.
13. Discuss the significance of mutation in PSO-based DSE.
14. Discuss the details of security–design cost trade-off fitness function.
15. Explain the impact of varying signatures on the probability of coincidence and tamper tolerance ability.
16. Analyze the impact of signature strength on the register count of different DSP applications.
17. Analyze the security algorithms in terms of hardware cost, embedded security constraints, and exploration time.
18. Discuss the process of computing the design latency and area.
19. Discuss and compare the impact of encoding algorithms for signature-based security algorithms.

References

15 nm open cell library. [Online], Available: https://si2.org/open-cell-library/, last accessed on January 2020.

Castillo, E., U. Meyer-Baese, A. García, L. Parrilla and A. Lloris 2007), "IPP@HDL: efficient intellectual property protection scheme for IP cores," *IEEE Transactions on VLSI Systems*, vol. 15, no. 5, pp. 578–591.

Chaurasia, R. and A. Sengupta (2022a), "Protecting Trojan secured DSP cores against IP piracy using facial biometrics," *in: Proceedings of IEEE 19th India Council International Conference (INDICON)*, Kochi, India, 2022, pp. 1–6, doi:10.1109/INDICON56171.2022.10039864

Chaurasia, R. and A. Sengupta (2022b), "Symmetrical protection of ownership right's for IP buyer and IP vendor using facial biometric pairing," in: *Proceedings of 8th IEEE International Symposium on Smart Electronic Systems (IEEE – iSES)*, India, Accepted.

Chaurasia, R. and A. Sengupta (2022c), "Crypto-genome signature for securing hardware accelerators," in: *Proceedings of 19th IEEE INDICON*, India, Accepted for publication, November.

Chaurasia, R. and A. Sengupta (2022d), "Security vs design cost of signature driven security methodologies for reusable hardware IP core," in: *2022 IEEE International Symposium on Smart Electronic Systems (iSES)*, Warangal, India. pp. 283–288, doi:10.1109/iSES54909.2022.00064

Colombier, B. and L. Bossuet (2014), "Survey of hardware protection of design data for integrated circuits and intellectual properties," *IET Computers & Digital Techniques*, vol. 8, no. 6, pp. 274–287.

Cui, A. and C. Chang (2007), "Watermarking for IP protection through template substitution at logic synthesis level," in: *Proceedings of ISCAS*, New Orleans, LA, pp. 3687–3690.

Express Benchmark Suite. Available: http://www.ece.ucsb.edu/EXPRESS/benchmark/. Accessed January 2022.

Hong, I. and M. Potkonjak (1999), "Behavioral synthesis techniques for intellectual property security," in: *Proceedings of DAC*, pp. 849–854.

Karmakar, R. and S. Chattopadhyay (2020), "Hardware IP protection using logic encryption and watermarking," in: *2020 IEEE International Test Conference (ITC)*, pp. 1–10.

Koushanfar, F., I. Hong and M. Potkonjak (2005), "Behavioral synthesis techniques for intellectual property protection", *ACM Transactions on Design Automation of Electronic Systems*, vol. 10, no. 3, pp. 523–545.

Koushanfar, F., S. Fazzari, C. McCants, W. *et al.* (2012), "Can EDA combat the rise of electronic counterfeiting?," in: *Proceedings of DAC*, San Francisco, CA, pp. 133–138.

Le Gal, B. and L. Bossuet (2012), "Automatic low-cost IP watermarking technique based on output mark insertions," *Design Automation for Embedded Systems*, vol. 16, no. 2, pp. 71–92.

Mahdiany, H.R., A. Hormati and S.M. Fakhraie (2001), "A hardware accelerator for DSP system design," in: *Proceedings of ICM*, Morocco, 2001, pp. 141–144.

Mishra, V. and A. Sengupta (2014), "MO-PSE: adaptive multi objective particle swarm optimization based design space exploration in architectural synthesis for application specific processor design," *Elsevier Journal of Advances in Software Engineering*, vol. 67, pp. 111–124, ISSN 0965-9978.

Newbould, R.D., J.D. Carothers and J.J. Rodriguez (2002), "Watermarking ICs for IP protection," *Electronics Letter*, vol. 38, no. 6, pp. 272–274.

Ni, M. and Z. Gao (2005), "Detector-based watermarking technique for soft IP core protection in high synthesis design level," in: *Proceedings of CCS*, Hong Kong, pp. 1348–1352.

Pilato, C., S. Garg, K. Wu, R. Karri and F. Regazzoni (2018), "Securing hardware accelerators: a new challenge for high-level synthesis," *IEEE Embedded Systems Letters*, vol. 10, no. 3, pp. 77–80.

Plaza, S.M. and I.L. Markov (2015), "Solving the third-shift problem in IC piracy with test-aware logic locking," *IEEE Transactions on Computer-Aided Design of Integrated Circuits and Systems*, vol. 34, no. 6, pp. 961–971.

Rathor, M. and A. Sengupta (2020), "IP core steganography using switch based key-driven hash-chaining and encoding for securing DSP kernels used in CE systems," *IEEE Transactions on Consumer Electronics*, vol. 66, no. 3, pp. 251–260.

Rathor, M., A. Anshul, K. Bharath, R. Chaurasia, and A. Sengupta (2023), "Quadruple phase watermarking during high level synthesis for securing reusable hardware IP cores," *Elsevier Journal of Computer and Electrical Engineering*, vol. 105, Article no. 108950.

Roy, D. and A. Sengupta (2019), "Multi-level watermark for protecting DSP kernel used in CE systems," *IEEE Consumer Electronics Magazine*, vol. 8, no.2, pp. 100–102.

Roy, J.A., F. Koushanfar and I.L. Markov (2008), "EPIC: ending piracy of integrated circuits," in: *Proceedings of DATE*, Munich, pp. 1069–1074.

Salivahanan, S. and A. Vallavaraj (2001), *Digital Signal Processing*, Mumbai: McGraw-Hill Education (India) Pvt Limited, ISBN: 9780074639962.

Schneiderman, R (2010), "DSPs evolving in consumer electronics applications," *IEEE Signal Processing Magazine*, vol. 27, no. 3, pp. 6–10.

Sengupta, A. and S. Bhadauria (2016), "Exploring low cost optimal watermark for reusable IP cores during high level synthesis," *IEEE Access*, vol. 4, pp. 2198–2215.

Sengupta, A. and R. Chaurasia (2022a), "Secured convolutional layer IP core in convolutional neural network using facial biometric," *IEEE Transactions on Consumer Electronics*, vol. 68, no. 3, pp. 291–306.

Sengupta, A. and R. Chaurasia (2022b), "Securing IP cores for DSP applications using structural obfuscation and chromosomal DNA impression," *IEEE Access*, vol. 10, pp. 50903–50913.

Sengupta, A. and S.P. Mohanty (2019), *IP Core Protection and Hardware-Assisted Security for Consumer Electronics*, The Institute of Engineering and Technology (IET).

Sengupta, A. and M. Rathor (2019a), "Crypto-based dual-phase hardware steganography for securing IP cores," *IEEE Letters of the Computer Society*, vol. 2, no. 4, pp. 32–35.

Sengupta, A. and M. Rathor (2019b), "IP core steganography for protecting DSP kernels used in CE systems," *IEEE Transactions on Consumer Electronics*, vol. 65, no. 4, pp. 506–515.

Sengupta, A. and M. Rathor (2020), "Securing hardware accelerators for CE systems using biometric fingerprinting," *IEEE Transactions on VLSI Systems*, vol. 28, no. 9, pp. 1979–1992.

Sengupta, A. and M. Rathor (2021), "Facial biometric for securing hardware accelerators," *IEEE Transactions on VLSI Systems*, vol. 29, no. 1, pp. 112–123.

Sengupta, A., D. Roy and S.P. Mohanty (2018), "Triple-phase watermarking for reusable IP core protection during architecture synthesis," *IEEE Transactions on Computer-Aided Design of Integrated Circuits and Systems*, vol. 37, no. 4, pp. 742–755.

Sengupta, A., E.R. Kumar and N.P. Chandra (2019), "Embedding digital signature using encrypted-hashing for protection of DSP cores in CE," *IEEE Transactions on Consumer Electronics*, vol. 65, no. 3, pp. 398–407.

Sengupta, A., R. Chaurasia and T. Reddy (2021), "Contact-less palmprint biometric for securing DSP coprocessors used in CE systems," *IEEE Transactions on Consumer Electronics*, vol. 67, no. 3, pp. 202–213.

Chapter 9

Taxonomy of hardware security methodologies: IP core protection and obfuscation

Anirban Sengupta[1] and Aditya Anshul[1]

In the modern electronic era, the increased usability and acceptability of reusable intellectual property (IP) cores induce the necessity to secure them from various hardware security threats and attacks. This chapter presents a taxonomy of hardware security methodologies for IP cores and a detailed design flow of hardware integrated circuits (ICs) along with vulnerability points where potential attacks/threats are possible. Trustworthy and untrustworthy regimes in the design flow have also been highlighted in the discussion. Further, a discussion of detective and preventive control-based hardware security approaches used for hardware IP cores have also been presented, including an analysis of prominent structural obfuscation, logic locking (logic encryption), and IP core protection (IPP) techniques. Each approach has been lucidly explained in terms of its threat model, algorithm, and security analysis. Finally, a security comparison of hardware IP obfuscation approaches in terms of strength of obfuscation security metric as well as security comparison of IPP approaches in terms of probability of coincidence security metric have also been introduced.

9.1 Introduction

Reusable application-specific integrated circuits (ASICs) [or digital integrated circuits (ICs)] are the central modules in various electronic and computing devices such as tablets, smartwatches, laptops, and mobile phones. These electronic and computing devices have found wide usability and applicability in the modern digital ecosphere, starting from intelligent kitchen appliances to complex embedded systems used in automobiles and aircraft. Some additional usages of ASICs include biometric fingerprinting, automation and advancement of military operations, robotics, automation at tolls, and medical imaging systems. ASICs are designed to achieve a particular type of functionality, generally data and computation-intensive tasks such as image pixel computation (compression and decompression of image data), digital data filtering, signal amplification and attenuation, and execution of the complex mathematical function in case of complex embedded systems. Therefore, it is crucial to

[1]Department of Computer Science and Engineering, Indian Institute of Technology Indore, India

efficiently design these ASICs by considering various parameters such as speed, area, power, security, and reliability. The complexity of ASICs in terms of size and number of transistors is huge. Therefore, it is efficient to design these ASICs through different layers of abstraction. The various abstraction layers provide easy error handling with a more detailed design environment. These digital ICs (ASICs) are also known as reusable intellectual property (IP) cores, and they are realized in the form of digital signal co-processors (DSPs). Some of the widely used DSP IPs are (a) discrete cosine transform (DCT), used in compression and decompression of video, image, and audio files; (b) Haar wavelet transform (HWT), used for performing compression on digital data; (c) finite and infinite impulse response filter (FIR and IIR), widely used to perform signal attenuation in audio and video systems; (d) fast Fourier transform (FFT), widely used for digital video broadcasting; (e) joint photographic expert group (JPEG) used for performing image and video compression in digital camera systems, etc. As discussed above, all these DSP applications are intended to perform a specific data and computation-intensive function. They are therefore designed as a dedicated reusable intellectual property (IP) cores from higher levels of abstraction using a high-level synthesis (HLS) framework (Pilato *et al.*, 2018).

The DSP IP core design process contains five different levels of abstraction, where the different layers perform different dedicated design functionalities. Figure 9.1 highlights the different abstraction levels of the DSP IP core design process. The five different abstraction levels are: (i) system level, (ii) algorithmic level, (iii) register transfer level (RTL), (iv) logic level, and (v) layout level, respectively. The system level represents the DSP IP core design at the highest level of abstraction. Here, the application (or design) is expressed in the form of system specifications such as input/outputs, functionality, speed, size, and power requirement. Next, the second abstraction level, also known as the algorithmic level, expresses the design description in terms of its behavior. Here, the design is expressed with the help of control data flow graph (CDFG), also referred to as an intermediate representation of the design.

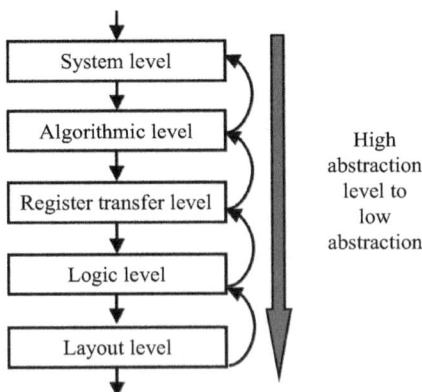

Figure 9.1   Top to bottom representation of different abstraction level used in digital ICs design process

The algorithmic level is also known as the electronic synthesis level (ESL) or behavioral level. Subsequently, the register transfer level expresses the design description in the form of an interconnection between different units, such as the arithmetic and logic unit (ALU), control unit, and storage hardware. Further, the design is expressed in terms of logic gates at the logic level. The final design layout is produced at the layout level and transferred to the fabrication center of the chips (or ASICs) fabrication and manufacturing. The output of the layout level is a layout-ready database file format, also known as a graphic data system (GDS) file. The GDS files are a collection of detailed binary information, such as the representation of text labels, planar geometric shapes, and other hierarchical data on the layout (Sengupta, 2016).

Further, the HLS process generates an optimized RTL datapath using IP vendor-specified design constraints, functional description of DSP application (in the form of transfer function or C/C++ code), and optimization algorithm. The detailed flow of the HLS process is shown in Figure 9.2. At first, the high-level specifications are specified, such as speed, power, energy, and area. Next, the target DSP application benchmark is selected, such as DCT, FIR, and FFT. Then, the selected DSP application is converted into its equivalent CDFG using its respective transfer function. Further, design space exploration is performed by considering the various high-level specifications. Various heuristic and meta-heuristic algorithms, such as particle swarm optimization (Mishra and Sengupta, 2014), ant colony optimization, genetic algorithm (Krishnan and Katkoori, 2006), and bacterial foraging algorithm (Sengupta and Bhadauria, 2014), are commonly used to perform design space exploration (DSE). The DSE prunes the search space according to the IP vendor-specified criteria (specifications) and yields an optimized architectural solution corresponding to the target DSP application. Finding an optimized solution corresponding to complex DSP applications is challenging, as they involve trade-offs between several orthogonal conditions. Therefore, DSE is a crucial phase of

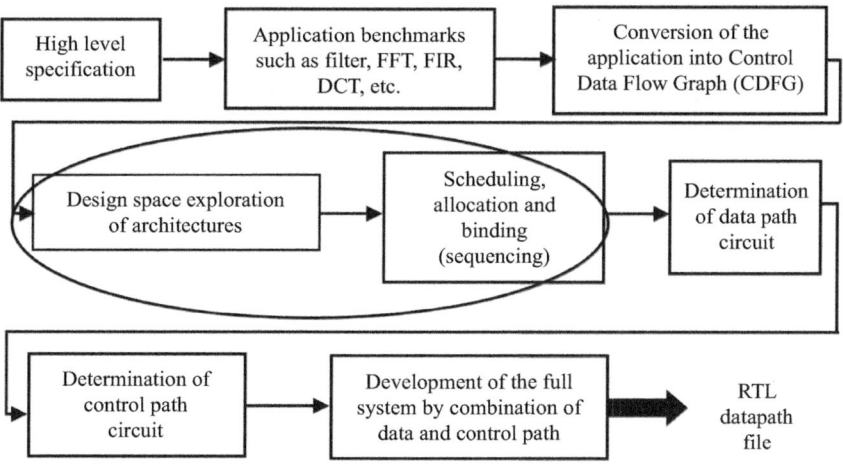

Figure 9.2   Flow of the HLs process

the HLS process. The obtained architectural solution (or resource constraints) is used to perform the scheduling, followed by allocation and binding. The three steps, scheduling, allocation, and binding, are the HLS process's core components. The CDFG is scheduled with the help of obtained architectural solutions and scheduling algorithms to produce a scheduled data flow graph (SDFG). There are various scheduling algorithms available in the literature, such as "as soon as possible," "as late as possible," "LIST scheduling," and "integer linear programming" (McFarland *et al.*, 1988). Post-scheduling hardware are allocated to the different operations. This step is known as allocation. Register allocation using a colored interval graph (CIG) framework is also integral to the hardware allocation. Further, the various operations of the SDFG are binded to common hardware using binding operation. The binding operation enables resource sharing among various operations of SDFG. To realize resource sharing, interconnects such as multiplexers and demultiplexers are used. After binding, an RTL datapath circuit and its corresponding control circuit are generated. The control circuit is required to maintain synchronization between different hardware used. The final output of HLS is an RTL datapath design, which is further subjected to RTL synthesis followed by gate-level synthesis and layout synthesis. The advantages of the HLS process are as follows: (a) it provides a shorter design cycle due to automation, (b) it leads to fewer errors as the design is expressed in terms of its behavior at a higher level, (c) integrates the ability to search the design space, and (d) since the decisions are made at a higher level, hence the decisions taken creates impacts at the subsequent lower levels of digital IC design process (Rathor *et al.*, 2022).

9.2   Possible hardware threats and attacks in the design flow of hardware IC

The design process of the digital ICs/system on chips (SoCs) can only be realized with the involvement of multiple third-party entities. The rapid globalization of the semiconductor supply chain industry has dramatically impacted the designing process of these IP cores. The participation of numerous third-party entities is essential as some can afford a cheaper technical workforce while some lower labor costs, besides aiming to meet design cost optimization and time-to-market demands. The involvement of multiple third-party entities (IP vendors) makes the IC design process vulnerable to various hardware threats and attacks, such as IP piracy (counterfeiting and cloning), insertion of malicious logic, reverse engineering (RE) attacks, overproduction, and fraudulent claim of ownership attack by an adversary. The most common type of hardware threat is hardware (IP) piracy. IP piracy can be broadly classified into two categories: (i) IP counterfeiting and (ii) IP cloning. Figure 9.3 depicts the different types of hardware threats and attacks in a nutshell during the digital IC design process. The adversary counterfeits the design in IP counterfeiting, selling under the same brand name. Here, it becomes difficult to demarcate between

S. No.		Security issues	Descriptions
1.		Intellectual property (IP) cloning:	Selling the cloned product under different brand name
2.		IP counterfeiting:	Selling the counterfeited product under same brand name
3.		Hardware trojan attack:	Insertion of malicious circuitry that damages the IP core functionality and trustworthiness
4.		Overproduction:	Production of IP core more than the IP vendor specified licencing limit
5.		False claim on IP ownership:	Fraudulent claim of IP ownership by an adversary

Figure 9.3   Different types of hardware security issues and their corresponding description

counterfeited and genuine products. On the other hand, an adversary clones the original design in IP cloning and sells it under their (different) brand name. Counterfeited IP may contain malicious logic or may not be rigorously tested (in terms of reliability checks), which can prove to be hazardous as they can (a) cause improper functioning of the device by affecting the computational and functional output (e.g., improper computation of image pixel leading to the wrong diagnosis of disease through medical imaging system), (b) cause excessive heating in the digital ICs, (c) leak sensitive information such as essential credentials, (d) cause performance degradation, and (d) cause loss of reputation and esteem of IP vendor. Therefore, it is essential to demarcate and isolate pirated IPs from genuine ones (Rathor *et al.*, 2022; Sengupta *et al.*, 2017a).

The second type of hardware threat and attack is known as RE attack and insertion of malicious logic at safe places inside the IP design file. In a reverse engineering attack, the attacker/adversary tries to understand the complete design functionality by carefully inspecting the design RTL or netlist file. This is a bottom-up approach where an adversary tries to decode the functionality by moving backward from the RTL or netlist design file. The careful inspection of the RTL or netlist design file also provides an edge to the adversary for inserting malicious logic at safe places, such that they cannot be discovered easily during the regular testing process. The added malicious logic (hardware Trojan) remains dormant until some particular triggering condition is achieved, thus easily evading the normal testing process (test vectors). This malicious logic can also

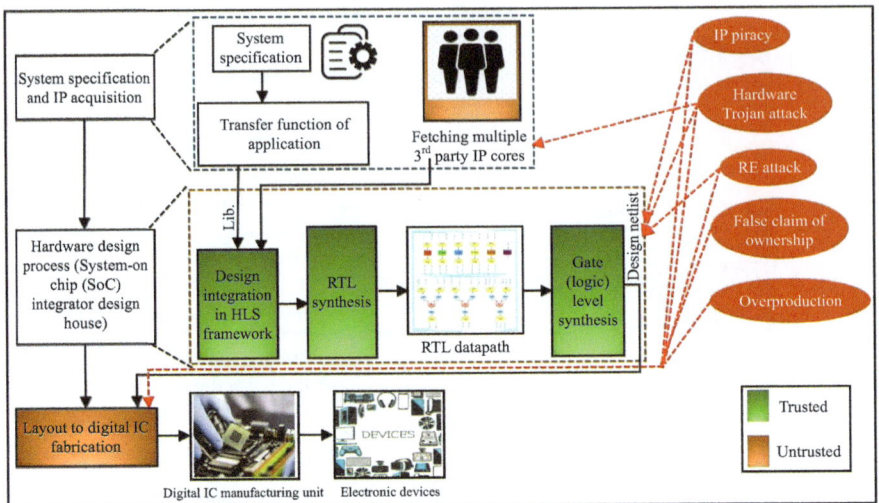

Figure 9.4   Possible hardware threats and attacks in the design flow of a digital IC

cause severe damage to IP vendors as well as end consumers, as explained above in this section. Further, the third type of hardware security threat is IP overproduction. Each IP core design file is associated with a licensing limit specified by the IP vendor or designer. Therefore, violating the authorized licensing limit of IP vendors leads to the overproduction of IP cores. Finally, fraudulent IP ownership claims are the fourth type of hardware threat and attack. Here, an adversary claims the ownership of the IP core fraudulently. IP core can be fraudulently claimed at two places, either by the SoC integrator or in the fabrication center. Figure 9.4 shows the possibilities of the various hardware threats and attacks in the complete design chain of an IP core (Becker *et al.*, 2017; Sengupta *et al.*, 2019).

9.3   Taxonomy representation of IP core protection methodologies

Threat model: The DSP IP cores imported from various third-party IP vendors may not always be trustworthy. Therefore, the demarcation and isolation of authentic DSP IP cores from the counterfeited ones are necessary before their integration into the electronic systems at the SoC level. Further, it might be possible that the counterfeited IP cores may contain malicious logic and may not be rigorously tested. Therefore, the presence of malicious logic can cause safety and reliability hazards. Further, another threat dimension is the fraudulent claim of IP ownership. Therefore, protecting the rights of IP vendors is also crucial in nullifying false claims of IP ownership (Sengupta *et al.*, 2017a).

9.3.1   Watermarking-based hardware security approach

HLS hardware watermarking is referred to as the art of embedding secret information in the hardware IP core design file. As depicted in Figure 9.5, hardware watermarking-based approach is broadly classified into two areas: (i) single-phase watermarking approach and (ii) multi-phase watermarking approach, respectively. Further, the single-phase watermarking approach is classified into two parts: (a) single-phase dual variable watermarking approach and (b) single-phase multi-variable watermarking approach. The single-phase watermarking approach embeds additional secret information into one phase known as the "register allocation" phase of the IP core design process (Koushanfar *et al.*, 2005). In contrast, multi-phase watermarking exploits three phases, "scheduling," "hardware allocation," and "register allocation" of the hardware design process to embed the additional secret information (Sengupta and Bhadauria, 2016). Further, the single-phase dual variable watermarking approach uses a two-variable ('0' and '1') signature encoding mechanism to embed additional secret information (in the form of additional edges in the CIG of the respective DSP application) (Koushanfar *et al.*, 2005). In contrast, the single phase multi variable watermarking approach uses four-variable ('i', 'I', 'T', and '!') signature encoding mechanism for hardware watermarking (Sengupta and Bhadauria, 2016). The IP vendor selected four variables, and their corresponding encoding mechanism (Sengupta and Bhadauria, 2016) are as follows: (i) 'i' = embedding an additional (artificial) edge between <prime, prime> storage variables node pair in the CIG of DSP application, (ii) 'I' = embedding an additional edge between <even, even> storage variables node pair, (iii) 'T' = embedding an additional edge between <odd, even> storage variable node pair, and (iv) '!' = embedding an additional edge between <0, any integer> storage variable node pair (Sengupta and Bhadauria, 2016). Similarly, an IP vendor-selected seven-digit multi-variable signature and their corresponding encoding mechanism in the triple (multi) phase watermarking approach (Sengupta *et al.*, 2017a) are as

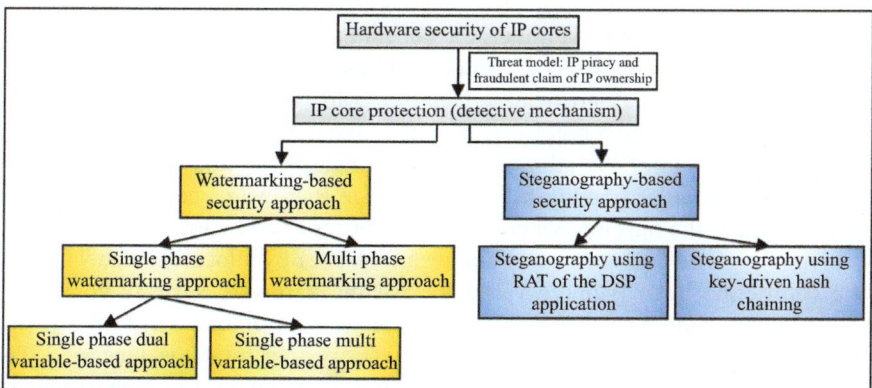

Figure 9.5   Taxonomy representation of IP core protection methodologies

follows: (i) 'α' = hardware of vendor type 1 are allocated to odd number operations, and hardware of vendor type 2 are allocated to even numbered operations in odd control steps, (ii) 'β' = hardware of vendor type 2 are allocated to odd numbered operations and hardware of vendor type 1 are allocated to even numbered operations on even control step, (iii) 'γ' = movement of non-critical path operation which is having greater mobility into its immediate next control step, (iv) 'i' = embedding an additional edge between <prime, prime> storage variables node pair, (v) 'I' = embedding an additional edge between <even, even> storage variables node pair, (vi) 'T' = embedding an additional edge between <odd, even> storage variable node pair, and (vi) '!' = embedding an additional edge between <0, any integer> storage variable node pair. All the above-mentioned IP vendor-selected multi-variable signature encoding mechanisms are used to generate the covert watermarking signature, which is further embedded into the design to secure the hardware from different threats and attacks (Sengupta *et al.*, 2017a).

The hardware watermarking methodology accepts the following as input: (i) transfer function (or C/C++ code) of the respective DSP application, (ii) IP vendor/designer selected architectural constraints (resource constraints); (iii) IP vendor/designer selected multi-variable signature combinations and their corresponding encoding mechanism. At first, the transfer function of the DSP application is converted into its equivalent CDFG graph. Next, the CDFG is scheduled using IP vendor-selected resource constraints and the LIST scheduling algorithm. An SDFG is generated as the result of the scheduling process. As mentioned above in the Introduction section, allocation and binding are further performed, which includes the allocation of storage variables (registers) using the CIG framework of the HLS process. Post allocation of storage variables, a RAT corresponding to the SDFG of the DSP application is generated. Now, for the single-phase watermarking approach, the generated signature and its corresponding hardware security constraints (generated using IP vendor-selected encoding mechanism corresponding to multi-variable signature) are embedded into the design using the CIG framework of the HLS process during the register allocation phase. A CIG is generated using the initial RAT (containing the information on register allocation in different control steps) of the DSP application. Further, the generated watermarking-based hardware security constraints are embedded into the CIG as additional artificial edges. Post embedding an additional artificial edge, the color of the register is changed accordingly in the case of a color conflict while satisfying the security constraints. If the color (register) of the adjacent nodes (storage variables) in the CIG is the same after embedding of additional artificial edge (hardware security constraint), then, in that case, the raised color conflict is either resolved by local alteration (i.e., swapping of registers between two storage variables) or by allocating a new register to the conflicting node (storage variable). All the hardware security constraints generated corresponding to the watermark signature are embedded similarly, and finally, a RAT (after complete embedding of security constraints) is generated. The final generated RTL datapath design containing the secret IP vendor-selected watermark is transferred to the next level of the IP design process, i.e., logic level synthesis, for further processing (Sengupta and Bhadauria, 2016).

On the other hand, the triple-phase watermarking approach uses the "scheduling" and "hardware allocation" phases in addition to the "register allocation" phase, as explained earlier. In the triple-phase watermarking methodology, first, a functional unit allocation table is generated using the SDFG of the DSP application with a non-critical operation timing table. If the movement/shifting of an operation into its immediate upward and downward control step does not violate the data dependency and functionality of the DSP IP core, then that particular operation is known as a non-critical operation. In the first phase of triple-phase watermarking, all the operations are initially marked and sorted in an increasing order in each control step. Then, all non-critical operations (beginning from control step one) are shifted to their immediate next control step for each occurrence of signature variable 'γ'. After shifting all non-critical operations in an SDFG, a modified non-critical timing table is generated. Subsequently, in the second phase of triple-phase watermarking, hardware re-allocation is performed based on the IP vendor-selected hardware allocation mechanisms 'α' and 'β'. Post performing hardware re-allocation, a modified hardware allocation table is generated. Further, the next step (i.e., the third phase of the triple-phase watermarking) is similar to the previously explained embedding of signature variables in the register allocation phase of single phase multi-variable watermarking mechanism. The IP vendor-selected multi variable signature ('i', 'I', 'T', and '!') are embedded in a similar fashion using CIG of the DSP application, and the IP vendor-selected embedding rule (explained above in this section). After embedding all seven variables of the IP vendor-selected signature in three different phases of the HLS framework, a final modified RAT (containing watermark signature) is generated (Sengupta *et al.*, 2017a). The embedded additional IP vendor-selected secret watermark information into the design of the hardware IP core facilitates easy and smooth detection of pirated IPs from authentic ones. Further, this watermark information also helps in resolving the fraudulent claim of IP ownership as the IP vendor can easily claim the ownership of the authentic IP based on the embedded covert watermark information (Sengupta *et al.*, 2017a). Figure 9.6 highlights the design flow of the triple-phase watermarking methodology.

The triple-phase watermarking approach (Sengupta *et al.*, 2017a) provides more robust security than the single-phase dual variable (Koushanfar *et al.*, 2005) and single-phase multi-variable (Sengupta and Bhadauria, 2016) watermarking approach due to the involvement of seven variable encoding mechanism, which leads to the generation of a larger number of security constraints leading to lower probability of coincidence (indicating stronger digital proof useful for piracy detection and nullifying false IP ownership claim). Further, the single-phase multi-variable (Sengupta and Bhadauria, 2016) watermarking approach surpasses the single-phase dual-variable watermarking approach (Koushanfar *et al.*, 2005) in terms of security strength due to the generation of a more significant number of hardware security constraints leading to lower probability of coincidence.

9.3.2 *Steganography-based hardware security approach*

Steganography is the art of concealing secret/confidential information in the hardware design file to protect it from hardware piracy and fraudulent claim of

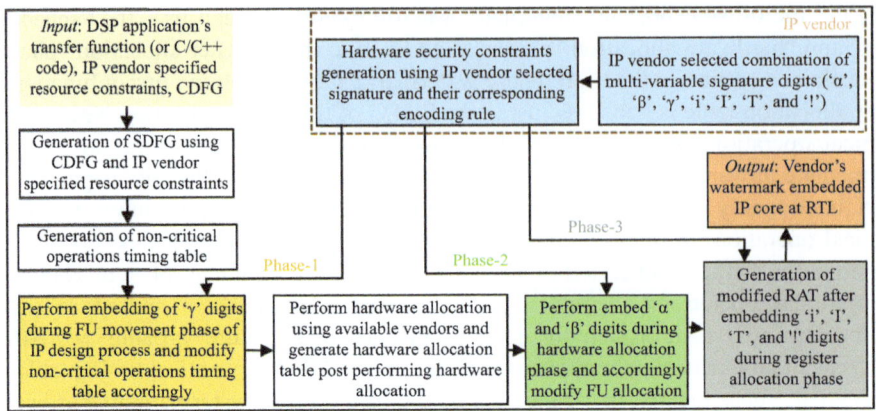

Figure 9.6   Flow diagram of triple-phase watermarking methodology

ownership. Figure 9.5 depicts the classification of hardware steganography into two categories: (a) hardware steganography using RAT of DSP application and (b) steganography using key-driven hash chaining.

Entropy thresholding-based IP steganography: The steganography using RAT of the DSP application (Sengupta and Rathor, 2019) accepts the transfer function of the DSP application, IP vendor-selected resource constraints, threshold entropy value, and module library (containing functional unit specifications such as area, delay, and power requirement). Initially, as explained in the previous watermarking-based security approach, the transfer function of the DSP application is converted to its corresponding CDFG. Further, the CDFG is scheduled using IP vendor-selected resource constraints to generate an SDFG. Post generating SDFG register allocation is performed using the CIG framework of the HLS process. After register allocation, a RAT is generated, which is further used to generate hardware security constraints in the steganography using the RAT of the DSP application. At first, all possible edges between the storage variables of similar color are listed. Then, the entropy value is computed corresponding to embedding each possible edge between the same colored storage variables. The entropy value is equivalent to the maximum number of color (register) transformations required corresponding to embedding each possible edge. After computing the entropy value, the final list of possible edges between storage variables of similar color is determined based on the IP vendor-selected threshold entropy value. Subsequently, all the generated possible edges between storage variables of similar color (hardware security constraint) are embedded into the design of the DSP IP core, followed by color transformations in the case of color conflict due to the embedding of additional artificial edges (security constraints). The IP vendor-selected threshold entropy value provides an edge to the designer to control the amount of additional information to be embedded into the design, to meet the design cost optimization objective. The complete details of the steganography using RAT of the DSP application are explained in Chapter 7.

Hash chaining-based hardware steganography: On the other hand, hardware steganography using key-driven hash chaining uses an IP vendor-selected encoding mechanism to generate a bitstream corresponding to the target DSP application. Further, this approach additionally uses the SHA-512 encryption algorithm to generate the hardware security constraints using generated bitstream, which are further embedded into the DSP IP design file to protect it from IP piracy and fraudulent claim of IP ownership. The details of the hardware steganography using key-driven hash chaining are depicted in Figure 9.7. The initial steps involving the generation of CDFG from the transfer function and its corresponding SDFG are similar to the previous explanation. After generating the SDFG of the target DSP application, all operations are marked accordingly. A bitstream is derived using IP vendor/designer-selected encoding rules and the DSP application's SDFG. The list of nine encoding rules, which are used to generate the initial bitstream are as follows: (a) 'E_1' = if the parity of control step number and operation number are even, then output is '0' otherwise '1', (b) 'E_2' = if the parity of control step number and operation number are same, then output is '0' otherwise '1', (c) 'E_3' = if the parity of control step number and operation number are odd, then output is '0' otherwise '1', (d) 'E_4' = if the parity of control step number and operation number are different, then output is '0' otherwise '1', (e) "E_5' = if the control step number and operation number are prime, then output is '0' otherwise '1', (f) 'E_6' = if the control step number and operation number are prime, then output is '1' otherwise '0', (g) 'E_7' = if the GCD of the control step number and operation number are '1', then output is '0' otherwise '1', (h) 'E_8' = if (operation number) mod (corresponding control step number) is '0', then output is '0' otherwise '1', and (i) 'E_9' = if the

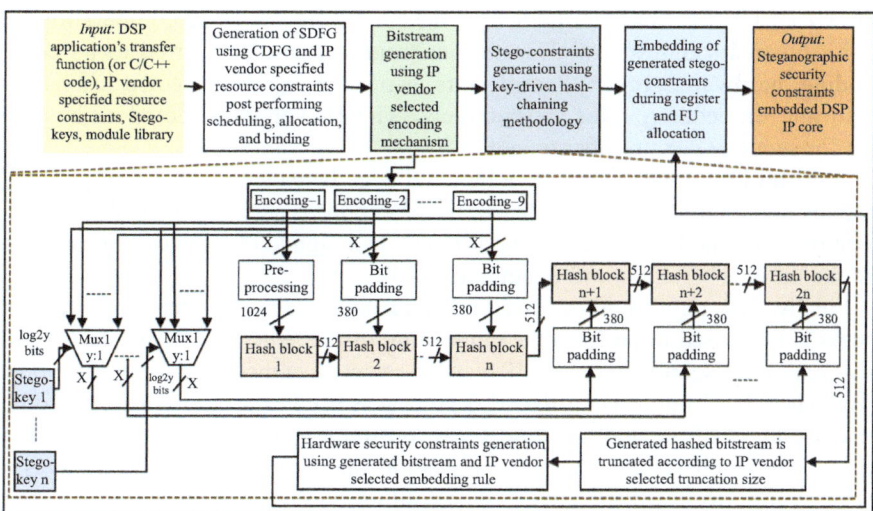

Figure 9.7   Flow diagram of steganography using key-driven hash training-based hardware security methodology

control step number is equal to the second odd sequence of operation number, then output bit is '0' otherwise '1'. The total length of the initially derived bitstream is the same as the total number of operations in the SDFG of the target DSP application (Rathor and Sengupta, 2020).

Further, the generated bitstream is fed as input to the SHA-512 encryption block to generate an encrypted 512-bit signature value. As the SHA-512 encryption block accepts input in 1024-bit format, the generated bitstream is converted into its equivalent 1024-bit format. To do so, 896-bit bitstream is produced by appending '1' and then '0' to the derived bitstream. Further, the generated 896-bit bitstream is again appended with the 128-bit representation of the derived bitstream to finally generate a 1024-bit bitstream value. After generating the 1024-bit bitstream, it is fed as input to the SHA-512 encryption block. Next, the obtained 512-bit output value from the first block is appended with "1000" and 380-bit padding block. Subsequently, the obtained 896-bit output is again appended with the 128-bit representation of obtained 512-bit digest from the first block. Now, the generated 1024-bit is fed as input to the second hash block, and this process continues till n-hash blocks. The extra padding bit added before the derived bitstream to generate a 380-bit bitstream remains completely unknown to the adversary. Moreover, the SHA-512 encryption block contains a round function computation block, which is designed to run for IP vendor-selected 'R' number of times. Next, the hardware steganography using key-driven hash chaining contains n-additional key-driven hash blocks apart from n regular hash blocks. The input to these additional hash blocks is similar to the previous ones, i.e., the derived bitstream using IP vendor-selected nine different encoding rules. The only difference is that the input to the additional n-hash blocks is controlled through IP vendor-selected stego-keys (as shown in Figure 9.7). The final output obtained at the end of the SHA-512 computation is a 512-bit bitstream digest, which is further truncated and converted to its corresponding hardware security constraints using IP vendor-selected truncation length and embedding rules, respectively. The IP vendor-selected embedding rules for security constraints are as follows:

If the bit is '0', embedding an additional edge between <even, even> storage variables node pair. Further, the hardware of vendor type 1 is assigned to all odd operations, and the hardware of vendor type 2 is assigned to all even operations in case the bit is '1' (Rathor and Sengupta, 2020).

Finally, the register allocation and hardware allocation phase of the HLS process is harnessed to perform the embedding of the generated hardware security constraints corresponding to the IP vendor-selected embedding rules mentioned above. The CIG framework of the HLS process is used to perform the embedding. The embedding of the additional security constraints (artificial edges) in the register allocation phase and hardware allocation phase, along with the changes in register colors (to avoid the raised color conflict due to embedding security constraints), is discussed in detail in the previously explained hardware watermarking approach. The generated secret steganographic security constraints embedded DSP IP core provide immunity from IP piracy and fraudulent claim of IP ownership threat. An authentic IP core can easily be demarcated and isolated based on the embedded secret steganographic IP vendor-specific information. Further, the same secret embedded information is used to prove the ownership of the IP core (Rathor and Sengupta, 2020).

9.4 Taxonomy representation of obfuscation methodologies

Threat model: The obfuscated/encrypted netlist files generated after HLS and logic level synthesis may be reverse engineered by an adversary. An adversary having access to the netlist file tries to understand the details and complete functionality of the respective IP core's design through reverse engineering followed by careful inspection (Sengupta *et al.*, 2017b). The thorough examination of the design file also provides an edge to the adversary to insert the malicious logic (hardware Trojan) covertly in safe locations in the design as well as induce RTL design alterations (tampering). The covert insertion of malicious logic in safe places makes its detection challenging during the testing process. The presence of hardware Trojan can cause safety and reliability hazards (as explained earlier in this chapter). Further, a complete understanding of the IP core design makes it easy for an adversary to counterfeit the design. Therefore, securing the IP core design file from reverse engineering attack is crucial (Sengupta *et al.*, 2019).

The classification of obfuscation and logic encryption-based hardware security approach is shown in Figure 9.8. As depicted in Figure 9.8, the obfuscation approach is broadly classified into two categories: (a) structural obfuscation and (b) functional obfuscation (also known as logic encryption). In general, hardware structural obfuscation is the art of performing modification (alteration) in the original design of the DSP IP core without affecting its functionality such that the modified design file does not reveal any meaningful information to the adversary. The motive for performing hardware obfuscation is to hinder the adversary from deploying an RE attack. As explained above, the primary aim of the adversary behind an RE attack is to understand the complete functionality of the design file through a thorough investigation of the design file, such as it also becomes feasible to covertly insert malicious logic into the design file as well as to induce RTL

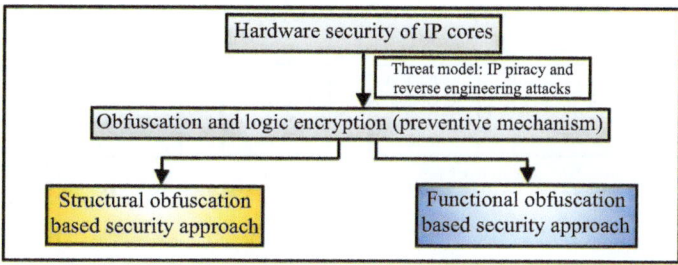

Figure 9.8   Taxonomy representation of obfuscation methodologies

design alterations. Therefore, making a design file unobvious (does not provide any meaningful information) to an adversary helps in hindering the adversary from deploying RE attack. Further, pirating an obfuscated design that does not provide meaningful information does not hold any value for the adversary (Sengupta *et al.*, 2017b).

9.4.1   Structural obfuscation-based security approach

Multi-layered high-level transformation-based structural obfuscation: Figure 9.9 illustrates the details of a low-cost structural obfuscation-based hardware security approach (Sengupta *et al.*, 2017b). The integration of particle swarm optimization-based design space exploration (PSO-DSE) helps in determining an optimized resource/architecture configuration corresponding to the structurally obfuscated DSP application designs. As discussed in the Introduction, DSE is an important phase of the HLS process, which prunes the design search space to determine an optimized resource configuration considering the IP vendor-specified high-level specifications (such as speed, area, power, and latency). The structural transformations used in the discussed obfuscation approach are the high-level transformations such as loop unrolling (LU), loop invariant code motion (LICM), redundant operation elimination (ROE), logic transformation (LT), and tree height transformation (THT). The input to the discussed structural obfuscation-based security approach are as follows: (i) DSP application's transfer function (or C/C++ code), (ii) module library, and (iii) IP vendor selected preprocessed unrolling factors (UF). Similarly, the input to the PSO-DSE block to generate a low-cost optimized architectural solution corresponding to obfuscated design are as follows: (i) IP vendor-selected resource constraints, (ii) maximum number of iterations, and (iii) control parameters (such as swarm size, acceleration coefficients, inertia weight, and iterations). Initially, the CDFG of the target DSP application is taken and is subjected to loop unrolling process using an IP vendor-selected preprocessed unrolled factor (UF). The loop body of the CDFG is repeated UF number of times.

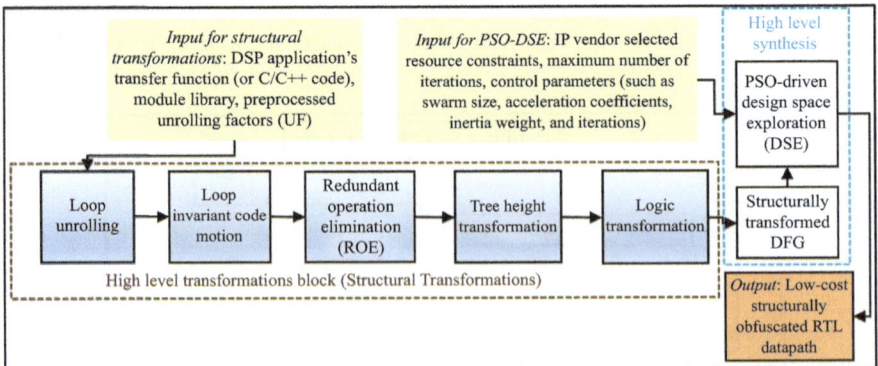

Figure 9.9   Flow diagram of low-cost structural obfuscation-based hardware

The repetition of the CDFG loop body increases the operation density, which in turn impacts (alters) the RTL circuit interconnectivity of the DSP design without changing its functionality. Next, LICM is executed on the loop unrolled CDFG of the DSP application. LICM involves the movement of loop-independent nodes out of the CDFG loop body. This movement of loop-independent node outside the CDFG loop body induces alterations (modifications) in the architecture of the RTL (size and type of used interconnect units) without affecting the end functionality of the DSP IP core. The higher modifications in the original DSP application RTL structure result in a stronger structural obfuscation (Sengupta *et al.*, 2017b).

After performing LICM, all the redundant nodes of the resulting CDFG are eliminated. The art of eliminating redundant nodes from the CDFG is called redundant operation elimination. The objective of performing ROE on the CDFG of the DSP application is to introduce greater modifications to the corresponding RTL design file without affecting the end functionality. The nodes in the CDFG having the same input with the same type of operation are called redundant nodes. To remove all redundant nodes, all such nodes are marked in increasing order first. Then, all marked redundant nodes except the least one are discarded. Further, the resulting RTL design connections are modified accordingly to preserve the DSP IP core's end functionality. Next, logic transformation is executed after performing ROE. The art of incorporating some kind of logical change into the original architectural design of the DSP application without affecting its end functionality is known as LT. Logical alteration in the DSP design file includes replacing functional units (such as adders and multipliers) with logically equivalent functional units such that the end functional output remains the same. The process of LT also induces structural change/alteration in the RTL design file. In the end, tree height transformation is performed on the DSP application's resulting CDFG. THT is the process of altering the height of the respective CDFG (primarily reducing the CDFG's height intentionally) without affecting the end functionality of the DSP IP core. THT also induces significant structural alterations in the DSP's RTL design file besides incorporating the delay optimization and parallel execution of sub-computations (Sengupta *et al.*, 2017b).

Further, the above-discussed high-level transformations result in greater RTL design file modifications when implemented sequentially. It is necessary to perform a feasibility check regarding implementing the above high-level transformation corresponding to any particular DSP application. All high-level transformations do not apply to every DSP application. Therefore, only the feasible transformation out of the discussed is used for a specific DSP application. Thus, the obtained structurally obfuscated RTL design file reveals no meaningful information to an adversary, making it challenging for him/her to deploy RE attack. Since the final generated structurally obfuscated IP core design file includes several structural alterations, therefore, it is essential to determine a low-cost optimized architectural solution corresponding to each structurally obfuscated design. The integration of the PSO-DSE block with the structural obfuscation methodology serves the objective of determining an optimized architectural solution (Sengupta

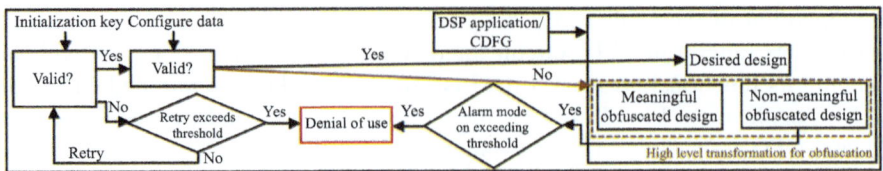

Figure 9.10   Flow diagram of key-controlled structural obfuscation-based hardware security approach

et al., 2017b). The details regarding the execution of PSO-DSE are explained in Chapter 7.

Key-controlled structural obfuscation: Furthermore, Figure 9.10 depicts a structural obfuscation approach with folding-based algorithmic transformation, controlled with the help of an IP vendor-selected initialization key (Lao and Parhi, 2015). To obtain the desired obfuscated design, the correct initialization key has to be provided as input. Folding is used to generate structurally obfuscated DSP designs. In folding, two similar operations are handled using single shared FU. For example, two addition operations on four different variables that were earlier performed using two adders will now be performed with only one adder on a sharing basis. This helps in a DSP IP design with lower area and power. As depicted in Figure 9.10, only a correct combination of initialization keys will yield the desired design. Wrong key-value entries will lead to either meaningful or non-meaningful obfuscated designs. If a person exceeds the IP vendor-selected retry attempt limit for initialization key entry, they are denied from using the DSP circuit (Lao and Parhi, 2015).

9.4.2   Functional obfuscation-based security approach

IP locking block-based functional obfuscation: Functional obfuscation employs the locking of the IP designer's netlist design file with the help of logic encryption. The implementation of logic encryption is performed using multi-key-gate logic. The locking of the design functionality restricts the adversary from its illegal use. To use the IP core, the adversary has to first unlock the locked design file with the help of the correct set of keys. Functional obfuscation is performed using locking blocks known as IP locking blocks (ILBs), which are combinations of various gates such as NOT, OR, AND, NAND, XNOR, and XOR gates. The advantages of ILBs are as follows: (i) prohibition of key gate isolation: no occurrence of isolated key gates in the inserted ILB structure, (ii) immunity from muting of key gates: the architecture of ILB restricts the adversary from muting any specific key gate, (iii) multi-pair wise security: an adversary has to perform decoding of eight key bits in order to sensitize one-bit input to the output (because of dependency of ILBs on multiple key bits), (iv) immune to run of key gates: it is impossible to substitute a run of key gates with a single key gate in the architecture of ILBs to sensitize the key value to the output. Figure 9.11 depicts a low-cost logic encryption technique to employ functional obfuscation of DSP IP cores (Sengupta *et al.*, 2019). The input

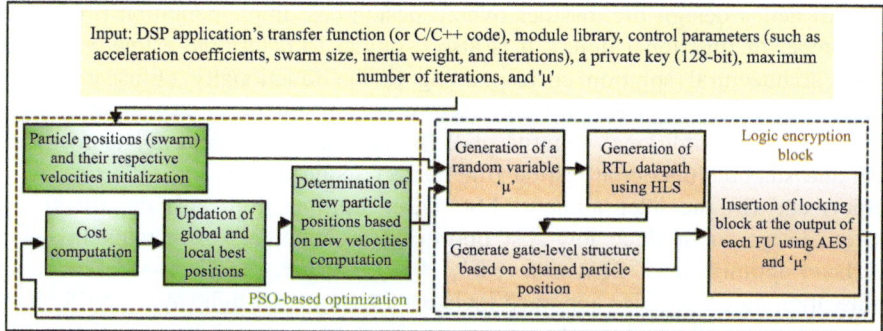

Figure 9.11 *Flow diagram of low-cost functional obfuscation-based hardware security approach*

to the discussed functional obfuscation-based security approach are as follows: (i) DSP application's transfer function (or C/C++ code), (ii) module library, (iii) IP vendor-selected resource constraints, (iv) maximum number of iterations, (v) control parameters (such as swarm size, acceleration coefficients, inertia weight, and iterations), (vi) 'μ' (controls the amount of ILBs to be inserted), and (vii) AES private key (128-bit). The value of 'μ' always lies between 1 and IP vendor/ designer-selected repeated patterns of ILBs (inserted into gate level design datapath). The discussed functional obfuscation-based security approach is employed in two blocks: (i) PSO-based optimization block and (ii) logic encryption block. As depicted in Figure 9.11, all the particle positions (swarm positions) are initialized along with their corresponding initial velocities, which are zero initially for all particle positions. Subsequently, high level and RTL synthesis is performed using transfer function/CDFG of DSP application and obtained resource configuration (particle position). The details of the HLS and RTL synthesis process are discussed in the Introduction section. Post HLS and RTL level synthesis, the generated RTL data path corresponding to the target DSP application is subjected to logic (gate) level synthesis. During gate level synthesis, the ILBs are inserted into the design at different appropriate locations (Sengupta *et al.*, 2019). Post insertion of ILBs, the design is resynthesized, and a functionally obfuscated (or logically encrypted) netlist design file is generated. Next, the design cost corresponding to obfuscated design is computed using the IP vendor-selected design cost function (used for PSO-based optimization), and local and global best particle positions are updated based on obtained design cost. Further, new velocities and their corresponding particle positions are computed, and again the same procedure is repeated as explained above (shown in Figure 9.11). The complete process is repeated till the generation of the final optimized low-cost architectural solution corresponding to the final functionally obfuscated DSP design, or IP vendor-specified stopping criterion is not met. The additional details corresponding to PSO-based optimization are explained in Chapter 7. The used design cost function (as discussed above) optimizes the overall gate count, delay, and power of the functionally obfuscated (locked) netlist design file, besides overall design cost implementation (including the number of functional units). It is also essential to determine a low-cost optimized architectural solution corresponding to the functionally obfuscated DSP design. Therefore, any arbitrary architecture may result in an expensive design. Moreover, some additional benefits of ILBs include (a) the capacity to facilitate easy and smooth functional testing (the correct sequence always leads to correct output) and (b) the integration of PSO-based optimization with the functional locked design to produce low-cost optimized architecture (also helps in delay overhead optimization) (Sengupta *et al.*, 2019). Further, the discussed functional obfuscation-based security approach includes a customized lightweight AES block (activates using IP vendor selected private key) to protect it from satisfiability (SAT) and removal attacks. The inputs of the ILBS depend on the output of the custom AES block. This makes the ILB structure reconfigurable and disables any detection via standard template during removal attack. Using custom lightweight AES (publicly unavailable architecture) helps in providing security from removal attack. It becomes tough for an adversary to locate and remove the custom AES logic as they are not publicly available. Moreover, the combination of custom lightweight AES with IP logic blocks results in an indistinguishable circuit design, making it complex and challenging for an adversary to locate and remove it (Sengupta *et al.*, 2019).

Key-based logic locking: Further, non-mutable XOR-XNOR key gate pairs are used to perform functional obfuscation (logic encryption) (Yasin *et al.*, 2016). This is a key-based logic-locking approach where the sensitization of the output is not possible by muting one of the key gates. The selection of the target location where the key gates are to be inserted is also a crucial parameter (Yasin *et al.*, 2016). The judicious selection and insertion of key gates at different locations hinder the adversary from identifying and removing the inserted key gates. Further, one-way random functions such as AES with a secret private key are used to generate locking block key values to protect it from SAT-based attacks (Pappu *et al.*, 2002). To decode the input key value of the logic locking block, the adversary must first decode the AES private key. As explained previously, AES's architecture is synthesized, resulting into a different gate level architecture (than the standard public AES) to provide immunity against removal attack. Moreover, the use of block ciphers such as AES encryption to secure inserted XOR-XNOR key gates from SAT attack has various benefits over the use of physically unclonable functions (PUFs) (Rührmair *et al.*, 2009) such as (i) block ciphers are immune to modeling attacks while PUFs are not, and (ii) block ciphers impart various well-known cryptographic properties and provide protection against cryptanalytic and modeling attacks (Yasin *et al.*, 2016).

9.5   Low-cost steganography-based hardware security approach

This section discusses the different security models required to perform a comparative analysis of discussed IP core protection and obfuscation methodologies.

The probability of coincidence (P_i) security metric is used to compare discussed IP core protection methodologies such as hardware watermarking and steganography. On the other hand, the strength of obfuscation (O_S) and power of obfuscation (O_p) security metrics are used to compare discussed security mechanisms, such as functional and structural obfuscation, respectively. P_i metric evaluates the probability of coincidentally detecting the implanted hardware security constraints (artificial edges) in an unsecured design. Further, the P_i security metric also provides insight into the presence of digital evidence inside a secured IP core design. The presence of digital evidence (IP vendor/designer specific) acts as a digital proof used to detect against IP piracy and verify the correct IP ownership. A lower value of P_i indicates stronger presence of digital evidence inside the IP core design with more robust security. Additionally, a lower value of P_i supports the signature generation with greater uniqueness, facilitating a definite, robust, and smooth differentiation between genuine and pirated IP cores during the detection process.

Further, the strength of obfuscation and power of obfuscation define the security in terms of the complexity faced by an adversary to deploy reverse engineering on design netlist file as well as against potential RTL alterations. Therefore, a higher value of strength and power of obfuscation indicates a greater complexity faced by an adversary in performing reverse engineering on the design netlist file as well as stronger resistance against potential RTL alterations. P_i of hardware security approaches (Koushanfar *et al.*, 2005; Sengupta and Bhadauria, 2016), P_i of security approach (Sengupta *et al.*, 2017a), are formulated as (9.1) and (9.2), respectively:

$$P_i = \left(1 - \frac{1}{s}\right)^h \tag{9.1}$$

where '*s*' represents the number of colors in the CIG of the DSP application, and '*h*' represents the total number of implanted secret security constraints:

$$P_i = \left(1 - \frac{1}{s \times \prod_{i=1}^{t} N(H_x)}\right)^d \tag{9.2}$$

where '*d*' represents the number of digits used for performing watermarking, '*s*' represents the total number of registers, '*t*' represents the different hardware types, and "$N(H_x)$" indicates the number of hardware of type '*x*'.

Further, the strength and power of obfuscation are formulated in (9.3) and (9.4), respectively, adopted from Sengupta *et al.* (2017, 2019):

$$O_S = (2)^z \tag{9.3}$$

The strength of obfuscation O_S indicates the logic encryption key space and '*z*' is a measure of the following $q*l*g$; where q = key bits per ILB, l = number of ILBs per FU, and g = number of FUs in the datapath:

$$O_P = \frac{\sum_{k=1}^{5} \frac{T_k}{T_N}}{T(HLT)} \tag{9.4}$$

where O_P represents the obfuscation robustness, measured in terms of obtained structural mismatch between the original design and the achieved obfuscated design while preserving the functionality, 'T_k' represents the total number of DFG nodes modified after employing 'kth' high-level transformation (HLT) technique, 'T_N' represents total nodes before employing 'kth' HLT, and "$T(HLT)$" represents total number employed HLT techniques for a specific DSP application.

The properties of the modified nodes are as follows:

(a) The primary input of a node or parent node is different from its original in an obfuscated CDFG.
(b) The child node is different from its original in an obfuscated CDFG.
(c) In an obfuscated CDFG, the operation type (such as addition and multiplication) of a node is changed.
(d) In an obfuscated CDFG, a node of original CDFG is non-existent.

9.6   Comparison between various hardware security methodologies

Figure 9.12 depicts a comparison of P_i between single-phase dual variable watermarking (Koushanfar *et al.*, 2005), single-phase multi-variable watermarking (Sengupta and Bhadauria, 2016), and multi-phase multi-variable watermarking (Sengupta *et al.*, 2017a) approaches. As evident from Figure 9.12, Sengupta *et al.* (2017a) surpass Sengupta and Bhadauria (2016) and Koushanfar *et al.* (2005) in terms of security strength (digital evidence). The generation of a higher number of hardware security constraints and involvement (usages) of a larger number of

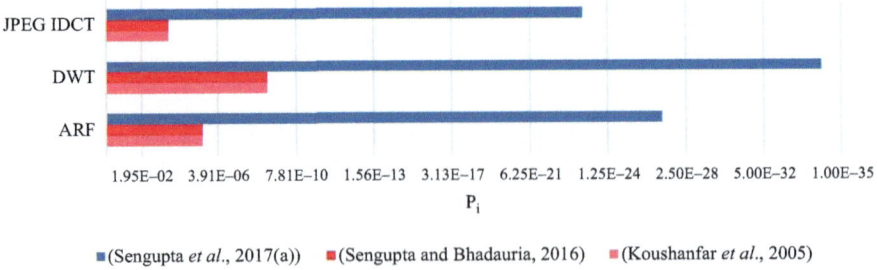

Figure 9.12   *Comparison of Pi between single-phase dual variable watermarking (Koushanfar et al., 2005), single-phase multiple variable watermarking (Sengupta and Bhadauria, 2016) and multi-phase multiple variable watermarking (Sengupta et al., 2017a)*

signature variables leads to a lower value of P_i with a more robust security strength corresponding to Sengupta *et al.* (2017a). The used signature variables in Koushanfar *et al.* (2005), Sengupta and Bhadauria (2016), and Sengupta *et al.* (2017a) are two, four, and seven, respectively. Next, Figure 9.13 compares Pi between multi-variable watermarking (Sengupta and Bhadauria, 2016) and hash chaining-based hardware steganography (Rathor and Sengupta, 2020). A lower value of P_i corresponding to (Rathor and Sengupta, 2020) indicates that the hash chaining-based hardware steganography approach is more robust than (Sengupta and Bhadauria, 2016). This is because Rathor and Sengupta (2020) lead to the generation and implanting of a larger number of hardware security constraints than Sengupta and Bhadauria (2016).

Further, a comparison of the strength of obfuscation between IP locking block-based functional obfuscation (Sengupta *et al.*, 2019) and key-based logic locking (Yasin *et al.*, 2016) is presented in Figure 9.14. The strength of obfuscation gives information regarding the number of encrypted key bits required to decipher the netlist. As shown in Figure 9.14, Sengupta *et al.* (2019) surpasses (Yasin *et al.*, 2016) in terms of the strength of obfuscation. This is because Sengupta *et al.* (2019)

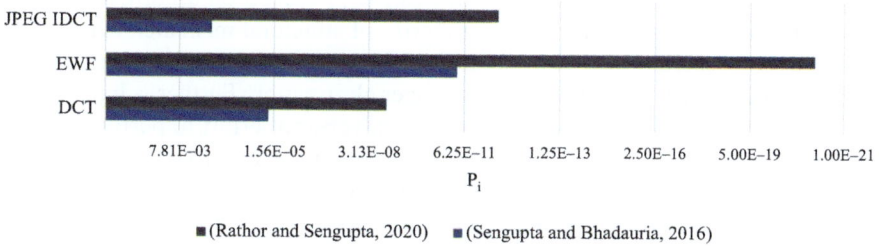

Figure 9.13   *Comparison of Pi between single-phase multi-variable watermarking (Sengupta and Bhadauria, 2016) and hash chaining-based hardware steganography (Rathor and Sengupta, 2020)*

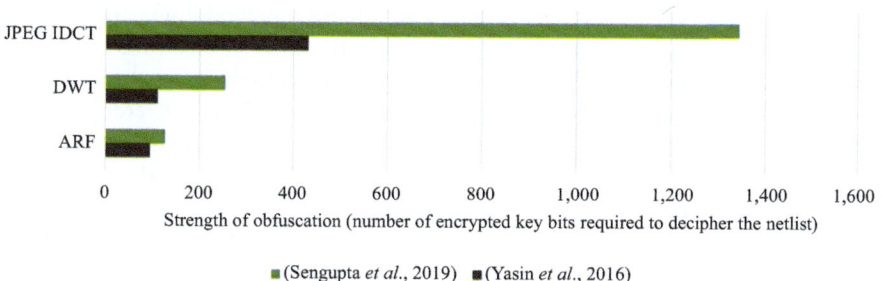

Figure 9.14   *Comparison of strength of obfuscation between IP locking block-based functional obfuscation (Sengupta et al., 2019) and key-based logic locking (Yasin et al., 2016)*

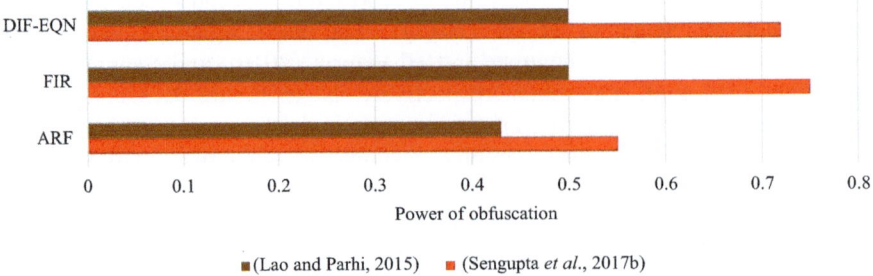

Figure 9.15   Comparison of power obfuscation between multi-layered high-level transformation-based structural obfuscation (Sengupta et al., 2017b) and key-controlled structural obfuscation (Lao and Parhi, 2015)

use a higher number of encoded key bits for functional obfuscation (logic locking) leading to larger key space. This results in significantly higher attack time using brute force attack. Figure 9.15 compares structural obfuscation-based security approaches in terms of the power of obfuscation presented on a normalized scale between 0 and 1 (indicating robustness in terms of structural mismatch between the original unobfuscated design and the achieved obfuscated design). A higher value of the power of obfuscation indicates a stronger obfuscation. Further, a higher value indicates an increased complexity in terms of adversarial effort to perform reverse engineering on the design netlist file. In Figure 9.15, multi-layered high-level transformation-based structural obfuscation (Sengupta *et al.*, 2017b) outperforms key-controlled structural obfuscation (Lao and Parhi, 2015) due to the involvement of multiple high-level transformation, ultimately leading to a greater number of modifications at the DFG nodal level, thereby subsequently creating modifications at the netlist level (without affecting the functionality).

9.7   Conclusion

This chapter discusses a taxonomy of hardware security methodologies for reusable IP cores. The increasing importance of reusable IP cores in electronic and computing devices in terms of the capability to perform complex and data-intensive functions has created an urge to secure these IP cores against different possible hardware security threats and attacks. The discussed IPP-based detective methodologies, such as hardware watermarking and steganography, facilitate the generation of a robust and unique signature template, which is converted and implanted into the IP core design as hardware security constraints (artificial edges in the register allocation framework of HLS). The implanted signature guards an IP core against IP (hardware) piracy, i.e., counterfeiting and cloning, and fraudulent claim of IP ownership. Further, the different preventive methodologies, such as structural and functional (logical) obfuscation, make a design unobvious and/or encrypted, thus hindering the reverse engineering attack by an adversary.

9.8 Questions and exercise

1. Describe the importance and usages of digital ICs in the current digital ecosphere.
2. Briefly describe the different levels of digital IC design process.
3. Describe the importance and HLS in the IC design process along with its different phases.
4. Define the different possible vulnerable points in the digital IC design process.
5. Describe the different possible hardware security issues and threats in details.
6. What is the difference between IP cloning and IP counterfeiting?
7. What do you understand from the hardware Trojan attack?
8. Draw a classification tree of the different detective and preventive hardware security mechanism discussed in this chapter.
9. Define the threat model of IP core protection-based security methodology.
10. What is the difference between hardware watermarking and hardware steganography?
11. Draw a comparison between single-phase dual variable, single-phase multi-variable, and multi-phase multi-variable hardware watermarking.
12. What are the benefits of using higher number of signature variables in multi-phase multi-variable hardware watermarking?
13. Define the different phases of the multi-phase multi-variable watermarking approach along with their importance in context of performing hardware watermarking?
14. What are the ways to resolve the color conflicts during embedding to hardware security constraints into the DSP IP core design?
15. Explain non-critical operations of the CDFG with a suitable example.
16. Explain the security methodologies of entropy threshold-based IP steganography and hash chaining-based hardware steganography.
17. Explain the process of hardware security constraints generation and their embedding into the design of a DSP IP core.
18. Define the threat model of IP core obfuscation-based security methodology.
19. What is the difference between structural and functional obfuscation?
20. What the different high-level transformation and how they are beneficial to generate structurally obfuscated design? Explain with a suitable example.
21. Explain multi-layered high-level transformation-based structural obfuscation and key-controlled structural obfuscation.
22. Explain IP locking block-based functional obfuscation and the importance of reconfigurable IP locking blocks.
23. Briefly explain how IP locking block-based functional obfuscation helps in providing security against SAT and removal attack.
24. What is key-based logic locking?
25. Define different security models corresponding to different discussed security approaches.

26. How multi-phase multi-variable watermarking provide more robust security than single-phase dual variable and single-phase multi variable watermarking?

References

Becker, G.T., M. Fyrbiak, and C. Kison (2017), "Hardware obfuscation: techniques and open challenges," in: Bossuet, L. and Torres, L. (eds.), *Foundations of Hardware IP Protection*, Springer, Cham.

Koushanfar, F., I. Hong, and M. Potkonjak (2005), "Behavioral synthesis techniques for intellectual property protection," *ACM Transactions on Design Automation of Electronic Systems*, vol. 10, no. 3, pp. 523–545.

Krishnan, V. and S. Katkoori (2006), "A genetic algorithm for the design space exploration of datapaths during high-level synthesis," *IEEE Transactions on Evolutionary Computation*, vol. 10, no. 3, pp. 213–229.

Lao, Y. and K.K. Parhi (2015), "Obfuscating DSP circuits via high-level transformations," *IEEE Transactions on Very Large Scale Integration (VLSI) Systems*, vol. 23, no. 5, pp. 819–830.

McFarland, M.C., A.C. Parker, and R. Camposano (1988), "Tutorial on high-level synthesis," in: *Proceedings of the 25th ACM/IEEE Design Automation Conference (DAC '88)*, Washington, DC: IEEE Computer Society Press, pp. 330–336.

Mishra, V.K. and A. Sengupta (2014), "MO-PSE: adaptive multi objective particle swarm optimization based design space exploration in architectural synthesis for application specific processor design," *Advances in Engineering Software*, vol. 67, pp. 111–124.

Pappu, R., B. Recht, J. Taylor, and N. Gershenfeld (2002), "Physical one-way functions," *Science*, vol. 297, no. 5589, pp. 2026–2030.

Pilato, C., S. Garg, K. Wu, R. Karri, and F. Regazzoni (2018), "Securing hardware accelerators: a new challenge for high-level synthesis," *IEEE Embedded Systems Letters*, vol. 10, no. 3, pp. 77–80.

Rathor, M. and A. Sengupta (2020), "IP core steganography using switch based key-driven hash-chaining and encoding for securing DSP kernels used in CE systems," *IEEE Transactions on Consumer Electronics*, vol. 66, no. 3, pp. 251–260.

Rathor, M., A. Sengupta, R. Chaurasia, and A. Anshul (2022), "Exploring handwritten signature image features for hardware security," in: *IEEE Transactions on Dependable and Secure Computing* Early access, https://doi.org/10.1109/TDSC.2022.3218506

Rührmair, U., J. Sölter, and F. Sehnke (2009), "On the foundations of physical unclonable functions," *IACR Cryptology*, vol. 2009, pp. 277.

Sengupta, A (2016), "Design flow of a digital IC for CE products," *IEEE Consumer Electronics*, vol. 5, no. 2, pp. 58–62.

Sengupta, A. and S. Bhadauria (2014), "Automated exploration of datapath in high level synthesis using temperature dependent bacterial foraging optimization algorithm," in: *2014 IEEE 27th Canadian Conference on Electrical and Computer Engineering (CCECE)*, pp. 1–5.

Sengupta, A. and S. Bhadauria (2016), "Exploring low cost optimal watermark for reusable IP cores during high level synthesis," *IEEE Access*, vol. 4, pp. 2198–2215.

Sengupta, A. and M. Rathor (2019), "IP core steganography for protecting DSP kernels used in CE systems," *IEEE Transactions on Consumer Electronics*, vol. 65, no. 4, pp. 506–515.

Sengupta, A., D. Kachave, and D. Roy (2019), "Low cost functional obfuscation of reusable IP ores used in CE hardware through robust locking," *IEEE Transactions on Computer-Aided Design of Integrated Circuits and Systems*, vol. 38, no. 4, pp. 604–616.

Sengupta, A., D. Roy and S.P. Mohanty (2017a), "Triple-phase watermarking for reusable IP core protection during architecture synthesis," *IEEE Transactions on Computer-Aided Design of Integrated Circuits and Systems*, vol. 37, no. 4, pp. 742–755.

Sengupta, A., D. Roy, S.P. Mohanty, and P. Corcoran (2017b), "DSP design protection in CE through algorithmic transformation based structural obfuscation," *IEEE Transactions on Consumer Electronics*, vol. 63, no. 4, pp. 467–476.

Yasin, M., J.J. Rajendran, O. Sinanoglu, and R. Karri (2016), "On improving the security of logic locking," *IEEE Transactions on Computer-Aided Design of Integrated Circuits and Systems*, vol. 35, no. 9, pp. 1411–1424.

Index

Printed and bound by CPI Group (UK) Ltd, Croydon, CR0 4YY

21/01/2025

01823630-0002